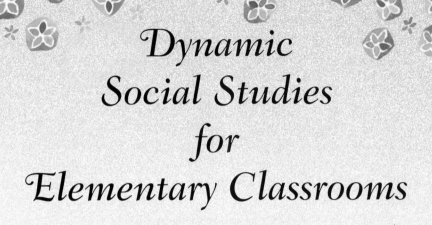

Dynamic
Social Studies
for
Elementary Classrooms

SEVENTH EDITION

George W. Maxim

West Chester University

Merrill
Prentice Hall

Upper Saddle River, New Jersey
Columbus, Ohio

Library of Congress Cataloging-in-Publication Data
Maxim, George W.
  Dynamic social studies for elementary classrooms / George W. Maxim.—7th ed.
    p.cm.
  Rev. ed. of: Social studies and the elementary school child. 6th ed. c1999.
  Includes bibliographical references and index.
  ISBN 0-13-048845-3
  1. Social sciences—Study and teaching (Elementary)—United States. 2.
Interdisciplinary approach in education—United States. 3. Effective teaching—United
States. I. George W. Maxim. Social studies and the elementary school child. II. Title.

LB1584 .M378 2003
372.83/044--dc21   2002075346

Vice President and Publisher: Jeffery W. Johnston
Editor: Linda Ashe Montgomery
Production Editor: Mary M. Irvin
Design Coordinator: Diane C. Lorenzo
Text Design: Carlisle Publishers Services
Cover Design: Ali Mohrman
Cover Art: "Winter Wonderland" by Katie S. Atkinson/Stock Illustration
Production Manager: Pamela D. Bennett
Director of Marketing: Ann Castel Davis
Marketing Manager: Krista Groshong
Marketing Coordinator: Tyra Cooper

This book was set in Wilke by Carlisle Communications, Ltd., and was printed and
bound by R.R. Donnelley & Sons Company. The cover was printed by The Lehigh Press, Inc.

Pearson Education Ltd.
Pearson Education Australia Pty. Limited
Pearson Education Singapore Pte. Ltd.
Pearson Education North Asia Ltd.
Pearson Education Canada, Ltd.
Pearson Educación de Mexico, S.A. de C.V.
Pearson Education—Japan
Pearson Education Malaysia Pte. Ltd.
Pearson Education, *Upper Saddle River, New Jersey*

Photo Credits: Scott Cunningham/Merrill, pp. 42, 80,
136, 176, 231, 244, 251, 257, 260, 307, 314, 406;
Laima Druskis/PH College, p. 358; Bruce Johnson/
Merrill, p. 61; KS Studios/Merrill, pp. 216, 391; Library
of Congress, p. 71; Anthony Magnacca/Merrill, pp. 2, 20,
34, 53, 77, 99, 111, 189, 252, 262, 300, 348; Elizabeth
Maxim, pp. 289, 367; Linda Peterson/Merrill, p. 227;
PH/College, p. 344; Barbara Schwartz/Merrill, pp. 12,
388, 396; Anne Vega/Merrill, pp. 89, 154, 161, 167,
181; Tom Watson/Merrill; p. 207.

Merrill
Prentice Hall

10 9 8 7 6 5 4 3 2
ISBN 0-13-048845-3

*Dedicated with affection to
Libby, Mike, and Jeff—
From whom I have learned*

Preface

Throughout the process of writing the seventh edition of *Dynamic Social Studies for Elementary Classrooms*, several times I found myself asking "Why am I doing this?" I didn't ask the question repeatedly because I was full of self-doubt, but because it helped me maintain focus for the project. For me, the answer each time was based on a conviction that social studies education has a lot to do with the development of informed, rational, and culturally responsive citizens. If I could help convince others of the essential role that social studies plays in bringing joy and meaning to the human condition, then the hours spent at the computer with this new edition were well spent.

This new edition was written with the conviction that subject matter is important, and that topics for instruction should be carefully selected and well taught. It is the subject matter that fuels young, inquiring minds. Subject matter should not be considered the only goal of social studies education, however; instead, it should be perceived as the vehicle through which the processes of real world inquiry and problem solving are carried out. Students who use content in authentic ways learn the subject matter well. Classrooms that teach social studies well become places of learning and wonder; they engage students in the subject matter in flexible and innovative ways.

That is the basic premise of *dynamic* social studies—using the subject matter of the social sciences as a vehicle for making knowledgeable decisions in students' daily lives and for the types of problem solving and critical thinking that may be considered the basic skills of the next generation. Therefore, students will be encouraged to expand their curiosity about things and events in the world around them, to develop a questioning attitude, and to become independent problem solvers. Because this edition gives great attention to capturing the spirit of wonder that all elementary school children naturally bring to learning, students will be referred to as "young social scientists." The young social scientists in our classrooms put subject matter to work to find and solve problems just like practicing social scientists do. This kind of teaching does not minimize the subject matter, for social scientists rarely find a powerful and interesting problem without substantial background knowledge. It simply suggests that subject matter can be put to work to stoke the curiosity in our students.

This is not a text steeped in research and theory, although research and theory are an important part of it. Nor is it a "cookbook" text full of delicious classroom recipes, although it does contain a wealth of teaching examples and suggested strategies. It does build bridges between theory and practice with the hope that future teachers understand that no single method of instruction, by itself, can help us achieve all the important goals of social studies instruction. The keys to effective teaching are variety and flexibility.

WHAT'S NEW TO THE SEVENTH EDITION?

This edition maintains the primary focus of previous editions, but has been thoroughly revised and updated. Each chapter begins with a *Classroom Scenario* that highlights the chapter's message. The scenarios, each having taken place in actual elementary school classrooms, become advance organizers that place the content into a meaningful context. In addition, a number of authentic scenarios called *Classroom Vignettes* are used within the chapters to help you understand and visualize how teachers have actually used recommended instructional strategies in their classrooms.

The *Classroom Scenarios* and *Classroom Vignettes* help exemplify concepts throughout the book as well as offer scripts for preservice teachers to follow as they plan classroom experiences. Embedded within these scenarios are practical classroom activities. Many of these activities appeared as separate boxed features in previous editions.

A number of fresh, new visual aids have been added throughout the book. Photographs, illustrations, figures, and tables help to illuminate and reinforce the information presented.

There is an increased attention to literature in this edition. Rather than devote a single chapter to literature in social studies, find suggestions I have integrated into all chapters.

One of the main changes in this edition is the reorganization of chapters. In some instances, two chapters were condensed into one. (The former Reading and Language Arts chapters were combined into a Literacy chapter, for example.) The chapters have been rearranged and brought together as thematic entities. The introductory chapter stands alone; Chapters 2 through 4 deal with children as young social scientists; Chapters 5 through 7 focus on general instructional considerations; and the remaining chapters focus on planning for instruction.

Finally, references have been updated throughout the text. New ideas from the social studies profession have been included, and appropriate citations have been made.

ACKNOWLEDGMENTS

Although the author's name is prominently displayed on the cover, the process of writing a book cannot be considered an individual accomplishment. This is especially true for this seventh edition. Many people have made important contributions to this book.

First, I must single out the contributions of my family. My wife Libby was as supportive as anyone can be. Her patience was pushed to the breaking point a number of times, but she rarely let me know. Whether it was a problem with the computer or the need to mow the lawn or walk Barkley and Betty (our basset hounds), she was always there to help me out. I would not have been able to write this new edition without her.

Next, I must acknowledge the important involvement of Mike and Jeff, my two sons. Although they may not be aware of how much they helped, the pride they showed in my work provided the motivation to keep me moving. In addition, their genuine energy and spirit never failed to pick me up whenever those "writing day blues" began to take over.

My parents, Rose and Stanley Maxim, although deceased, deserve special recognition. Their honorable work ethic instilled in me the value of sticking with a job, especially one as tough as writing a book. Their love of parenthood was an important model and inspiration for me throughout my career and life.

On a professional level, I am indebted to the outstanding professional team at Merrill/Prentice Hall, headed by Linda Montgomery. I am greatly impressed with Linda's knowledge of, and insights for, quality social studies education. Her expertise as Editor for Curriculum and Instruction is unparalleled. I am fortunate to have benefited from her wisdom and patience. Mary Irvin, the production editor, expertly coordinated all aspects of the book's development. And, Lea Baranowski of Carlisle Publishers Services did a remarkable job with the text design. She was a pleasure to work with—personally and professionally.

I wish to mention a few of my colleagues at West Chester University for their help and inspiration: Tony Johnson, Dean of the School of Education; Mary Ann Maggitti, Associate Dean; Martha Drobnak, Chair of the Elementary Education Department; and faculty engaged in teaching elementary school social studies methods courses—Martha Drobnak, A. Scott Dunlap, Gil Minacci, Ruby Peters, and Fran Slostad. A special word of thanks is extended to Dan Darigan for his inspiration to explore more deeply the literacy–social studies connection.

The form and format of this seventh edition was greatly influenced by the judicious suggestions and critical commentary offered by the reviewers: Minerva Caples, Central Washington University; Marjorie Krebs, Bowling Green State University; Duane M. Giannangelo; and Sally Beisser, Drake University. I am grateful for their expert recommendations.

Discover the Companion Website Accompanying This Book

THE PRENTICE HALL COMPANION WEBSITE: A VIRTUAL LEARNING ENVIRONMENT

Technology is a constantly growing and changing aspect of our field that is creating a need for content and resources. To address this emerging need, Prentice Hall has developed an online learning environment for students and professors alike—Companion Websites—to support our textbooks.

In creating a Companion Website, our goal is to build on and enhance what the textbook already offers. For this reason, the content for each user-friendly website is organized by topic and provides the professor and student with a variety of meaningful resources. Common features of a Companion Website include:

FOR THE PROFESSOR

Every Companion Website integrates **Syllabus Manager™,** an online syllabus creation and management utility.

- **Syllabus Manager™** provides you, the instructor, with an easy, step-by-step process to create and revise syllabi, with direct links into Companion Website and other online content without having to learn HTML.
- Students may logon to your syllabus during any study session. All they need to know is the web address for the Companion Website and the password you've assigned to your syllabus.
- After you have created a syllabus using **Syllabus Manager™,** students may enter the syllabus for their course section from any point in the Companion Website.

- Clicking on a date, the student is shown the list of activities for the assignment. The activities for each assignment are linked directly to actual content, saving time for students.
- Adding assignments consists of clicking on the desired due date, then filling in the details of the assignment—name of the assignment, instructions, and whether or not it is a one-time or repeating assignment.
- In addition, links to other activities can be created easily. If the activity is online, a URL can be entered in the space provided, and it will be linked automatically in the final syllabus.
- Your completed syllabus is hosted on our servers, allowing convenient updates from any computer on the Internet. Changes you make to your syllabus are immediately available to your students at their next logon.

FOR THE STUDENT

- **Introductions**—General information about each topic covered on the website
- **Organizations**—Links to various pertinent organizations within each topic on the website
- **General Resources**—Links to useful and meaningful websites that allow you to access current Internet information and resources to support what you are learning about instruction
- **Additional Resources**—An additional bank of links found in selected topics that are more specifically broken down
- **Lesson Plans**—Links to various sites that help you incorporate theories and practices into the classroom. Some examples include virtual field trips and webquests
- **Assessments**—Links to various means of assessing your students' progress in the classroom
- **Internet Safety**—Information found in each topic to help you keep your students safe on the Internet while in the classroom
- **Learning Network**—A link to Pearson Education's Learning Network, which provides additional information in all fields of teaching
- **Electronic Bluebook**—Send homework or essays directly to your instructor's email with this paperless form
- **Message Board**—Serves as a virtual bulletin board to post—or respond to—questions or comments to/from a national audience
- **Chat**—Real-time chat with anyone who is using the text anywhere in the country—ideal for discussion and study groups, class projects, etc.

To take advantage of these and other resources, please visit the *Dynamic Social Studies for Elementary Classrooms*, Seventh Edition, Companion Website at
www.prenhall.com/maxim

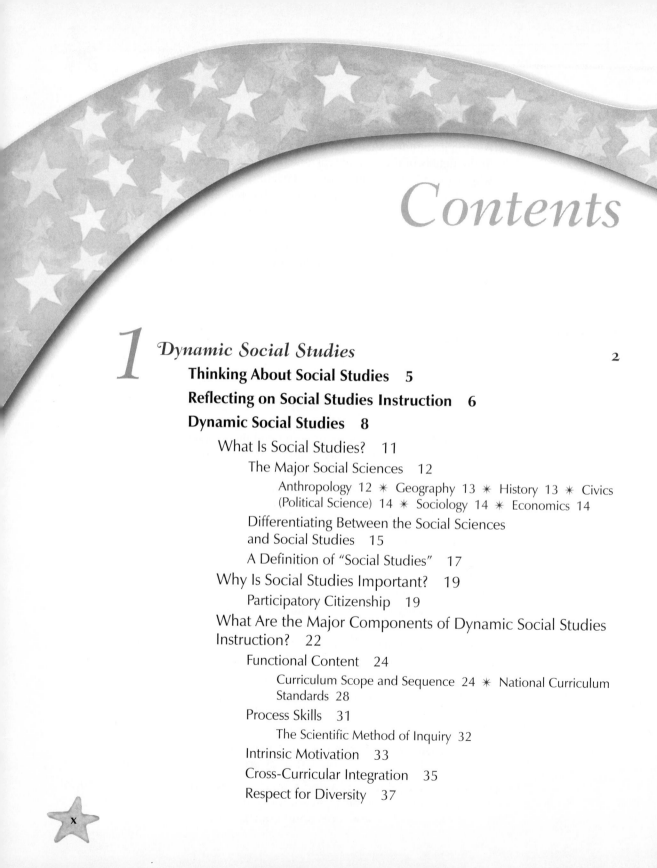

Contents

5 *Direct Instruction*   *176*

Task and Concept Analysis   182

Introduction   190

Lesson Presentation   200

Learner Response   204

9 *Literacy-Based Social Studies Instruction*   300

Reading in Dynamic Social Studies Classrooms   303

Writing in Dynamic Social Studies Classrooms   318

Dynamic
Social Studies
for
Elementary Classrooms

1

Dynamic
Social Studies

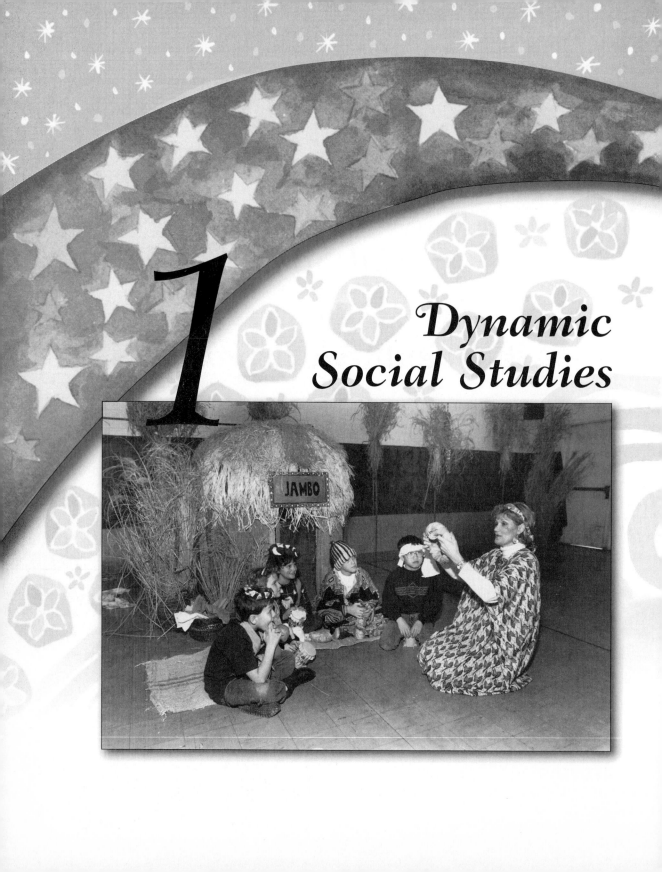

Mrs. Holzwarth's fourth graders in Upper Darby, Pennsylvania, were about to wind up a thematic unit on their state when Naisha brought in a newspaper story about Maryland having recently adopted the monarch butterfly as its state insect. The idea of a state insect captured the students' interest and provoked them to ask, "Does Pennsylvania have a state insect, too?" That was all it took to launch Mrs. Holzwarth's class into one of the most rewarding social studies learning adventures it had ever tackled.

The students got the ball rolling by looking up information in various books and pamphlets; they found a state flower, a state song, a state tree, a state nickname, and various other official state symbols, but no official "state bug." The class suggested that they should write to the president of the United States to see if they could have one, but Mrs. Holzwarth explained that since this was a state matter they should direct their inquiry to their district legislators in Harrisburg (the state capital).

Before they did so, however, the children decided to conduct a regular democratic election to determine what the state insect should be. Several insects were nominated; each nominee became the subject of careful study. The students explored the pros and cons of such bugs as the praying mantis, dragonfly, ladybug, and grasshopper. After weighing the advantages and disadvantages of each, a class vote settled the matter: the firefly was their choice.

Why? One reason was that the scientific name, *Photuris pennsylvanica,* closely resembled the name of their state, Pennsylvania. Students also liked the fact that summer evenings often reflected the soft glow of hundreds of these insects; Mrs. Holzwarth's youngsters spent many a summer night catching "lightning bugs." (The children had been taught that to catch a firefly and let it walk all over their hands for a while is "okay" but to hurt the insect in any way is very wrong.)

After the children had democratically selected the firefly as their choice for state insect, they decided to write a letter to their two state representatives informing them of their dream. Both lawmakers were impressed with the children's civic venture and decided to visit Mrs. Holzwarth's classroom. They discussed the process of introducing a law in the state legislature and advised the students how to proceed. The children learned that their next step would be to persuade other legislators to support their cause. Undaunted, these 26 children wrote over 250 letters—203 to the House, 50 to the Senate, and 2 to the governor and his wife. The children also learned that they needed support from voters in their area, so they circulated a petition and obtained more than 2,100 signatures.

To interest others in their work, the students printed over 600 luminous bumper stickers proclaiming "Firefly for State Insect." They also kept up their letter-writing campaign, asking legislators to vote YES when the bill came onto the floor. The children were invited to Harrisburg for the House Government Committee hearings on their bill, and they went to Harrisburg armed with banners on the side of the bus and singing an original song they wrote especially for this occasion:

Oh firefly! Oh firefly!
Please be our state bug.
Photuris pennsylvanica,
You'll fly forever above.
Oh firefly! Oh firefly!
You light up so bright.
It's fun to see such a pretty sight.
Oh firefly! Oh firefly!

Imagine the thoughts of the children as they arrived in Harrisburg to be met by television crews and reporters from the major newspaper wire services. The hearing itself was held according to established decorum, the children testifying with all their knowledge about fireflies for a period of about 2 hours. The committee reported its unanimous support of the bill to the House of Representatives, and eventually the bill passed the House by a vote of 156 to 22. Next, the Senate passed the firefly bill by an overwhelming vote of 37 to 11. When the governor finally signed the bill (Act 59), the children were again in Harrisburg to watch the establishment of a new state law. *Photuris pennsylvanica* officially took its place alongside the whitetail deer, ruffed grouse, and Great Dane as official state animals.

For Mrs. Holzwarth's class, this "happening" was much more than an exercise in choosing a state insect. It was a purposeful problem-solving episode where the children took direct political action and participated in meaningful legislative processes. They learned about petitioning and writing letters to their representatives, and they saw firsthand how government works. One child noted, "Now we have something to tell our grandchildren." Another, when asked if she would like to get another law passed, blurted, "Darn right! I'd like a law against homework. Homework gives you pimples!"

THINKING ABOUT
SOCIAL STUDIES

Social studies can be the single most exciting subject in the elementary school, for no other subject helps guide children through the captivating world around them with such strength and power. Children come to school ever curious about their surroundings, and social studies provides the perfect medium to capture their instinctive drive to explore and investigate. Social studies teachers understand that elementary school children are the most eager of all learners and that there seems to be a "natural social scientist" residing within each. Therefore, their social studies classrooms are richer in variety, richer in stimulation, and richer in challenge than we might find for any other subject.

Good social studies teachers—the "Mrs. Holzwarths" of our world—bring an aura of distinction to their social studies classrooms; they obviously enjoy their work and value the lives they touch. There are no "secret recipes" or "die-cast molds" that might help us duplicate these special individuals; each is one of a kind. They know and love our nation and hold bright hopes for its future. Their sense of democratic values influences everything they do in their classrooms. These teachers know that young children are our future and the way they live and learn today becomes the way they will live and learn tomorrow. These teachers expertly handle with keen insight and skill all the subtle professional tasks of a social studies program, and their instructional choices are based on a maze of complicated decisions.

The most vital ingredient of good social studies programs is an informed, caring teacher. The physical setting is important, as are the materials used, but a teacher's professional skills and personality are of major importance. Personal and professional behaviors dictate the tone of the environment and make a lasting impact on the students, on their families, and on society. Few individuals are more significant in the lives of elementary school children than their parents, close relatives, and teachers. Teachers of elementary school children should be among the finest people we can imagine, but being a fine person does not in itself guarantee success in teaching. A superior teacher must also possess a set of professional skills founded on sound theoretical and research-based principles.

By working hard to become well informed, you will be bound for success. Successful teachers welcome the challenge of creating superb social studies classrooms. These teachers look to sound theory as the basis for their instruction. They deliberately build their programs on their best knowledge of whom they are teaching, what they are teaching, and how they are teaching. Successful social studies teachers have what it takes, in spirit and skill, to tailor-make their classrooms to fit the children who come to them. They demonstrate, without timidity, that they would rather be challenged than safe and bored.

It's a good idea to acquire this can-do spirit early in your career, for succeeding in risky situations helps identify potential greatness more clearly than any other factor. Those who take risks have a high degree of self-confidence, a distinctive quality of outstanding social studies teachers. As I once heard, having a strongly positive sense of self-worth is worth 50 IQ points. So work hard, dream a lot, and muster up the grit to establish a point of view. However, risks cannot, and should not, be taken unless your fundamentals are solid. Good social studies teachers never take risks blindly; their decisions are based on a strong foundation of knowledge and skill. Build that foundation in social studies education and take your risks there, for it is the one area of the elementary school curriculum that most openly invites the ideas and dreams of adventurous and creative teachers.

REFLECTING ON
SOCIAL STUDIES INSTRUCTION

Do you remember a "Mrs. Holzwarth" from your elementary school days? Think back to when you were an elementary school child. What memories do you have of a social studies teacher doing the things you liked, but also the things you didn't like? Make a list of both kinds of experiences. Share your list with your classmates, perhaps by constructing a group chart listing positive and negative experiences under their appropriate headings. What kinds of memories did most students have? I enjoy doing this activity on the first day of a semester with my classes. It is instructive both for my students and me. Although I hesitate to describe this category first, the "dislike" category usually includes such memories as reading pages 79 to 81 in the text and writing answers to the questions at the end of the section (while the teacher corrected the weekly spelling tests), listening to the teacher drone on about how a cotton gin works (without benefit of a picture or model), being required to memorize facts about the early explorers of North America (where they came from, when they left their homeland, the date they arrived here, and where they explored), or copying "research reports" directly from the encyclopedia. After generating a list of 15 to 20 "dislikes" such as these, I ask the students to suggest one-word descriptive labels that best capsulate those types of instructional practices. "Boring," "deadly," "dull," "unexciting," and "humdrum" are some of the most frequently suggested labels.

Unfortunately, when their elementary school days are filled with experiences like these, individuals tend to underestimate the hard work that goes into effective social studies instruction: "Is that all there is to it? Why, anybody can teach social studies to elementary school kids! Who can't direct them to take out their textbooks? Read a few pages? Answer the questions at the end of the section? Why does anyone need to take a college methods course to learn to do something so simple?" When faced with such an accusation, the best way to cope is to admit its validity. Anybody *can* tell children to take out their textbooks to read a few pages. And, yes, it's true that anybody *can* ask them the questions printed at the end of a reading selection. The accusations are true, but there is one thing wrong—they miss the whole point of elementary school social studies education. Social studies is not meant to be taught that way; if it were, it's indeed true that there is no point in taking this course.

In contrast to the "dislike" category, the "like" category usually includes memories such as: "Taking a field trip to the seashore to study the shells that washed up," "Cooking venison stew as we read the book, *Sign of the Beaver*," "Hearing a Peace Corps volunteer tell of her experiences in Ethiopia," "Making a large mural of the rainforest," "Taking food and clothing to a homeless shelter," "Learning about the history and origin of the families of everyone in the classroom," and "Making a large dragon as part of a Chinese New Year celebration." When asked to suggest one-word descriptive labels to best capsulate these experiences, students regularly come up with words like "fun," "exciting," "interesting," "lively," "rewarding," "active," and "instructive." I bring closure to the activity by asking my students to think deeply about these questions: "Which set of words would you want others to use while describing *your* social studies program?" "What will you need to know or do in order for that to happen?"

The resulting discussion usually elicits questions about the kinds of professional expertise necessary to carry out social studies programs that are fun, exciting, interesting, lively, rewarding, and active, but at the same time, instructive. Some worry that using "fun-type" programs all the time can create serious classroom management problems: "I'd like my social studies class to be fun and exciting, but I'm worried that the children will get out of control." "Won't children think of the 'fun' activities as 'play time' and just fool around in class, not learning anything?" They are worried that a "fun-type" program downgrades social studies classes to a series of inconsequential experiences that serve only to create hard-to-handle behavior problems. Others are concerned that being strictly serious about the content will do just the opposite—result in a dull and boring social studies program: "How do you get across the important social studies content without being run-of-the-mill or ordinary?" "How can you teach content without communicating to the children that we think they're basically incompetent?" They fear that "serious-type" social studies programs can become trivial and tedious for both the teacher and the children and are prone to provoke resentment and hostility in the children. In essence, the question becomes, "How can I make social studies fun and still maintain control over what the children do and understand?" One of the most significant challenges faced by social studies teachers is, on one hand, helping children acquire the knowledge, skills, and values that help prepare them for constructive participation in a democratic society, and, on the other hand, organizing and conducting lessons that offer a blend of pleasure, intrigue, variety, active involvement, and excitement.

Back in 1933, John Dewey addressed this dilemma and offered some sage advice that remains relevant even today. In speaking to the serious–fun dichotomy of social studies instruction, Dewey (1933) said that when either is used exclusively, "Play degenerates into fooling, and work into drudgery" (p. 286). Dewey recommended a program in which a balance between seriousness and fun is maintained to help promote learning. That is, teachers can easily build a participative approach that takes the best elements from all sources and blends them to their needs and those of their students. The philosophy of this text echoes Dewey's thoughts; when learning is fun, students become more interested and open to acquiring new knowledge. As a result, they see social studies as an important and fulfilling part of their lives and will strive for serious learning.

DYNAMIC
SOCIAL STUDIES

I like to call this approach *dynamic social studies* largely because it is characterized as an active, engaging, enjoyable, and rewarding system of instruction that results in meaningful and substantial learning. Regardless of what we call this approach, however, its major purpose is to help children discover their scientific selves, and to rediscover the "natural social scientist" within. Like adult social scientists who look at life around them with an overpowering sense of wonder, taking nothing for granted within their own fields, children are captivated by spectacular phenomena of their social world. We encourage and support their sense of wonder when we open their minds to the creative spirit that floods the social sciences. I find it best to explain the concept of an unfolding social scientist in every child by telling a story about Heinrich Schliemann.

Classroom Vignette

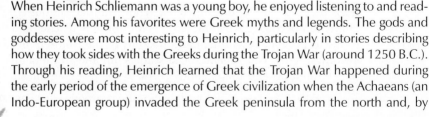

When Heinrich Schliemann was a young boy, he enjoyed listening to and reading stories. Among his favorites were Greek myths and legends. The gods and goddesses were most interesting to Heinrich, particularly in stories describing how they took sides with the Greeks during the Trojan War (around 1250 B.C.). Through his reading, Heinrich learned that the Trojan War happened during the early period of the emergence of Greek civilization when the Achaeans (an Indo-European group) invaded the Greek peninsula from the north and, by

FIGURE 1–1
Geography of Ancient Greece

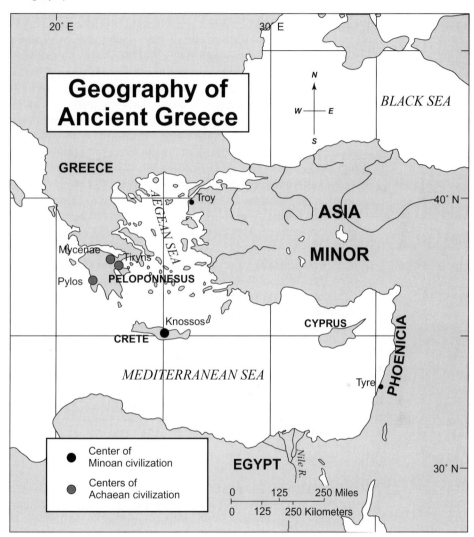

about 1400 B.C., controlled both what is now Greece and the Aegean Sea (see Figure 1–1). Because the Aegean was crucial to the development of early Achaean civilization, the Achaeans banded together to attack Troy, a rival power on the Greek peninsula that controlled trade routes between the Aegean and Black Seas. After a long and bitter war, the Achaeans emerged victorious. But victory did not come without its devastating cost. Invaders quickly swarmed into Greece, easily defeating their severely weakened Achaean rivals.

The invaders plundered the rich cities and disrupted trade; advances in art, architecture, and writing were largely forgotten. As a consequence, the glory of Achaean civilization vanished and Greek history entered a troubled period called the Dark Age.

Historians first learned about the Trojan War from the *Iliad* and the *Odyssey,* two of the best-known epic poems of all time. Homer, a blind Greek poet, is thought to have composed these poems about 750 B.C., about 500 years after the fall of Troy. It is said that he based his poems on stories that had been passed down through the generations.

When Schliemann was growing up, most historians believed that Homer's story of Troy's burning was just a myth. Because the story mixed the exploits of gods and goddesses with those of human heroes, it seemed to have no realistic historical base. But Heinrich Schliemann was not convinced; he thought Troy really existed. Schliemann had an opportunity to test his convictions when, as an adult and amateur archaeologist, he visited a location in northwestern Asia Minor to investigate stories of buried cities that matched Homer's description of Troy. After much searching, he eventually found a huge mound of earth close to the Aegean Sea—an area where the ancient city of Troy was believed to have once stood. Schliemann dug into the earth and discovered that the site had been the location of several cities, each built upon the other, as the previous city had been destroyed—nine separate cities altogether. Archaeologists rushed to the site and used their special methods and equipment to study the area. To Schliemann's surprise and joy, they found that charred wood and other evidence suggested that the city of Troy actually did stand at this location. Archaeologists believed that the Troy made famous by Homer was the city found in the seventh layer of Schliemann's mound. By pursuing a lifelong interest that began during his childhood, then, Schliemann showed that Homer's great epic poems were more than fiction; they have a real historical base.

Heinrich Schliemann created a link to the past by acting on a powerful childhood interest. Not all children will grow up to be "Heinrich Schliemanns," but just think about the sense of wonder to all of life they display as they act like "natural social scientists" on their world: a "geographer" bends down to study the effect of sand sifting through his or her fingers; an "economist" helps determine how the class will obtain the money necessary to buy a sapling for the school playground; a "political scientist" petitions the principal for a new piece of playground equipment; an "anthropologist" watches the way a mother interacts with her baby during feeding time; an "historian" watches and listens in awe as a senior citizen augments stories of World War II with fascinating memorabilia. Like these children, Schliemann did not attain the educational credentials required of practicing social scientists. However, he was curious about his social world and he acted on his curiosity. That is the basic premise of this text—children are curious about their world and strive to seek answers to their questions so they can obtain knowledge about their wondrous social environment. Observing children in their "child's world" leads anyone to conclude that they are doers and thinkers; in other words, natural scholars. I recall one first grader who was asked by his teacher what it would be like to know everything. "Awful," the little boy replied.

"Why?" his surprised teacher asked.

"'Cause then there would be nothing to wonder about," he responded.

We build this same spirit when we honor students' ideas and offer them opportunities to create their own understandings. This is not a traditional notion of social studies instruction, one that simply views learning as a one-way transfusion of information from the teacher to the students. A dynamic social studies classroom helps children look on their world as a never-ending mystery. For many of you, facilitating such a program may require a radical shift in thinking. However, to start you on your way to acquiring a sound plan for teaching dynamic social studies, it is important to carefully examine these fundamental questions:

What is social studies?

Why is social studies important?

What are the major components of dynamic social studies instruction?

What Is Social Studies?

Our first responsibility in acquiring the level of competence needed for making content and methodological decisions in dynamic social studies is to establish a clear idea of the subject we will be teaching. Although social studies has been part of the elementary school curriculum for decades, it is not an easy subject to define. What do *you* think of when you hear the words "social studies"?

A definition of social studies begins by examining the difference between two potentially confusing terms: *social science* and *social studies.* The word *science,* derived from the Latin word *scientia,* means *knowledge.* So, putting *social* and *science* together, we can define a *social science* as any of several fields that seek to understand and explain the social realm of human existence. The story of Heinrich Schliemann largely described the work associated with a single social science— archaeology (a branch of *anthropology*). In addition to anthropology, 5 other major social sciences contribute knowledge and processes that help us explain and understand the social realm of human existence; including *geography, history, political science, sociology,* and *economics.* Research is the major activity for most social scientists; they use various methods to assemble facts and construct theories, such as carrying out field investigations, living and working among people being studied, examining historical documents and records, preparing and interpreting maps, administering tests and questionnaires, and conducting interviews and surveys. The

*Young children learn about
social studies by living it.*

work associated with each major specialty varies greatly, but specialists in one discipline often find their research overlaps with work being done in another.

The Major Social Sciences

Anthropology

Anthropologists study people to find out about their physical, social, and cultural development. They examine the total pattern of human behavior and its products particular to a special group (language, tools, beliefs, social forms, art, law, customs, traditions, religion, superstitions, morals, occupations, and so on). Anthropologists usually concentrate in one of four specialties—*archaeology, sociocultural anthropology, linguistic anthropology,* or *biological-physical anthropology. Archaeology* is the scientific study of earlier civilizations carried out by recovering and examining material evidence of the past, such as skeletal remains, fossils, ruins, implements, tools, monuments, and other items from past human cultures, in order to determine the history, customs, and living habits. *Sociocultural anthropology* is the study of customs, cultures, and social lives in settings that vary from non-industrialized societies to modern urban centers. *Linguistic anthropology* looks at the role of language in various cultures. *Biological-physical anthropology* studies the evolution of the human body and analyzes how culture and biology influence one another. Most anthropologists specialize in one region of the world. Because of this immense scope of study, anthropology has often been described as a universal discipline, one that comprehensively studies cultures by looking at all aspects of their existence.

Geography

Geographers study people and places, the natural environment, and the capacity of the earth to support life. They ask questions about places on the earth and their relationship to the people who live in them. The first task in geography is to locate places, describing and explaining their physical characteristics (climate, vegetation, soil, and landforms) and their implications for human activity. Geographic inquiry continues by exploring the relationships that develop as people respond to and shape their physical and natural environments. It permits us to compare, contrast, and comprehend the regions of the world and their various physical and human features and patterns. This knowledge helps us to manage the world's resources and to analyze a host of other significant problems in terms of the spaces they occupy and how these spaces interact with each other on the earth's surface.

History

Historians systematically research, analyze, and interpret the past. They use many sources of information in their research, including interviews, government and institutional records, newspapers and other periodicals, photographs, film, diaries, letters, and photographs. Historians usually specialize in a certain country or region or a specific time period. Some function as biographers, collecting detailed information on individuals. Others are genealogists, interested in tracing family histories. Some help preserve or study archival materials such as artifacts or historic buildings and sites.

History is a principal discipline contributing to social studies but is not itself considered by many to be a true social science because the processes of empirical research are not utilized rigorously in historical study. This is not to fault the efforts of historians, but only to indicate that they are handicapped in their abilities to control or reproduce the phenomena they are studying, as scientists do. Instead, historians must reconstruct events of the past from surviving evidence, and resulting interpretations are not always accurate. For example, interpretations change as historians develop more sophisticated techniques of examining evidence and accumulate different kinds of evidence. Consider that, in the 1920s, many historians argued that the first large constructions in the world were the Egyptian pyramids. This contention was disproved in the 1960s with the introduction of the technique of carbon-14 dating, which indicated that northern European monuments such as Stonehenge were constructed earlier than most of the pyramids. As we enter the 21st century, powerful new methodologies, many using computers, may offer historians the rigorous processes that a true science demands. But for now, history is treated as a discipline wavering somewhere in the murky region between the sciences and the humanities. This is why you may frequently see references to social studies as "history and the social sciences."

Civics (Political Science)

Political scientists study the origin, development, and operation of political systems and public policy. The emphasis of civics inquiry has been to examine the structure and functions of government—how people get power, what their duties are, and how they carry out their duties. Political scientists conduct research on a wide range of subjects such as the relationship between the United States and other countries, the political life of nations, the politics of small towns or a large metropolis, or the decisions of the U.S. Supreme Court. Political scientists approach their subject through analyses of political institutions or policies, and by means of detailed examinations of the day-to-day workings of contemporary governments. Depending on the topic, political scientists take polls, conduct public opinion surveys, analyze election results, interview public officials, examine the speeches of politicians, study the actions of legislators and judges, and probe the beliefs and personalities of political leaders.

Sociology

Sociologists study society and social behavior by examining the groups and social institutions people form, such as the family, government, religion, business, or school. They also study the behaviors and interactions of groups, analyzing the influence of group activities on individual members. Sociologists investigate the values and norms of groups to discover why group members behave as they do. They study how groups form, how they operate, and how they change. Sociologists organize their study of groups around many questions, such as, "What kinds of groups of people form in any given society?" "What are the expectations of each group member?" "What problems do the group members face?" "How does the group control its members?" To answer these questions, sociologists may visit a particular group, observe what the people in that group do, interview group members, or even live with a group for a short time to more completely understand its nature. The results of a sociologist's research aid educators, lawmakers, administrators, and others interested in resolving social problems and formulating public policy.

Economics

Economists study the production, distribution, and consumption of goods and services. They examine the ways economics affects all of our lives. From youngsters who save their allowance for a special toy, to college students who must scrape together enough money for tuition, through newlyweds who apply for a mortgage as they buy their first home, all people face situations where they attempt to satisfy unlimited wants with limited resources. Referred to as the *scarcity* concept, it is from this idea that a family of economics emerges. Because of scarcity, humans have attempted to find ways to produce more in less time with less material, by which *specialization* of labor was developed. From specialization has emerged the idea of *interdependence,* a reliance of people upon one another that necessitates monetary, transportation, and communication systems. From interactions of these factors, a *market* system developed through which buyers and sellers produce and exchange goods or services. Finally, *governments,* responsible for controlling segments of the market system, ensure the welfare of all their citizens. Information about the economy, including the study of taxation, consumer economics, and economic policy, helps one assess pressing issues of the day.

Differentiating Between the Social Sciences and Social Studies

To differentiate between a *social science* and *social studies,* we must travel back in time to 1916 when waves of immigration and a newly industrialized economy were propelling our nation into unprecedented social change. Responding to this change, the National Education Association (NEA) searched for new approaches to teaching that could best meet society's unfolding needs. Up to then, the major purpose of elementary school education was to foster competencies in the basic skills of reading, writing, and arithmetic. Content that we would associate with social studies today was drawn from three major social sciences—geography, history, and political science. These were taught primarily as separate high school subjects with an emphasis on memorizing facts. In the elementary school, teachers read stories from a limited supply of books that stressed such citizenship virtues as courage, honesty, fairness, and obedience. *The Power of Kindness* is a sample of a history selection a teacher might use.

The Power of Kindness

William Penn, the founder of Philadelphia, always treated the Indians with justice and kindness. The founders of other colonies have too often trampled on the rights of the natives, and seized their lands by force; however, this was not the method of Penn. He bought their lands, and paid for them. He made a treaty with the Indians, and faithfully kept it. He always treated them as men.

After his first purchase was made, Penn became desirous of obtaining another portion of their lands, and offered to buy it. They returned an answer that they had no wish to sell the spot where their fathers were buried; but to please their father Onas, as they named Penn, they said they would sell him a part of it.

A bargain was thereby accorded that, in return for a certain amount of English goods, Penn should have as much land as a young man could travel around in one day.

The story of William Penn was obviously intended to instill such virtues as honesty and fairness. Similarly, stories of George Washington and the cherry tree, Ben

Franklin and the kite, and other exemplary stories were read to the children to instill democratic virtues and patriotic pride.

Information contained in these stories was to be "learned by heart and, when forgotten, learned again." The following statement, written by two historians at the turn of the century, capsulated the educational methodology of the time:

> In its effect on the mind, American history is distinctly to be commended. The principal reasons for the study of history are that it trains the memory . . . exercises the judgment and sets before the students' minds a high standard of character. In all these respects, American history is inferior to that of no other country. The events which are studied and should be kept in memory are interesting in themselves and important to the world development. (Channing & Hart, 1903, p. 1)

The National Education Association (NEA) established various study committees to examine whether this strong emphasis on memorizing facts should remain the central priority of schooling in a "new age," or whether the voice of new, "progressive" educators should be recognized. Notice how the following proposal for change in history instruction differs from what you have just read:

> The one thing that [history] ought to do, and has not yet effectively done, is to help us to understand ourselves and our fellows and the problems and prospects of mankind. It is this most significant form of history's usefulness that has been most commonly neglected. . . . It is high time we set to work boldly . . . to bring our education into the closest possible relation with the actual life and future duties of . . . those who fill our public schools. (Robinson, 1912, p. 134)

The NEA established a special committee to study which of these contrasting philosophies might be more useful in addressing the dual processes of assimilating immigrants into a new society as well as preparing an industrial workforce. After careful reflection, the committee suggested that the school curriculum and teaching strategies should be brought more in tune with contemporary society. Supported by the work of leading educators such as Francis Parker and John Dewey, the NEA proposed that the work of our schools was to guarantee *social efficiency* through educating students to understand and resolve social problems; readying students for the job market in a changing world of business and industry; and developing in students practical skills related to health, hygiene, and nutrition. The committee recommended that school subjects should be practical:

> In their lives and work and thought, people do not need simply to be able to recall facts or preset procedures in response to specific stimuli. They need to be able to plan courses of action, weigh alternatives, think about problems and issues in new ways, converse with others about what they know and why, and transform and create new knowledge for themselves; they need, in short, to be able "to make sense" and "to learn." (Peterson & Knapp, 1993, p. 136)

To help achieve meaningful learning, the special NEA committee proposed a new school subject—*social studies*—as the major vehicle to help accomplish these social education goals. Making this proposal was a bold move, since never before had a school subject called social studies existed—at least, not in the United States. However, this completely new subject would be responsible for promoting the idea of "social efficiency" in a rapidly changing society. The NEA (1916) envisioned social studies as "the subject matter related directly to the organization and development of human society, and to individuals as members of social groups" (p. 5). The "subject matter" for this new school subject was to be drawn from the most influential social sciences of the time—history, geography, and civics—and blended together as one school subject for the purpose of helping children understand our American heritage and acquire the skills and sensitivities basic to constructive participation in our nation's democratic society.

A Definition of "Social Studies"

What could prevent such a deceptively simple idea from becoming immediately embraced by the educational community? An answer to that question was as difficult to arrive at in 1916 as it is today, for over the years splinter groups have presented contradictory views of what this new subject called social studies should be. According to critics, their efforts have come up empty; they have constructed nothing more than an endless maze of ambiguity, inconsistency, and contradiction. Finn (1988) expounds on this point: "The great dismal swamp of today's school curriculum is not reading or writing, not math or science, not even foreign language study. It is social studies, a field that has been getting slimier and more tangled ever since it changed its name from history around 1916" (p. 16).

Students in social studies methods courses often become confused when they learn of the years of deep internal conflict: "After all these years, why can't social studies educators still agree on the nature of this school subject?" To gain some perspective on this question, you must understand that, as with every sensitive educational issue, attempts to reshape social studies are sparked by experts holding strong opinions about how students learn best and what knowledge, skills, and values are most important for active citizenship participation. Such disagreements cannot be quickly resolved; years of controversy, disagreement, and debate are to be expected (see Figure 1–2). After decades of carefully and patiently gazing at a clouded tangle of puzzle pieces, however, an intelligible, practical image of social studies seems to have appeared. Out of the hazy mist, razor-sharp patterns are taking shape. The leading professional association for social studies educators—the National Council for the Social Studies (NCSS) (1993b)—helped bring about harmony by aggressively addressing the nature of social studies and ultimately approving the following definition:

> Social studies is the integrated study of the social sciences and humanities to promote civic competence. Within the school program, social studies provides coordinated, systematic study drawing upon such disciplines as

FIGURE 1–2

What to Include in the Social Studies Curriculum? The Search for Identity

anthropology, archeology, economics, geography, history, law, philosophy, political science, psychology, religion, and sociology, as well as appropriate content from the humanities, mathematics, and natural sciences. The primary purpose of social studies is to help young people develop the ability to make informed and reasoned decisions for the public good as citizens of a culturally diverse, democratic society in an interdependent world. (p. 3)

With this NCSS initiative, social studies can now be defined and set apart as a viable field of study. Clearly, the definition has affirmed citizenship education as the primary purpose of social studies instruction, presented an argument for integrated content from the social sciences and other disciplines (expanded from political science, history, and geography), emphasized the need for higher-order thinking skills as opposed to memorization of facts, focused on the need for multicultural understandings, and stressed decision making and personal responsibility as important characteristics of good citizens. We will thoroughly examine these fundamental attributes throughout the remainder of this chapter.

Why Is Social Studies Important?

From the start of public schooling in the United States, the major purpose of elementary education has been to teach children the basic skills of reading, writing, and arithmetic. But, following the NEA's 1916 initiative, social studies took its important place in the curriculum for the primary purpose of helping prepare our nation's youth for constructive participation in society. Today, it is accepted that our schools must not only help prepare children with competencies in the "three Rs" but, to become effective citizens, they must also possess the knowledge, skills, and civic virtues necessary to become active participants in the world of the 21st century. The survival of our democratic society depends upon a citizenry who care for their country and for humanity itself. Will our students be delivered an education that cultivates the moral grit to stand up and cry out for the protection of our nation's future, or one that minimizes civic responsibility and respect for the legacies and principles of our families, neighborhoods, and national communities? How will our future citizens measure the purpose and quality of life in our nation? The Children's Defense Fund, an organization with a solid reputation for social advocacy, commented on the impact of schools in responding to our nation's increasingly diverse society:

> A thousand years from now, will civilization remain and humankind survive? Will America's dream be alive, be remembered, and be *worth* remembering? Will the United States be a blip or a beacon in history? Can our founding principle "that all men are created equal" and "are endowed by their Creator with certain inalienable rights" withstand the test of time, the tempests of politics, and become deed and not just creed for *every* child? Is America's dream big enough for every fifth child who is poor, every sixth child who is Black, every seventh child who is Latino, and every eighth child who is mentally or physically challenged? (1996, p. xi)

Participatory Citizenship

The previous questions are not new; social studies educators have asked variations of these questions for years. From the first colony at Plymouth Rock to our bustling communities today, educators and community leaders have debated over the best way to fulfill their responsibility to educate America's young citizens. Every generation has experienced deep concern about how to place in its children's collective hands the trust and understanding required to protect peace and freedom and how to teach its youngsters tolerance, cooperation, and the skills of living together in a diverse, democratic society. As a nation, we have situated social studies instruction at the core of this responsibility because we are a proud people—a democratic republic of over 250 million citizens, each of whom is part of a unique political venture. We prize our political processes, our institutions, our shared heritage, and our freedom. To preserve and protect this prized inheritance, we call upon our schools to fully prepare our youngsters as *good citizens*—individuals possessing the knowledge, skills, and attitudes required to participate in and maintain our democratic nation. We want our future citizens to respect our past as a democratic society and recognize the rich contributions of all groups who have made modern America a free and powerful nation. We want students to take active roles as champions of freedom who stand up and cry out for the rights and responsibilities of citizenship. We want students to speak with their voices, actions, and votes for the improvement of the quality of life in our families, neighborhoods, nation, and world. The Curriculum Standards Task Force of the National Council for the Social Studies (1994) recommends that the overriding purpose of social studies should be to provide students with the education required to guard and protect their cultural inheritance, to ready students for the most important post they will be entrusted with in the new millennium—Jefferson's concept of *the office of citizen:*

> Citizens who take this office seriously are in touch with the cultural heritage of the nation. They possess knowledge of the economic, political, and so-

Helping clean up a local park offers these students an opportunity to engage in participatory citizenship.

cial factors that make up the human ecosystem in which all must function, and they understand its relationship to natural systems. They understand the principles of rule of law, legal limits to freedom, and majority rule with protection for minority rights. They have informed spatial, temporal, and cultural perspectives. They possess the attitudes and behaviors that support fair play and cooperation. Without a conscious effort to teach these ideals, a free republic will not long endure. (p. xx)

The preservation of our democracy depends upon a system of education that prepares its young people to accept responsibility in society, but today's social commentators say we are failing that test. They sadly lament the growing apathy of today's adult citizens, describing our nation as being comprised of "civic couch potatoes" (Tyack, 1997). Submerged in public apathy, they reject such basic citizenship responsibilities as volunteering in civic or public arenas (PTA, town council, organizations that feed and clothe the poor); helping solve such real problems as eliminating drug abuse and homelessness, writing letters to the editor, deliberating and debating public policy, working for public interest groups, and voting. *The fundamental purpose of social studies education is to* **prepare future citizens** *to be trustees of our democratic community.* As teachers of future citizens, how can we help young adults resist social indifference and accept full responsibility as a stakeholder in our democratic nation?

Preparing students to participate in the political, social, and economic affairs of our democracy and to live together as good citizens in a drastically different world of the future requires a dynamic social studies program with citizenship education as its heart. A program centering on participatory citizenship must be made a regular part of school life to help students establish the sense of personal responsibility and accountability that forms a solid foundation to participate as adults in public life. Teachers must provide opportunities for students to go beyond the simple acquisition of facts; they must offer situations where learners use what they know to reason, to analyze, to create, and to evaluate. The NCSS Task Force on Early Childhood/ Elementary Social Studies (1989) suggests:

> For children to develop citizenship skills appropriate to a democracy, they must be capable of thinking critically about complex societal problems and global problems. Teachers must arrange the classroom environment to promote data gathering, discussion, and critical reasoning by students. Another important aspect of citizenship is that of decision maker. Children must acquire the skills of decision-making, but also study the process that occurs as groups make decisions. . . . Children need to be equipped with the skills to cope with change. (p. 16)

A significant responsibility of social studies teachers, then, is to offer a sound program with proper balance. Several professional organizations have issued position statements describing precisely how that might be accomplished, but the "Essentials of Exemplary Social Studies Programs," formulated by the National Council for the

FIGURE 1–3

Essentials of Exemplary Social Studies Programs

Knowledge

Students need knowledge of the world at large and the world at hand, the world of individuals and the world of institutions, the world past, the world present and future. . . . From this knowledge base, exemplary programs teach skills, concepts, and generalizations that can help students understand the sweep of human affairs and ways of managing conflict consistent with democratic procedures.

Democratic Beliefs

Fundamental beliefs drawn from the Declaration of Independence and the United States Constitution with its Bill of Rights form the basic principles of our democratic constitutional order. . . . These democratic beliefs depend upon such practices as due process, equal protection, and civic participation, and are rooted in the concepts of justice, equality, responsibility, freedom, diversity, and privacy.

Thinking Skills

It is important that students connect knowledge with beliefs and actions. To do that, thinking skills can be developed through constant systematic practice throughout the years of formal schooling.

In addition to strengthening reading and computation, there is a wide variety of thinking skills essential to the social studies which can be grouped into four major categories: [*Data-Gathering Skills, Intellectual Skills, Decision-Making Skills,* and *Interpersonal Skills.*]

Participation Skills

As a civic participant, the individual uses the knowledge, beliefs, and skills learned in the school, the social studies classroom, the community, and the family as the basis for action.

Source: National Council for the Social Studies. (1981). Essentials of the social studies. *Social Education, 45* (3), 163–164.

Social Studies (1981), appears to present as clearly as any other source the major goals of social studies instruction today. These goals are described in Figure 1–3. The rest of this text will be directed toward helping you acquire skills to carry out each of the four major areas of teaching responsibility: *knowledge, democratic beliefs, thinking skills,* and *participation skills.*

What Are the Major Components of Dynamic Social Studies Instruction?

In a dynamic social studies program, our major mission is to help advance the four major goals of social studies education by encouraging students to function as

FIGURE 1–4
Dynamic Social Studies Model

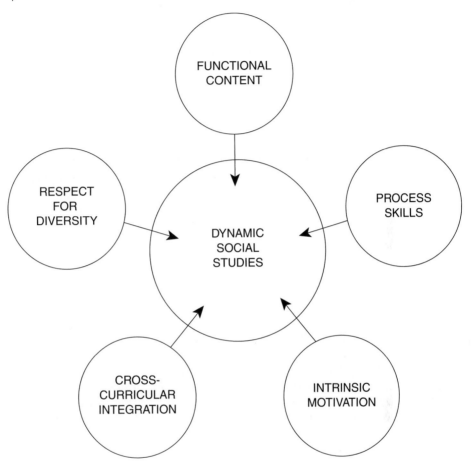

young "social scientists." What is the teacher's role in that process? The model that will play a major role in describing the teacher's role as well as in shaping the remainder of this text is shown in Figure 1–4. It is simple, but the model brings together in a clear way the major components of dynamic social studies instruction. The model has five components, each of which must be in place for effective learning to take place: *functional content, process skills, intrinsic motivation, cross-curricular integration,* and *respect for diversity.*

Functional Content

Social scientists must be knowledgeable, for social science investigations do not spring forth in a vacuum. Social scientists launch their investigations from a complete and organized knowledge base related to the problems they want to solve. It is when they encounter a new phenomenon that cannot be explained in terms of their organized knowledge base ("trusty framework") that social scientists become energized to seek new information that might alter or replace their "trusty framework." Think about how this works: In the case of Heinrich Schliemann, the inconsistencies in his knowledge base brought on by conflicting accounts of the Trojan War were the basis for his questions and subsequent research. Likewise, the unanticipated discovery of what was going on with Maryland's state insect fueled Mrs. Holzwarth's students' strong interest in learning how to propose one for Pennsylvania. In neither case were individuals satisfied with simply knowing something at face value. Content knowledge in dynamic social studies programs is not treated merely as received knowledge to be accepted and memorized, but as a means through which questions or problems are explored and confronted. If students are satisfied to merely store the information and retrieve it for others they could not be considered "social scientists." If children are to become explorers and questioners rather than passive acceptors of information, they must know enough about something to raise questions, discover something new, or solve a problem. Therefore, dynamic social studies instruction does not stop with the "what" of the social sciences, but continues on with the "how" of the social science disciplines. Students who are learning about geography, for example, must know not only the facts about geography, but also experience how geography works and what geographers do.

The hardest lesson for beginning teachers, then, is to learn that telling is not teaching; told is not taught. Those who consider teaching as information transmission ignore the paramount role of the learner in learning; students will not learn content if they cannot attach meaning to it. Therefore, if children are going to develop the skills appropriate for active citizenship in a democracy, they must be capable of thinking about complex social problems in a classroom environment that promotes decision-making and problem solving. Thinking and content are clearly inseparable in dynamic social studies programs.

The most basic responsibility of dynamic social studies teachers, then, is to select "functional content," or a meaningful knowledge core that furnishes the information necessary to become aware of, identify, and solve problems. This component can be thought of as the background of subject area content that spurs the questioning mind and allows one to solve problems or do original work. A firm knowledge background spurs questions, and these questions, in turn, provide the context for obtaining more knowledge.

Curriculum Scope and Sequence

Think for a moment about all the social studies knowledge you have learned over the years. Some, like the name of the 17th president of the United States or the major exports of Bolivia, may have been long forgotten, but there is probably more that you remember to this day and feel every child should know, too. What five specific facts or understandings do you feel are indispensable for today's youth? Write them down. Compare your list with your classmates. Should children learn about important *people* like Benjamin Franklin, Confucius, or Sarah Tubman; or important *places* like the Gobi Desert, the Fertile Crescent, or the rainforests of the Amazon? What about important *things* like pueblos, railroads, or the Great Wall of China; or *events* such as the Battle of Bull Run, the rise of Christianity, or the discovery of Mohenjo-Daro? Did anyone list significant *ideas* such as Henry Ford's assembly line, Hinduism, or the Bill of Rights?

When we ask the question, "*What* content from the social science disciplines should be selected for our social studies program?" we are dealing with the *scope* of the program, or what will be taught. This is but one critical question when considering the makeup of the social studies curriculum. The second important question is, "*When* is the most developmentally appropriate time to teach the selected content?" Answers to this question create the *sequence*, or order, in which the content is treated.

There are a number of possible ways to arrange the scope and sequence but, in elementary school social studies, these two important elements of curriculum design have been traditionally organized according to the *expanding environment approach.* No one knows exactly how this system came into existence, but beginning in the 1930s, Lucy Sprague Mitchell (1934) started the ball rolling by proposing that children's knowledge of their world developed through stages, beginning with the awareness of what is near at hand to that which is far away—beginning with the *here* and *now* of their lives and gradually expanding their environments. Mitchell described growth in understanding one's social environment much like the ripples that radiate out after throwing a pebble into a pond. The first ripple can be considered knowledge gained about one's body and the physical characteristics of the immediate surroundings (crib, play area, room, house); the next ripple includes knowledge of the home, family, friends, relatives, and other significant people in the environment; the next focuses on the classroom, school, and other familiar environments such as the street, neighborhood and the homes of friends and relatives; and, as the child matures, constant investigations open up new environments, so succeeding ripples spread out to the community, the state, the United States, the hemisphere, and so on. Mitchell suggested that social studies content should be organized with respect to this natural developmental progression—begin with the self, home, and school in grade 1, then widen to the neighborhood in grade 2, to the community in grade 3, and systematically outward to the state in grade 4, United States in grade 5, and the Western hemisphere or world in grade 6. In the 1930s, when the new social studies curricula were becoming firmly entrenched in our nation's school districts, the overwhelming pattern of content selection became the expanding environment approach. Likewise, when textbook companies began to publish their first social studies textbooks, the overwhelming scope and sequence pattern was the expanding environment approach. Paul Hanna (1963) reaffirmed this concept of expanding environments in an article so influential that his name is most closely associated with the expanding environment approach today.

In 1983, the NCSS organized a Task Force on Scope and Sequence to study whether the expanding environment approach remained suitable for contemporary times. The task force reaffirmed the merit of the expanding environment approach (shown in Figure 1– 5), and it remains the overwhelming favorite among social studies curriculum developers and textbook publishers today.

A dynamic social studies program, then, must provide for consistent and cumulative learning from kindergarten through the middle school years. Content must be extended outward, away from a self-centric focus, and, at each grade level, students should build upon knowledge and skills already learned while receiving preparation for that which is yet to come.

Although the expanding environment approach withstood serious challenges during the early 1980s, considerable controversy continued to surround the question of what specific core of essential knowledge should be incorporated into the instructional program at each grade level. This controversy was not confined to social studies, however, as education in general was placed under the microscope in response to the publication of reports like *A Nation at Risk* (National Commission on Excellence in Education, 1983). *A Nation at Risk* fired volleys of criticism at our nation's schools with searing commentaries such as, "If an unfriendly foreign power had attempted to impose on America the mediocre instructional performance that exists today, we might well have viewed it as an act of war" (p. 9). For social studies in particular, standardized test scores of students from the United States were compared with students from other industrialized nations. These reports indicated that America's students did not possess the knowledge required to make informed decisions about issues that affect them, their families, and their communities. One report specifically dealing with social studies was titled *What Do Our 17-Year-Olds Know?* The co-authors, Diane Ravitch and Chester E. Finn, Jr. (1987), reported the results of the first nationwide academic assessment of American 17-year-olds. In history, the national average for correct answers to basic history questions was 54.5 percent. The authors pointed out that if we approach this percentage from the commonly accepted view that 60 percent is the dividing line between passing and failing, our American students are in serious trouble. (More recently, the Colonial Williamsburg Foundation commissioned a basic history test to assess the knowledge of our nation's teens. Some surprising results showed that 14 percent of America's teens identified Abraham Lincoln as our country's first president. The same percentage said that our country celebrates its independence from France each July 4, 11 percent named John Adams, our second president (succeeding Abe Lincoln?), as the composer of *The Star-Spangled Banner,* while 9 percent believed it was Betsy Ross!) In response to these criticisms, some argued that it was the responsibility of social studies programs to move "back to the basics" and teach essential elements of lasting knowledge children need to do well on achievement tests.

FIGURE 1–5
Recommendations of the Task Force on Scope and Sequence

Kindergarten—Awareness of Self in a Social Setting

Providing socialization experiences that help children bridge their home life with the group life of school.

Grade 1—The Individual in Primary Social Groups: Understanding School and Family Life

Continuing the socialization process begun in kindergarten, but extending to studies of families (variations in the ways families live, the need for rules and laws).

Grade 2—Meeting Basic Needs in Nearby Social Groups: The Neighborhood

Studying social functions such as education, production, consumption, communication, and transportation in a neighborhood setting.

Grade 3—Sharing Earth-Space With Others: The Community

Focusing on the community in a global setting, stressing social functions such as production, transportation, communication, distribution, and government.

Grade 4—Human Life in Varied Environments: The Region

Emphasizing the region, an area of the earth defined for a specific reason; the home state is studied as a political region where state regulations require it.

Grade 5—People of the Americas: The United States and its Close Neighbors

Centering on the development of the United States as a nation in the Western Hemisphere, with particular emphasis on developing affective attachments to the principles on which the nation was founded; Canada and Mexico also studied.

Grade 6—People and Cultures: The Eastern Hemisphere

Focusing on selected people and cultures of the Eastern Hemisphere, directed toward an understanding and appreciation of other people through development of such concepts as language, technology, institutions, and belief systems.

Grade 7—A Changing World of Many Nations: A Global View

Providing an opportunity to broaden the concept of humanity within a global context; focus is on the world as the home of many different people who strive to deal with the forces that shape their lives.

Grade 8—Building a Strong and Free Nation: The United States

Studying the "epic of America," the development of the United States as a strong and free nation; emphasis is on social history and economic development, including cultural and aesthetic dimensions of the American experience.

Source: Task Force on Scope and Sequence. (1984). In search for a scope and sequence for social studies. *Social Education, 48* (4), 376–385.

National Curriculum Standards

In 1990, reacting to this considerable educational faultfinding, but largely unsubstantial proposals for improvement, President Bush brought our nation's governors together for the first time in history to discuss national educational policy and to determine what might be needed to improve our schools. Their discussions were summarized in the well-known *America 2000*—six national goals for public education. Goal 3, the goal most specifically related to social studies education, stated, "By the year 2000, American students will leave grades 4, 8, and 12 having demonstrated competency over challenging subject matter including English, mathematics, science, history, and geography; and every school in America will ensure that all students use their minds well, so they may be prepared for responsible citizenship, further learning, and productive employment in our modern society" (U.S. Department of Education, 1991, pp. 5–6). A great deal of attention was given to *America 2000,* prompting our nation's leading professional organizations to develop instructional standards (statements of what all students should know and be able to do at the completion of their education) in various subject areas for their fields (the arts, civics and government, economics, English, foreign language, geography, history, mathematics, physical education, science, and vocational education). The NCSS (1994) observed these developments with keen interest, and then expressed its distress: "Congress . . . passed . . . the *Goals 2000: Educate America Act,* codifying educational goals and sanctioning the development of national educational standards as a means of encouraging and evaluating student achievement. While that act included the disciplines named above, *it omitted social studies*" (emphasis added) (p. viii). Although bringing up the rear of the standards bandwagon, in 1993 the NCSS successfully argued for the annexation of social studies to the list of professional groups developing national standards and named a task force to define what students should know and when they should know it. The central questions guiding the NCSS task force were: "What will students be taught?" "How will students be taught?" "How will student achievement be evaluated?" Using the questions as a guide, the task force worked for over one year before publishing *Curriculum Standards for Social Studies: Expectations of Excellence* (NCSS, 1994).

The social studies curriculum standards are expressed as 10 thematic statements that begin: *Social studies programs should include experiences that provide for the study of . . . .* For example, the first thematic strand reads, "Social studies programs should include experiences that provide for the study of culture and cultural diversity." When presented in list form, this thematic strand is preceded by a roman numeral and shortened thusly, I Culture. Similarly, each of the 10 themes that serve as organizing strands for the social studies curriculum are:

    I Culture
   II Time, Continuity, and Change
  III People, Places, and Environments

IV Individual Development and Identity
  V Individuals, Groups, and Institutions
 VI Power, Authority, and Governance
 VII Production, Distribution, and Consumption
VIII Science, Technology, and Society
 IX Global Connections
  X Civic Ideals and Practices

Each of these thematic strands is defined and explained in a separate chapter of the standards document (see Figure 1–6 for a condensed version). Then, in separate chapters for each school level (early grades, middle grades, and high school), performance expectations and two or three examples of classroom activities are put forward. For example, the performance expectations for the first theme, "Culture," are as follows (NCSS, 1994):

> Social studies programs should include experiences that provide for the study of *culture and cultural diversity,* so that the learner can:
>
> a. explore and describe similarities and differences in the ways groups, societies, and cultures address similar human needs and concerns;
> b. give examples of how experiences may be interpreted differently by people from diverse cultural perspectives and frames of reference;
> c. describe ways in which language, stories, folktales, music, and artistic creations serve as expressions of culture and influence behavior of people living in a particular culture;
> d. compare ways in which people from different cultures think about and deal with their physical environment and social conditions; and
> e. give examples and describe the importance of cultural unity and diversity within and across groups (p. xiii).

Following a presentation of each standard and associated performance expectations, the chapters describe classroom activities to illustrate how the standards can be applied. For example, to meet performance expectations a, b, and d, the experiences of Carlene Jackson are recounted. Before the first day of school, Jackson examined her class list and inferred from the children's surnames that her class was a rich mix of cultural backgrounds—Mexican, Vietnamese, Korean, African American, and European American. By the end of the first month of school, Jackson and her students decided to study and compare how families meet their basic needs of food, clothing, and shelter in five places: their community; Juarez, Mexico; Hanoi, Vietnam; Lagos, Nigeria; and Frankfurt, Germany. Throughout the unit of study, Jackson and her students read books, looked at photos and slides, watched videos, and talked to speakers from their designated cities. The students honed their reading, writing, speaking, and map-reading skills. They created a chart summarizing the data they collected. You can request a copy of the Curriculum Standards for Social Studies (Bulletin 89) by

FIGURE 1–6
Curriculum Standards for Social Studies

Ten Thematic Strands

Ten themes serves as organizing strands for the social studies curriculum at every school level (early ,middle, and high school); they are interrelated and draw from all of the social science disciplines and other related disciplines and fields of scholarly study to build a framework for social studies curriculum design.

I Culture

Human beings create, learn, and adapt culture. Human cultures are dynamic systems of beliefs, values, and traditions that exhibit both commonalities and differences. Understanding culture helps us understand ourselves and others.

II Time, Continuity, and Change

Human beings seek to understand their historic roots and to locate themselves in time. Such understanding involves knowing what things were like in the past and how things change and develop—allowing us to develop historic perspective and answer important questions about our current condition.

III People, Places, and Environment

Technological advancements have ensured that students are aware of the world beyond their personal locations. As students study content related to this theme, they create their spatial views and geographic perspectives of the world; social, cultural, economic, and civic demands mean that students will need such knowledge, skills, and understandings to make informed and critical decisions about the relationship between human beings and their environment.

IV Individual Development and Identity

Personal identity is shaped by one's culture, by groups, and by institutional influences. Examination of various forms of human behavior enhances understanding of the relationships between social norms and emerging personal identities, the social processes which influence identity formation, and the ethical principles underlying individual action.

V Individuals, Groups, and Institutions

Institutions exert enormous influence over us. Institutions are organizational embodiments to further the core social values of those who comprise them. It is important for students to know how institutions are formed, what controls and influences them, how they control and influence individuals and culture, and how institutions can be maintained or changed.

VI Power, Authority, and Governance

Understanding of the historic development of structures of power, authority, and governance and their evolving functions in contemporary society is essential for the emergence of civic competence.

VII Production, Distribution, and Consumption

Decisions about exchange, trade, and economic policy and well-being are global in scope and the role of government in policy making varies over time and from place to place. The systematic study of an interdependent world economy and the role of technology in economic decision making is essential.

VIII Science, Technology, and Society

Technology is as old as the first crude tool invented by prehistoric humans, and modern life as we know it would be impossible without technology and the science which supports it. Today's technology forms the basis for some of our most difficult social choices.

IX Global Connections

The realities of global interdependence require understanding of the increasingly important and diverse global connections among world societies before there can be analysis leading to the development of possible solutions to persisting and emerging global issues.

X Civic Ideals and Practices

All people have a stake in examining civic ideals and practices across time, in diverse societies, as well as in determining how to close the gap between present practices and the ideals upon which our democracy is based. An understanding of civic ideals and practices of citizenship is critical to full participation in society.

Source: Nickell, P. (1995). Pullout feature: Thematically organized social studies. *Social Studies & The Young Learner, 8,* 1–8. © National Council for the Social Studies. Reprinted by permission.

writing or phoning the National Council for the Social Studies or visiting the NCSS Website:

National Council for the Social Studies
8555 Sixteenth Street
Suite 500
Silver Spring, MD 20910
(301) 588-1800
http://www.ncss.org/links/home.html

As with most breaks from precedent, the standards movement has elicited both arguments of support as well as blistering criticisms. In speaking of the benefits of standards, Resnick says "standards and assessments [will] help bring about better student outcomes—a different quality and higher level of student achievement" (O'Neil, 1993, p. 17). Standards become "images of excellence" to their defenders.

Ochoa-Becker (2001) recognized the challenges associated with designing curriculum standards for the social studies, but criticized them with the following concerns:

1. The NCSS Standards minimize and all but ignore the importance of controversial issues for citizens of democracy.
2. The NCSS Standards give insufficient attention to important intellectual processes such as critical thinking, data analysis, evaluation of evidence, consideration of values, and decision making that are vital if individuals are to participate effectively in a democratic society.
3. The NCSS Standards give insufficient attention to citizen participation, social action, and service learning. Without such abilities, citizens of a democracy are rendered mute and self-governance is seriously compromised (p. 165).

As a final point, it must be emphasized that social studies standards are not federal mandates for what should be taught in our nation's schools, nor are they an attempt to establish a national social studies curriculum. Instead, they are simply brought forward in an attempt to offer a professionally sanctioned, systematic vision of what all students should know and be able to do as a result of their schooling.

Process Skills

Content knowledge is but one component of dynamic social studies—a necessary component, but not sufficient in itself. Knowing how the social sciences work is the "something extra" of dynamic social studies instruction. Historians, for example, must not only understand the historical information uncovered by others, but also know where the information comes from, what kinds of questions to ask about the information, and how to investigate their questions. Bruner (1960) advised:

To instruct someone in [a] discipline is not a matter of getting him to commit results to mind. Rather, it is to teach him to participate in the process that makes possible the establishment of knowledge. We teach a subject not to produce little living libraries on the subject, but rather to get a student to think . . . for himself, to consider matters as a historian does, to take part in the process of knowledge-getting. Knowledge is a process, not a product. (p. 72)

Without consideration of the processes as well as the content of the disciplines, students are left with the impression that anything they are told or read about in their textbooks is true, a sad commentary for a subject area responsible for preparing students to assume the office of citizen in the United States.

The Scientific Method of Inquiry

Because the fundamental nature of their job is asking questions about people, things, places, and ideas, intellectual curiosity and creativity are indispensable traits of social scientists. They must be able to think logically and methodologically; objectivity, fairness, open-mindedness, perseverance, and systematic work habits are important in all kinds of social science research. Sometimes, scientists use what is called the *scientific method* of inquiry. You might recall this as a process of investigation that includes such steps as generating a hypothesis, collecting data, or forming a conclusion. You certainly learned the scientific method in elementary school and high school, probably coming away with the notion that social scientists never face a problem until they put on their horn-rimmed glasses, sharpen their pencils, pull out a checklist, and mark off each step as it is completed. However, in the real world, social science problems aren't all solved that way. Sometimes social scientists follow the scientific method deliberately, but often they work in a rather haphazard way, too, grubbing around and searching for answers through trial-and-error strategies. In actuality, social scientists are normally somewhere between those two extremes. Does this appear to be the way children approach problems and strive to uncover new knowledge, too? It is. Children, by their very nature, are curious about their world—just like social scientists.

Regardless of how rigidly we adhere to the scientific process in the elementary school classroom, it is what sets the social sciences apart from all other ways of knowing. Therefore, if we are to facilitate dynamic social studies learning, we must teach the strategies, habits, abilities, and attitudes that are typically thought of as the scientific process (skills for gathering, organizing, and analyzing information). Although these will be more fully discussed in Chapter 6, inquiry episodes usually contain the following features:

1. *Start with a question or problem.* It may be one the teacher suggests or the students themselves ask.
2. *Make hypotheses.* Students make a reasonable statement (or inference) about the question that is based on past experience. (If an inference is stated in such a way that allows it to be tested, it is called a *hypothesis.*)
3. *Collect data.* This simply means gathering information to answer the question. Each discipline has its own data collection procedures (review the so-

cial science descriptions), so there are many different ways information may be collected.

4. *Interpret the data.* Here the teacher guides the students to record their data, think about it, and make it public. Usually, this is done with such techniques as discussion, graphs, illustrations, written reports, or creative skits.

5. *Form a conclusion.* Closure occurs with the inquiry process when students generalize about the results on the basis of the data.

Although social scientists differ in their interpretation of the exact steps of the inquiry process, educators starting with John Dewey in the 1930s until today have advocated the investigation of real problems. But the steps previously highlighted are included in virtually every list. The important thing to remember, however, is that teaching dynamic social studies entails a community of learners in a classroom who are provided with meaningful content and invited to use processes that allow them to investigate the social science disciplines.

Intrinsic Motivation

The third component of the dynamic social studies model is intrinsic motivation, or an internal impulse that provokes us to action or keeps us absorbed in certain activities. For example, in recent months I have learned how to play fantasy sports on the Internet. I love drafting teams, trading players, and fighting for league championships (yes, I've won a few). I participate in fantasy sports not because someone has told me I must, but because it brings me pleasure. Certainly, the T-shirts that come with league championships are a special reward for doing well, but I enjoy the competition whether or not it results in a championship or a T-shirt. To me, the activity is pleasurable and worthwhile in and of itself. What activities do you engage in on a regular basis simply because you enjoy doing them? Intrinsic motivation is what leads you to action and keeps you engaged in those activities.

There are some activities I engage in that are not so enjoyable, but I do them anyway because they bring me things I do enjoy. Washing my car is an example. It's not something I would freely choose to do during my free time because I don't particularly like doing it. However, my family and I do appreciate a clean, shiny automobile. The motivation to wash my car is the same as it is for mowing my lawn or shoveling the snow from my driveway. It is called extrinsic motivation because the drive to do these chores does not come from within; the activities themselves are not enjoyable, but the rewards are—a shiny car, a clean yard, and a clear driveway. Likewise, a child who searches a number of trade books to find out how it might have felt to leave home at the age of 12 to work as an apprentice in Colonial America because he or she is genuinely interested in finding out is intrinsically motivated, while the child who researches the topic just to get a good grade is extrinsically motivated. The interest, enjoyment, and satisfaction is in the work itself when one is intrinsically motivated; when one is extrinsically motivated, the driving force is outside pressure or reward.

Dynamic social studies reveals how to move children from a passive to an active role in learning.

No activity in itself, however, is intrinsically motivating. Fantasy sports are important to me, but I'm sure there are some of you who would rather watch paint dry than play fantasy sports. The same is true for any other activity; it can be motivating only to a particular person at a particular time. What are some of the factors influencing whether we enjoy doing something, or whether students will become intrinsically motivated to learn? Following are several suggestions that can guide you in your efforts to promote student motivation in dynamic social studies programs (Amabile, 1989; Ormrod, 1999):

- *Interest.* Students are more motivated by something that has captured their interest than by something with no perceived value.
- *Pleasure.* Students become more involved when they are enthusiastic and excited about the subject matter or classroom activities.

- *Competence.* Students will seek out activities and persist longer at them if they feel they are capable of accomplishing tasks successfully and when they feel in control of their lives. The more confident children feel, the more likely they are to begin a task and see it through to completion.
- *Self-determination.* Students will become more highly intrinsically motivated when they are working on something for their own reasons, not someone else's.

However, having children work for an expected reward appears to make students less likely to take learning risks or approach a task with an experimental or pleasurable attitude. In addition, setting up competitive situations where students lock horns with one another for some desirable reward or other form of recognition tends to undermine intrinsic drive. Lastly, although classroom assessment can serve as an intrinsic motivator when it is perceived as a valid measure of course objectives, all too often the focus is turned away from the intrinsically enjoyable aspects of learning. As a result, students often become overly tense and suffer test anxiety.

To summarize, students are more likely to exhibit high intrinsic motivation when they find the subject matter they are studying to be interesting, when they like what they're doing, when they feel they are capable of accomplishing a task, and when they believe they have some control over the learning situation. Ormrod (1999) suggests, however, that it is important to note that students' intrinsic motivation for social studies learning emerges slowly over a period of time, especially if students have previously been accustomed to receiving extrinsic reinforcement for their efforts. In these situations, she advises, you might want to slowly increase their focus on the intrinsic rewards of learning while gradually weaning your students from an overdose of external reinforcement.

Cross-Curricular Integration

In an effort to make their students' education more authentic and worthwhile, dynamic social studies teachers challenge the traditional practice of teaching each elementary school subject separately; in other words, isolating reading, spelling, writing, math, science, and social studies into separate blocks of time with very little connection to one another. Dynamic social studies teachers have become particularly critical of this arrangement since the limited number of hours in the school day prevents all subjects from receiving equal treatment. In fact, most teachers spend the bulk of their day teaching the basic skills subjects such as reading, writing, spelling, and math. The content subjects (science and social studies) have traditionally been the disciplines most likely to be shoved to the end of the day and, if time runs out, postponed or forgotten about altogether. If social studies is somehow squeezed into the end of the day, the children are often led through a quick oral reading of a textbook section and a brief question-answer recitation period, "just to get it in." To remedy these lopsided practices, dynamic social studies teachers are seeking ways to connect, or integrate, all the subjects commonly taught in the elementary school curriculum. By bringing together the various subject areas and relating the content to a central theme, teachers are not only able to effectively and efficiently use the allotted amount of time during a busy school day, they are also able to create interesting and challenging learning opportunities. To that end, dynamic social studies brings together, or integrates, the various areas of the school curriculum into a wholly unified program where learning experiences cut across *all* subjects. Dynamic social studies provides the obvious connection between the arts, humanities, and the physical sciences. For example, consider all the possible cross-curricular experiences possible on a trip to the dairy farm:

> Read *The Milk Makers* by Gail Gibbons (New York: Simon & Schuster, 1985), *Milk from Cow to Carton* by Aliki (New York: HarperCollins, 1992), *What a Wonderful Day to Be a Cow* by Carolyn Lesser (New York: Knopf, 1995) and *Cows in the Parlor: A Visit to a Dairy Farm* by Cynthia McFarland (Colchester, CT: Atheneum, 1990). (Literature)
>
> Talk about the trip before, during, and after taking it. (Language Arts)
>
> Sing *Old MacDonald Had a Farm* and *The Farmer in the Dell.* (Music)
>
> Establish rules for the trip. (Political Science)
>
> Talk about appropriate group behavior. (Sociology)
>
> Figure out how much the trip will cost. (Math/Economics)
>
> Illustrate the two types of cows seen on the farm—Holsteins and Jerseys. (Art)
>
> Construct an experience story of the field trip. (Reading/Writing)
>
> Construct an informational chart about cows with illustrations and appropriate captions—"Cows are milked twice a day," "Cows have four stomachs," and so on. (Science)
>
> Observe daily activity on the dairy farm. (Geography)
>
> Arrange clothes and props for a dairy farmer in a dramatic play area. (Creative Dramatics)
>
> Make cottage cheese and/or butter. (Science/Health/Nutrition)
>
> Print the caption "Foods Made From Milk" across the top of a large sheet of paper. Have the children display appropriate foods by cutting pictures from magazines or drawing their own. (Science)
>
> Collect different types of food containers. Place all in a large box. The children sort the containers representing dairy products from the other types of food containers. (Math)

All of these concepts, skills, and strategies (and more) emerged within the context of a trip to the dairy farm, yet the whole experience was labeled social studies. That is because social studies is usually a central element in the development of cross-curricular learning experiences and serves as the "core" to pull everything together.

Elementary schools around the country are moving toward the idea of integrating the curriculum, and dynamic social studies appears to be at the center. Berg (1988) answers the question, "Where does social studies fit into cross-curricular integration?" with this perceptive response:

> Right in the middle! A major goal of the social studies is to help students understand the myriad interactions of people on this planet—past, present, and future. Making sense of the world requires using skills that allow one to read about the many people and places that are scattered about the globe; to use literature to understand the richness of past events and the people who are a part of them; to apply math concepts to more fully understand how numbers have enabled people to numerically manage the complexity of their world. The story of humankind well told requires drawing from all areas of the curriculum. (unnumbered pull-out feature)

The NCSS (1993a) has supported the integrative aspect of social studies instruction with this statement from its influential position statement, *A Vision of Powerful Teaching and Learning in the Social Studies:*

> . . . *social studies teaching integrates across the curriculum.* It provides opportunities for students to read and study text materials, appreciate art and literature, communicate orally and in writing, observe and take measurements, develop and display data, and in various other ways to conduct inquiry and synthesize findings using knowledge and skills taught in all school subjects. . . . Particularly in elementary and middle schools, instruction can feature social studies as the core around which the rest of the curriculum is built. (p. 217)

By integrating subject areas with themes having a social studies focus, students become involved in activities and experiences that are both purposeful and meaningful.

Respect for Diversity

Have you ever been involved in a situation where you had more difficulty doing something than anyone else? Dancing? Playing a musical instrument? Dribbling a basketball? Speaking a foreign language? Ice skating? Maybe you were the only one who had problems. How did you feel? How would you have felt if every day when you came to school you were the only one who couldn't perform certain tasks that could be completed by others with relative ease? What kind of "treatment" or "special help" would you need to keep you coming to school each day?

Now consider if you have ever been involved in a situation where you were able to do something more easily and quickly than anyone else. How did you feel then? What could be done in school to keep you from becoming miserably bored? What could be done to best help you work toward fulfilling your unlimited potential?

Like any of us who have been in these situations, all children come to school with varied strengths and limitations and form unique collections of distinctive talents and abilities. Some children stand out because they are exceptional performers; others face certain challenges that require special services to help them reach their potential. Teachers achieve quality in the dynamic social studies program when they deliver the best for each youngster and make the most of their time with everyone in their classrooms. Effective teachers adapt instruction to meet the special needs, talents, and interests of all their students; the quality of their dynamic social studies programs is distinguished by a keen awareness and consideration of each youngster as a distinct individual, *including those whose backgrounds or exhibited needs are not shared by most others.* These children may exhibit specific developmental disabilities, speak a home language that does not match the school's, come from diverse cultural or ethnic backgrounds, or possess unique gifts and talents; whatever the circumstance, the field of elementary school social studies education should be consistently responsive by offering appropriate experiences to fully develop their native capabilities.

At the heart of this topic is a concern about equity and fair treatment for groups that have traditionally experienced discrimination because of race or ethnicity, language, gender, or exceptionality. Because of the wide range of diversity in contemporary society, all teachers must become instructionally effective with diverse groups of students. There are at least three arguments why this charge is especially meaningful for social studies teachers (Winitzky, 1991):

> The first is that these issues should be of major concern to every citizen, that it is incumbent upon us as citizens to work toward the public good by trying to ameliorate these problems. . . . The second argument is that Americans have a strong belief in the power of education as the route to later success in life—economically, politically, and culturally. . . . Finally, many believe that we really have no choice. We simply live in a [diverse] world, and our schools should reflect that aspect of modern life. (pp. 126–127)

Children in today's schools come from an enormous range of backgrounds, languages, and abilities. To meet their educational needs, dynamic social studies teachers must be able to work effectively with *all* students.

AFTERWORD

At the roots of a democracy are knowledgeable and thoughtful citizens. Of course, they have many other qualities, too, but high on the list of behavior for democratic citizens is *thinking for themselves.* Democracy requires individuals who are able to search for and examine the facts whenever they must make up their minds about important issues. These issues might relate to one's personal life or to complex international concerns; regardless, the protection of our freedoms lies in the hands of rational people. Such skills must be learned during the early years with a dynamic social studies curriculum that offers meaningful experiences to all. All youngsters must find something to excite their interest and stimulate their thinking. A one-dimensional approach to social studies instruction cannot do this. We fail our children with our narrowness; if our myopic view of teaching has caused them to feel stupid or to be bored, we have lost. Learning for an informed citizenry is too important to be thought of as something that everyone must be able to do in any single way. The danger to our future is great when we restrict the adventuresome, "can-do" spirit of childhood. Therefore, dynamic social studies programs must be those that employ various teaching strategies and promote functional thinking skills. The probing, wondering mind of childhood must be freed. Our society of tomorrow starts in your classroom today.

The complexities of the 21st century may dictate a quality and quantity of education far different than we can currently imagine. You must take your emerging view of what social studies is and constantly search for ideas that help construct new roles for teachers in a new century. In addition to what we have considered up to this point, one of the most helpful ways of keeping up with current trends is to review the activities and publications of professional organizations.

The largest and most influential professional organization for social studies educators is the National Council for the Social Studies (NCSS). The council publishes several publications of interest to social studies teachers: *Social Education,* the primary journal, focuses on philosophical, theoretical, and practical classroom-application articles involved with K–12 instruction; *Social Studies and the Young Learner,* a separate journal for elementary teachers, offers articles primarily concerned with teaching strategies. The NCSS also periodically publishes "how-to" pamphlets that offer in-depth suggestions for implementing specific instructional responsibilities (such as using creative dramatics or current affairs strategies) in the social studies classroom. You should also become familiar with *The Social Studies,* a journal not associated with any particular professional organization. It deals with classroom practices on the K–12 level and contains a wealth of articles describing ideas for classroom use and stimulating thought on philosophical issues. The relevant addresses follow.

National Council for the Social Studies
8555 Sixteenth Street
Silver Spring, MD 20910
(301)588-1800

The Social Studies
1319 Eighteenth Street, NW
Washington, DC 20036

References

Amabile, T. M. (1989). *Growing up creative.* New York: Crown.

Berg, M. (1988). Integrating ideas for social studies. *Social Studies and the Young Learner, 1,* unnumbered pull-out feature.

Bruner, J. (1960). *The process of education.* New York: Vintage.

Channing, E. & Hart, A. B. (1903). *Guide to the study of American history.* Boston: Ginn and Company, 1.

Children's Defense Fund. (1996). *The state of America's children: Yearbook 1996.* Washington, DC: Author.

Clark, D. C., & Cutler, B. C. (1990). *Teaching.* New York: Harcourt Brace Jovanovich.

Dewey, J. (1933). *How we think.* Boston: D.C. Heath & Company.

Finn, C. E. (1988). The social studies debacle among the educationaloids. *The American Spectator,* (May 1988), 15–16.

Hanna, P. R. (1963). Revising the social studies: What is needed? *Social Education, 27,* 190–196.

Hirsch, Jr., E. D. (1987). *Cultural literacy: What every American needs to know.* Boston: Houghton Mifflin.

Mitchell, L. S. (1934). *Young geographers.* New York: John Day.

National Commission on Excellence in Education. (1983). *A nation at risk: The imperative for educational reform.* Washington, DC: Author.

National Council for the Social Studies. (1981). Essentials of the Social Studies. *Social Education, 45* (3), 163–164.

National Council for the Social Studies. (1993a). A vision of powerful teaching and learning in the social studies: Building social understanding and civic efficacy. *Social Education, 57,* 213–223.

National Council for the Social Studies. (1993b). Definition approved. *The Social Studies Professional, 114* (January/February 1993), 3.

National Council for the Social Studies. (1994). *Curriculum standards for social studies: Expectations of excellence.* (Bulletin 89). Washington, DC: National Council for the Social Studies.

National Council for the Social Studies Task Force on Early Childhood/Elementary Social Studies. (1989). Social studies for early childhood and elementary school children preparing for the 21st century. *Social Education, 54,* 16.

National Education Association. (1916). *The social studies in secondary education.* Report of the Committee on Social Studies, Bulletin 28. Washington, DC: Bureau of Education.

Ochoa-Becker, A. S. (2001). A critique of the NCSS curriculum standards. *Social Education, 65,* 165–168.

O'Neil, J. (1993). On the new standards project: A conversation with Lauren Resnick and Warren Simmons. *Educational Leadership, 50,* 17–23.

Ormrod, J. E. (1999). *Human learning.* Columbus, OH: Merrill.

Peterson, P., & Knapp, N. (1993). Inventing and reinventing ideas: Constructivist teaching and learning in mathematics. In G. Cawelti (Ed.), *Challenges and achievements of American education* (pp. 134–157). Alexandria, VA: Association for Supervision and Curriculum Development.

Ravitch, D., & Finn, C. (1987). *What do our 17-year-olds know?* New York: Harper & Row.

Robinson, J. H. (1912). *The new history.* New York: The Macmillan Company. 17–18, 134.

Tyack, D. (1997). Civic education—What roles for citizens? *Educational Leadership, 54,* 22–24.

U.S. Department of Education (1991). *America 2000: An education strategy.* Washington, DC: Author.

Winitzky, N. (1991). Multicultural and mainstreamed classrooms. In R. I. Arends, *Learning to teach* (pp. 125–156). New York: McGraw-Hill.

2

Young Historians

O
n a clear, crisp day just before Halloween, students from Dawn Roe's sixth-grade class learned there's more to cemeteries than cold gravestones and chilling ghost stories. The scene is Oaklands Cemetery, a venerable setting where young imaginations don't require much prompting to take off. Ms. Roe has been taking her students on field trips to Oaklands Cemetery for the past three autumns, always around Halloween; her students' motivation to explore the graveyard, for some mysterious reason, seems to peak during that spooky season.

Ms. Roe prepares her students for the trip by reading the picture book, *Saying Good-Bye to Grandma* by Jane Resh Thomas (New York: Clarion, 1988). The book is a tender story of Suzie, a young girl who faces myriad emotions during her extended family's gathering for her grandmother's funeral. Suzie is consoled at the end of the story by family members who tenderly help her understand that funerals can be times for reflection and reaffirmation. Afterwards, Ms. Roe displays a model gravestone that she constructed out of cardboard for Suzie's grandmother, complete with epitaph and simulated carving. She then launches a series of cemetery-based learning activities that her students are eager to pursue.

To start, Ms. Roe listens to the children's questions and comments as they examine her simulated gravestone for Suzie's grandmother: "What's an epitaph? Why do we have headstones with epitaphs? How else might headstones be decorated?"

Ms. Roe then presents a series of questions or tasks that her young historians would use to guide their cemetery investigation, always leaving open the opportunity for students to add their own:

- Record the ages at death for any 20 men and 20 women. Determine the average for each group. Which group lived longer? When you get back to the classroom, look through the material on the library table (books, electronic media, and magazine articles) and find as many reasons for this as you can.
- Record the average age at death for any 10 men and 10 women who died during each of the following periods: 1800 to 1849, 1850 to 1899, 1900 to 1949, and 1950 to present. During which period did they live longest? Think of some reasons why this happened. Again, check the resources at this table to verify your predictions.
- Look at the gravestones for epitaphs. Record the longest, shortest, funniest, most interesting, most religious, and so on.
- Examine the writing on the gravestones. Do any of the words or letters seem peculiar to you? List the ones that do.
- How are the gravestones of the past like those of the present? How are they different? What changes can you predict for gravestones in the future?
- Make a list of the most popular names on the gravestones from the 1800s. Are they popular today? What ethnic groups seemed most prevalent during that time? Why?

When the class arrives at the cemetery, they are greeted by Sam Linton, a caretaker who leads them on a walking tour past obelisks, shrines, sarcophagi, and ornate gravestones. Some were barely legible, worn away by the wind and rain. Some were cut deep enough into slabs of granite that even hundreds of years couldn't take them away. "Look at the names; they're sure different than names today!" commented Herschel. Mr. Linton picked up on Herschel's comment and explained how several of these people had helped shaped their community's history. As they walked, Mr. Linton pointed out some of the cemetery's most interesting features. The students discovered that some epitaphs were just records, "In memory of Elijah Fahnestock. He was born September 15 AD 1801 and died on February 18 AD 1859." Others immortalized people with glowing accolades ("Honorably recognized for his gallant and meritorious conduct in the War with Mexico 1849.") or even denials of death itself: "She is not gone. She is just gone away." The cemetery stroll even revealed a wealth of information for those persistent enough to try to read between the lines. For instance, one simple grave marker was etched, "A little refugee from S.C. who died in 1865." Some students asked, "What kinds of refugees came to our town from South Carolina in 1865? Why didn't this marker have a name?" Sam Linton explained that this was the grave of a slave brought here by the Hastings family: "You can see that she was buried along with several Hastings family members. This presents evidence of whites and blacks being buried together before it was commonly thought." "Some of the earliest monuments date back to the Civil War," explained Mr. Linton. A bronze statue of a Union soldier sitting on his horse guarded the cemetery's fenced-in Civil War-era section. The students counted 24 soldiers buried there. "Here's a Union soldier who was only 15 years old when he got killed," commented Kendra. "That's so young to die in battle."

"We used to plant tulips at the base of the statue guarding this section of the cemetery, but the deer and squirrels ate them all," explains Mr. Linton. "We're looking into something safer that could be planted."

After this interesting introduction by Mr. Linton, each young historian was assigned a partner. The dyads were assigned to a different section of the cemetery to carry out their investigations. The students carefully filled out their observation sheets with responses to the questions and tasks they talked about in the classroom, and made dozens of gravestone rubbings by placing large sheets of newsprint against the gravestones and carefully rubbing crayons over the paper. Everything on the gravestone (names, dates, epitaphs) transferred to the paper and provided excellent research material that could be taken back to the classroom.

Returning to the classroom, Ms. Roe's young historians gathered and analyzed their data and shared their findings. Perhaps the most heartwarming outcome of the entire experience happened when several of the students expressed concern about the tulip bulbs that were eaten by the wildlife. "I wish there was something we could do," lamented Bryce. "Perhaps there is," suggested Ms. Roe. "We can check with a nursery to see if they can suggest something that the animals won't bother."

The class checked with a local nursery that day and were informed that deer, squirrels, and other wildlife love to eat tulips, but that they passionately dislike daffodil bulbs. The students mounted a fundraising drive, earning enough money to buy 700 daffodil bulbs. Each member of the class planted 30 bulbs that year, and as a result the base of the Civil War monument became awash with bright yellow flowers each spring.

Ms. Roe doesn't bother to quiz her children on these cemetery adventures. "That's one story I'm sure they'll remember," she proclaims proudly.

Ms. Roe bases her overall approach to teaching history on a conviction that history is not confined to memorizing "who did what to whom, when, and where" or to an outline of events commonly found in textbooks. She believes history can be found in cemeteries, museums, or newspapers. It can be found in a box of old receipts; in games children play; in stories people tell; in paintings, clothes, tools, furniture, books, letters, and diaries. Wherever we look, we can find clues to our past. Ms. Roe wants her students to know that the past is not just a list of names and dates, but a story that tells of life in other times.

Unfortunately, not all children experience history like Ms. Roe's students. When asked to share their feelings about history, many children echo Henry Ford's opinion that history is "bunk"; they don't like history because they say it is "boring." Ms. Roe, however, avoided this aversion to history by accomplishing the goals of historical study in a much more creative way. Her *young historians* learn history, but the part they like best is the "story" of history—Ms. Roe's history is today's history.

The *National Standards for History* (National Center for History in the Schools, 1996) reinforces Ms. Roe's convictions. This influential document proclaims history—along with literature and the arts—as the *most enriching studies* in which elementary school students can be engaged, because "History connects each child with his or her roots and develops a sense of personal belonging in the great sweep of human experience" (p. 2).

WHAT IS HISTORY?

Simply put, *history* is considered everything we know about the past. Anything that has a past has a history. People who study the past to determine what happened, how it happened, and why it happened are called *historians.* Connecting these two thoughts, I like to refer to elementary school students engaged in solving the mysteries of the past as *young historians.* Young historians, like all historians, do not just collect facts—they look at the past as a puzzle to be solved. They ask questions of the past and look for answers in the records of people who lived before them. Many people compare good historians with good detectives; detectives investigate crimes by looking for evidence and interpreting the clues they uncover. Historians search for evidence and interpret the clues they find, too, but for the purpose of explaining the past.

To gather clues about the past, historians seek out evidence from many possible sources, both written and nonwritten. *Written sources* include books, almanacs, letters, diaries, songs, speeches, poems, court records, and campaign slogans. They include the gravestones investigated by Ms. Roe's class, old calendars, posters, maps—anything with writing on it. However, written sources go back to only about 3000 B.C. when the Sumerians developed a pictographic type of writing known as *cuneiform* (meaning "wedge shaped"). Writing made it possible for people to keep records, write poems and stories, or inscribe a treaty—all potential sources of information for historians. History before the development of writing is called *prehistory.*

Nonwritten sources include such artifacts as jewelry, coins, tools, toys, masks, weapons, utensils, furniture, monuments, buildings, clothing, photographs, statues, or cookware. Nonwritten evidence also includes a culture's oral tradition—stories such as folktales and myths that have been passed down from generation to generation by word of mouth. Written and nonwritten clues that were produced at the same time that the event took place are called *primary sources.*

Clues that were produced by people who studied and attempted to explain primary sources are called *secondary sources.* For example, a videotape of Martin Luther King delivering his famous "I Have a Dream" speech is a primary resource while someone describing the speech, even though he or she may have actually been there at the time it was delivered, is a secondary source. Obviously, primary sources are much more valuable to historians than secondary sources because they are generally more accurate.

To function constructively in dynamic social studies classrooms, young historians must know how historians do their job.

WHY IS HISTORY IMPORTANT?

One of the most encouraging signs in social studies education during the past decade has been the widespread and growing support for more history in our schools. Not the kind that demands memorizing a bunch of facts about the entire history of the world, but the history that makes the past seem real—the history that captivates young historians and activates them to weave together various pieces of information in the best tradition of a storyteller. Taking on the role of young historian helps students recognize their place in history, realizing that their lives will be part of a yet unrecorded history.

These are important reasons to support more and better history in our schools, but the National Center for History in the Schools (1996) adds that none is more important to a democratic society than this: *Knowledge of history is the precondition for political intelligence:* "Without history, a society shares no common memory of where it has been, of what its core values are, or of what decisions of the past account for our present circumstances. Without history, one cannot undertake any sensible inquiry into the political, social, or moral issues in society. And without historical knowledge and the inquiry it supports, one cannot move to the informed, discriminating citizenship essential in the democratic processes of governance and the fulfillment for all our citizens of the nation's democratic ideals" (p. 1).

WHAT SHOULD STUDENTS KNOW OR BE ABLE TO DO?

History must be an important part of the educational experience for every American child, and returning history to the elementary school classroom is one of this decade's major movements for educational improvement. No longer a lifeless subject filled with empty dates and useless names, history has become an investigation into everything around us, and a search for answers to the question, "How did it get that way?" History is a quest to understand the complexities of the Constitution as well as the dynamics of the Revolutionary War. It is the study of immigration as well as of the struggles against slavery and for civil rights. Certainly, American history is a major part of historical study in the elementary school, but students also encounter the origins of diverse cultural heritages of people around the world—their political, religious, and social experiences as well as their struggles for conquest and survival.

Today's historical study focuses on broad, significant themes and questions rather than the memorization of facts without a frame of reference. Heather Junker, a fifth grade teacher, recounted, "We used to teach about Western migration by giving the students a lot of facts and dates and battles. We also built a lot of model wagons and made a lot of quilt squares. We immersed the students in pioneer life so they could understand what life was like for the pioneers." This year, Ms. Junker put away the quilt squares and began in a small and simple way. She asked her students to ask their parents about ancestors who had come to the United States. She is focusing not on facts about the pioneers of the 1800s, but on the theme of migration that cuts through several eras. This approach isolates themes—such as migration, patriotism, or revolution—that exist over several periods and regions. "What we do," explains Ms. Junker, "is more along the lines of getting students to think critically about why people do what they do and less along the lines of memorizing hundreds of facts about a single time period. Once we identify a theme and the topic is clearly defined, we must carefully select high interest experiences that help students look at the world of the past in new ways and 'turn them on' to history. With themes, we don't have to rush to get through *The Gold Rush* by the middle of June."

Although the memorization of names and dates has little place in such a curriculum, important facts and concepts must be carefully selected so that students acquire a meaningful understanding of the past, from local and regional history to national, Western, and world history. Historians need comprehensive historical content—we can't teach our children *how* to think as historians without making sure they have something worth thinking about. The argument that we can teach students *how* to understand the world of the past without conveying to them the events and ideas that have brought it into existence is a weak argument.

Perhaps the most influential source of content recommendations for elementary school history is the *National Standards for History* (National Center for History in the Schools, 1996). The standards have made explicit the knowledge goals that all students should have the opportunity to acquire through the K–4 social studies curriculum. The Standards Committee recommends that, to bring history alive, historical studies should be centered on four topics under which eight standards are distributed (see Figure 2–1).

In General, How Should History Be Taught?

Although the *National Standards for History* pinpoints the content for a K–4 social studies program, the National Center for History in the Schools (1996) does not ignore the important matter of instructional methodology. The *National Standards for*

FIGURE 2–1
Standards in History for Grades K–4

Overview

Topic 1: **Living and Working Together in Families and Communities Now and Long Ago**

Standard 1: Family Life Now and in the Recent Past; Family Life in Various Places Long Ago

Standard 2: History of Students' Local Community and How Communities in North America Varied Long Ago

Topic 2: **The History of the Students' Own State or Region**

Standard 3: The People, Events, Problems, and Ideas That Created the History of Their State

Topic 3: **The History of the United States: Democratic Principles and Values and the Peoples From Many Cultures Who Contributed to Its Cultural, Economic, and Political Heritage**

Standard 4: How Democratic Values Came to Be, and How They Have Been Exemplified by People, Events, and Symbols

Standard 5: The Causes and Nature of Various Movements of Large Groups of People Into and Within the United States, Now and Long Ago

Standard 6: Regional Folklore and Cultural Contributions That Helped to Form Our National Heritage

Topic 4: **The History of Peoples of Many Cultures Around the World**

Standard 7: Selected Attributes and Historical Developments of Various Societies in Africa, the Americas, Asia, and Europe

Standard 8: Major Discoveries in Science and Technology, Their Social and Economic Effects, and the Scientists and Inventors Responsible for Them

Source: National Center for History in the Schools. (1996). *National standards for history.* Los Angeles, CA: Author.

History stresses that teachers should bring history alive primarily by using "stories, myths, legends, and biographies that capture children's imaginations and immerse them in times and cultures of the recent and long-ago past" (p. 3). The Standards further recommend that, "In addition to stories, children should be introduced to a wide variety of historical artifacts, illustrations, and records that open to them first-hand glimpses into the lives of people in the past: family photos; letters, diaries, and other accounts of the past obtained from family records, local newspapers, libraries, and museums; field trips to historical sites in their neighborhood and community; and visits to "living museums" where actors reenact life long ago" (p. 3).

The history curriculum engages students with a mixture of learning resources. All the resources should be used imaginatively to help young historians raise questions and build historical thinking, such as the ability to process information, solve problems, compare and contrast past and present, and create historical narratives of their own. Much has been written over the years about the drawbacks of textbook-focused social studies instruction. All good social studies teachers are aware of these difficulties and avoid overreliance on all-inclusive textbook usage, in spite of how good the textbook is. Even when the textbook *must* be the sole source of instruction, it is still possible to help the students appreciate and benefit from your teaching. The following classroom episode illustrates how Lillian Collado was able to engage her children's historical thinking, even when the sole source of content was a social studies textbook.

Classroom Vignette

The purpose of Ms. Collado's lesson was to help students understand the major factors that transformed many of the colonists from loyal British subjects to dissidents on the verge of revolutionary war. She began by announcing to the class that the school district budget just bottomed out and very little money was available to purchase the supplies necessary to finish out the school year. A committee of teachers had met to study the problem and decided that a good source of revenue would be to have students pay a small charge each time they put something into the wastebasket or sharpened their pencils. Ms. Collado asked the students if they thought this was a fair solution to the money problem, especially since they had never before been required to pay such a "tax." She also raised the issue of whether teachers had the right to levy such a tax on students. Ms. Collado engaged the students in a discussion of the options available to them (avoid paying the fee, protesting the plan, complaining to others) and the consequences of their actions.

Ms. Collado then asked the students to recall that the Seven Years' War drained the treasury of Great Britain, so the government was in desperate need to raise money. Britain began to do something it had never done before—it decided to tax the colonies. The class drew parallels between the British taxation scheme to their tongue-in-cheek classroom "tax." Then, to begin today's textbook reading assignment, Ms. Collado invited predictions about what the British might tax in the colonies and how the colonists might react to those British taxes.

Ms. Collado next directed the students to read a section in their textbooks to find out (1) what some of those taxes were, (2) the colonial reaction to each, and (3) the British response to the colonists' actions. After a short discussion of the reading selection, the students summarized the three British taxes described in the textbook selection: (1) the Stamp Act, (2) the Townshend Duties,

FIGURE 2–2
Taxation Graphic
Organizer

The Tea Act	Colonial Reaction	British Reply
Parliament taxed the tea sold to the colonies by the British East India Company.	Dumped tea from three ships waiting to unload tea chests in Boston Harbor. The event was known as the Boston Tea Party.	Passed the "Intolerable Acts," which closed Boston Harbor and limited the power of the colonists.

and (3) the Tea Act (see Figure 2–2). Ms. Collado directed the students to chart each on a graphic organizer. (Note how the organizer directly relates to the stated purposes for reading.)

Despite the fact that even textbook-focused instruction can be enriched with imaginative instructional strategies, history is much more enjoyable and effective when teachers enrich instruction with a wide selection of resources and activities. These special experiences pump life into the curriculum and deepen the students' understanding of history.

A Literature-Based Approach

As we just read in the preceding section, it is recommended that a literature-centered approach to history can be one of the most satisfying and enjoyable ways to study past events in the elementary school. This approach focuses instruction each year on inspiring selections from many historical periods, and then expanding those stories by exploring more deeply the historical times they bring to life. The pattern is described as an elementary school version of a "Great Books" approach to the humanities where literature is used as a context for thinking and learning about a variety of historical eras and cultures. Historical fiction, biographies, and folk literature dramatize and humanize the sterile facts of history. They transport young historians to the past and enable them to more clearly understand that today's way of life is a continuation of what people did in the past; that the present will influence the way people live in the future. A history textbook tells; a quality piece of literature not only tells but also has the power to evoke emotion.

Textbooks can tell children what racism is and that racism is wrong, but one need not be told these things while reading a book such as *Marching to Freedom: The Story of Martin Luther King, Jr.* by Joyce Milton (1987). Readers experience the atrocities of racism as they are carried off on the wings of Milton's words to the seats of a bus where they enter a new dimension of imagination to discover racism's horrors. Milton describes an event in 1943, when Martin Luther King, Jr. (then known as M. L.), was only 14 and had an encounter with "Jim Crow" laws that he would never forget. M. L. and his teacher (Mrs. Bradley) had boarded a bus that was pulling off the road to make a local stop:

> Suddenly, the bus driver started yelling at them. Some new passengers were boarding the bus—white passengers. "Get up, you two," the driver shouted. "The white folks want to sit down." Mrs. Bradley started to gather her belongings. But she didn't move fast enough to suit the driver. He began to curse. "You black _____ _____. You git up. Now!" M. L. sat frozen in his seat. No one had ever talked to him that way before. Ever so slowly M. L. stood up. . . .
>
> "That night will never leave my memory," he said years later. "It was the angriest I have ever been in my life." (pp. 4–5)

Textbooks keep students in the classroom; compelling stories from good books transport them to new and exciting worlds filled with individual heroism and epic events, evoking emotions rarely found in textbook passages—compassion, humanness, misfortune, happiness, awe, and grief. Darigan, Tunnell, and Jacobs (2002) elaborate: "If history is indeed the story of ourselves, then [a] limitation of history textbooks is that people are missing! The best one-word definition of history is, in fact, 'people.' Without human beings, whose emotions and actions influence the times, there is no history" (p. 271). Jean Fritz, in her autobiographical novel, *Homesick: My Own Story* (1982), recalls her first disappointing experience with history:

> Miss Crofts put a bunch of history books on the first desk of each row so that they could be passed back, student to student. I was glad to see that we'd be studying the history of Pennsylvania. Since both my mother's and father's families had helped settle Washington County, I was interested to know how they and the other pioneers had fared. Opening the book to the first chapter, "From Forest to Farmland," I skimmed through the pages but I couldn't find any mention of people at all. There was talk about dates and square miles and cultivation and population growth and immigration and the Westward movement, but it was as if the forest had lain down and given way to farmland without anyone being brave or scared or tired or sad, without babies being born, without people dying. Well, I thought, maybe that would come later.

Sadly for Jean, it never did. And, in many elementary school classrooms throughout the country, it never will. If we truly want our students to deepen their understanding of people and to bring to life the historical periods in which they lived, good books are called for. Three types (genres) of literature seem particularly adaptable to the social studies program: *historical fiction, biographies,* and *folk literature.*

Historical Fiction

Historical fiction is a category of realistic stories that are set in the past; the facts are accurate but the characters are fictional (although they sometimes interact with actual historical figures). Historical fiction offers children opportunities to vicariously

Children love to read and listen to stories about noteworthy people and events from the past. To share what they have read about, these students will simulate a television interview during which their classmates will appear as famous historical figures.

experience the past by entering into a convincingly true-to-life world of people who have lived before them. By being transported to the past through the vehicle of literature, students enter into the lives of the characters and, through mental imagery, become inspired to think as well as to feel about their condition. As Darigan, Tunnell, and Jacobs (2002) stress, "Historical fiction . . . expands and brings to life the landmark as well as seemingly mundane events of the past, accenting the true 'stories' in history" (p. 270).

It is easy to see how good historical fiction makes history erupt into life. Note how Patricia Beatty brings a young boy's struggles to life in her book, *Charley Skedaddle* (1987). The story is set during the Civil War when 12-year-old Charley Quinn, a member of the Bowery Boys (the toughest street gang in New York City), vows revenge for his brother's death at Gettysburg and plans to enlist in the Union Army. One day, he sees Con, an old friend, as Con is marching along a street in New York City with other soldiers of the Union Army:

"Take me along with you, Con, *please!*"

For a moment the man's face twisted and thought, then he boomed, "Sure, Charley. Why not? I got five day's cooked rations with me. Ye can share 'em."

> A short man with brown sideburns and a dark face in the ranks next to Con said, "He's only a shirttail kid, Con."
>
> "No matter, Jem Miller. The Army's got uses for kids. Ye keep up with us men, Charley, ye hear. Ye let go of me now."
>
> "I'll keep up, you bet." As Charley trotted past the Bowery Boys, he waved his hat and shouted, "I'm off to join the Army!"

Unfortunately, when the horrors of war became a reality and he sees his best friend killed, Charley is terrified. He "skedaddles" away from the Union Army. Eventually, though, while still a deserter, an act of courage far from the battlefields proves to Charley that he's not the coward he thought himself to be. Beatty brings the Civil War to life for elementary school students by telling about its horrors through the experiences of someone who thinks of life the way they, as a 12-year-olds, do.

If written well, historical fiction offers children fuller understandings of human problems and human conditions. To be considered for social studies programs, well-written historical fiction should be selected by using the following guidelines:

1. *It must tell an interesting story.* The book should "set the reader on fire."
2. *It should be accurate and authentic.* The historical period should be so precisely described that the people within the book "walk right into the room."
3. *It should reflect the spirit and values of the times.* The stories cannot be made to conform to today's ethical values (such as contemporary points of view concerning women and minorities).
4. *It should contain authentic language.* The spoken words should give the flavor of how people actually talked.
5. *It should provide insight into today's problems.* Putting the conditions of women and minorities into historical context is a prime example.

Biographies

Biographies are much like historical fiction in that they are based on historical facts that can be documented. Rather than telling stories of fictional characters, however, biographies are carefully researched accounts of the lives of real people. The work of Esther Forbes offers an excellent comparison of a biographical story and historical fiction. In 1942, Forbes won the Pulitzer Prize for her adult biography, *Paul Revere and the World He Lived In.* While she was researching the book, she uncovered many fascinating stories about the duties of Boston's young apprentices. As a result, she wrote a book for children, *Johnny Tremain* (Houghton Mifflin); it won the Newbery Medal in 1944. The book is about a fictional silversmith's apprentice to Paul Revere who lived in Boston in the days leading to the American Revolution. A skilled craftsperson, Johnny becomes one of the best at his trade. But a practical joke backfires, and Johnny's hand becomes maimed for life. Johnny then becomes involved in pre-Revolutionary War activities. The book remains one of the most popular works of historical fiction to this day.

Children are introduced to Deborah Sampson, Hammurabi, General Santa Ana, Neil Armstrong, Sojourner Truth, Abraham Lincoln, Molly Brown, Lewis and Clark, Mary McLeod Bethune, Hiawatha, or Frederick Douglass in their social studies texts, but these figures seem to spring to life as the biographer chooses and presents details in ways that engage the minds of the readers. Consider this account of a portion of the ritual, led by his tribe's shaman (spiritual leader), that transformed Sitting Bull from a boy into manhood:

> Following Moon Dreamer's [the shaman's] instructions, Sitting Bull entered the [sweat lodge] and sat by the [hot] stones for what seemed like an eternity. Soon his skin was hot and throbbing, and sweat poured down his back and sides. Finally, when the boy felt as if his body were on fire, the shaman ordered him back outside. There he told Sitting Bull to jump into a freezing creek.
>
> Without questioning the shaman's command, Sitting Bull raced across the ground and leaped into icy water, gasping at the shock. As he climbed up out of the creek, he saw that Moon Dreamer was getting ready to say farewell and leave. Naked and without food or water, Sitting Bull remained alone on the hill [for several days]. (Eisenberg, 1991, pp. 30–31)

Well-written biographies personalize historical subject matter with a degree of vividness unattainable with the straight reporting style of most textbooks. A "master biographer" whose works serve as excellent resources for elementary school students is Russell Freedman. Freedman reacted strongly against the "I cannot tell a lie" make-believe stories that were previously popular as biographies and presents a more objective treatment of historical figures in his biographies. In *Lincoln: A Photobiography* (1989), for example, Freedman contrasts the physical differences between Lincoln and Douglas as they prepared for one of their famous debates:

> The striking contrast between Douglas and Lincoln—The Little Giant and Long Abe, as reporters called them—added color and excitement to the contests. Douglas was Lincoln's opposite in every way. Barely five feet four inches tall, he had a huge round head planted on massive shoulders, a booming voice, and an aggressive, self-confident manner. He appeared on the speakers' platform dressed "plantation style"—a navy coat and light trousers, a ruffled shirt, a wide-brimmed felt hat. Lincoln, tall and gangly, seemed plain in his rumpled suit, carrying his notes and speeches in an old carpetbag, sitting on the platform with his bony knees jutting into the air. (pp. 58–59)

Biographies impress children with a sense of historical reality. Even early primary-grade children become interested in biographies if you read them picture biographies such as those by Ingri and Edgar Parin d'Aulaire (Abraham Lincoln, Benjamin Franklin, Pocahontas, and others). Gradually, you can move on to many excellent biographies suitable for children at each age level: Fernando Monjo's *The One Bad Thing About Father* gives a son's view of Teddy Roosevelt; James T. DeKay's *Meeting Martin Luther King, Jr.* is a picture essay describing King's early childhood and growth into adulthood; Dan D'Amelio's *Taller Than Bandia Mountain: The Story of Hideyo Noguchi* tells of a Japanese doctor's efforts to combat serious obstacles while achieving success in bacteriological research; and Evelyn Lampman's *Wheels West: The Story of Tabitha Brown* describes a 66-year-old woman's wagon train trip to Oregon, where she became a famed educational pioneer.

The same guidelines for selecting historical fiction apply to biographies. Be sure the story is a well-researched, carefully documented account of the person's life with fast-moving narrative and a clear, readable writing style.

Folk Literature

Fables, myths, legends, and folktales belong to the great component of literature we refer to as *folklore*. In essence, these are the stories that began with illiterate people and were handed down by storytellers for generations. Originating wherever people gathered—in marketplaces, during tasks such as weaving or sewing, in taverns, or around the hearth—the stories were told not only for the entertainment of the listener but often as an expression of philosophies and living conditions. The rich oral tradition of these stories was kept alive by generations of storytellers; the tales eventually found their place within printed literature.

Because folktales have been retold from generation to generation within every culture, they clearly reflect those cultures' beliefs, values, lifestyles, and histories. An authentic tale from Chinese culture, for example, will include references to the land on which the people lived, their food, their homes, their customs, and their beliefs. One particularly entertaining and informative folktale is the Chinese pourquoi (tells how or why something came to be) story *Tikki Tikki Tembo* by Arlene Mosel (Holt). The tale tells of the family's "first and honored" son who was proudly given the distinguished, long name of Tikki-Tikki-Tembo-No-Sa-Rembo-Chari-Bari-Ruchi-Pip-Peri-Pembo. Unfortunately, as a little boy, he falls into a well and it takes so long for his brother Chang to tell someone about his plight that the elder son nearly drowns. That is why, to this day, all Chinese have short names. Young children love this story and enjoy repeating the elder son's name.

As you can see, folktales reflect a culture so plainly that it is nearly impossible to confuse a folktale from a village in China with a folktale from central Africa. This is a major reason folktales belong in social studies classrooms; they help children understand a culture's past through its values, beliefs, and customs.

A cross-cultural study of folk literature also adds an extra dimension to helping children discover humankind's past. For instance, the theme of reward for a good, generous person and punishment for a greedy, disobedient one seems to be universal. In our culture, Cinderella illustrates the "good" on the good–bad scale. In addition, she receives her rewards for her goodness in various ways from culture to culture throughout the world. To demonstrate this to her young historians, Mary Gilland, a third-grade teacher, used such a cross-cultural study to enrich her unit on China.

Classroom Vignette

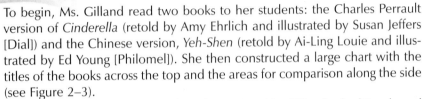

To begin, Ms. Gilland read two books to her students: the Charles Perrault version of *Cinderella* (retold by Amy Ehrlich and illustrated by Susan Jeffers [Dial]) and the Chinese version, *Yeh-Shen* (retold by Ai-Ling Louie and illustrated by Ed Young [Philomel]). She then constructed a large chart with the titles of the books across the top and the areas for comparison along the side (see Figure 2–3).

To help the children fill in the information, Ms. Gilland asked timely and appropriate questions: "Who were the main characters in the stories? How were they alike? Different? Where do the stories take place? When? How are the settings alike? Different? What problems did Cinderella face? Yeh-Shen?" Ms. Gilland's overall goal was to help the children understand that, despite many differences in cultures around the world, similar problems often have motivated people for generations. Ms. Gilland brought this phase of her China unit to closure by having her students dramatize the events of both stories.

Folk literature has deep roots in all cultures. Through this genre, students broaden their understandings of those cultures, as well as sense the common bonds that have linked together cultures for centuries. Countless books have memorable impact and may serve as the substance around which you could plan multiple learning opportunities in history.

To ensure that these important benefits actually emerge from the use of historical literature in your classroom, you must carefully research the story content so that you offer an accurate context for the students. I recall observing a field experience student read Elizabeth George Speare's *The Sign of the Beaver* to a group of fourth graders. During the introductory phase, she told the students that the story took place long ago in the Maine wilderness, where Matt is left by his father to tend a new cabin while he returned to Massachusetts for the rest of the family. The students located Maine and Massachusetts on a map and traced the probable

FIGURE 2–3
Chart Comparing Cinderella Variants

TITLE AND AUTHOR	<u>YEH – SHEN</u> BY AI-LING LOUIE	THE EGYPTIAN CINDERELLA BY SHIRLEY CLIMO	<u>CINDERELLA</u> BY AMY ERLICH
CULTURE OF ORIGIN			FRANCE
SETTING (TIME/PLACE)			• CINDERELLA'S HOUSE • A CASTLE
CHARACTERS AND CHARACTER TRAITS			• CINDERELLA—SWEET, HARD WORKING • FAIRY GODMOTHER—KIND, GENEROUS • STEPSISTERS—CRUEL, SELFISH • STEPMOTHER—SELF-CENTERED, MEAN • PRINCE—HANDSOME, CHARMING, WEALTHY
PROBLEM			CINDERELLA WANTS TO GO TO THE BALL, BUT HAS NOTHING TO WEAR. SHE HAS TO STAY HOME AND CLEAN WHILE HER MOTHER AND STEP-SISTERS GO.
SOLUTION			FAIRY GODMOTHER HELPS CINDERELLA WITH MAGIC. SHE AND THE PRINCE FALL IN LOVE AND GET MARRIED.
STORY PURPOSE			TO TEACH THE LESSON THAT GOOD IS BETTER THAN EVIL.

route. That was good, but to further establish the story background, the field experience student displayed a large study print depicting Native Americans of the past (because their interactions with Matt are central to the story). The story is wonderful, but it is an Eastern Woodlands story. The study print, however, depicted a buffalo hunting camp of the Cherokee (Plains Indians) with its cone-shaped tepees and campfire. The student complicated her error by referring to the shelters in the study print as "wigwams." As the regular classroom teacher stepped in to help correct the error-filled introduction, I couldn't help wonder about how pre-service teachers might be convinced of the need to confirm the accuracy of what they are teaching.

You must understand that the many benefits of historical literature can be realized only if you place the story content in an accurate context. Historical literature is based on a strong background of fact and should be used as a vehicle to enlighten, not confuse. What follows is a classroom story about a teacher who properly used children's literature to enrich understandings of the associated historical era.

Classroom Vignette

Gary Schuck's fifth-grade social studies curriculum focuses on the major regions of the United States, with major emphasis on their geography and history. The current region under study is the Great Plains. Students have been enthralled with stories of how settlers pushed west into the Great Plains during the 1860s and settled what are now Kansas, Nebraska, South Dakota, and North Dakota. They learned that wheat grew very well in the rich soil of the Great Plains and became the most important crop of the region. However, they also found out that early farmers faced many challenges, such as storms, dry spells, insects, blizzards, and floods, which often ruined or damaged their crops. Nevertheless, the pioneers were relentless and worked hard to build successful farms throughout the region.

One of the major resources for the study of this region was the collection of "Little House" books by Laura Ingalls Wilder. Most recently, the class had been reading *Little House on the Prairie* (Harper and Row, 1935), a story filled with adventures about the pioneering Ingalls family moving to the wild Kansas country. Today, Mr. Schuck will help his students explore one aspect of pioneer life—the kinds of buildings the Ingalls family would have found on a trip from their farm to a small prairie town: "Imagine that you are going with the Ingalls family on a trip to a small prairie town. What do you expect to find there?" After a brief discussion, Mr. Schuck divides the students randomly into teams of two and each team is assigned to a center: one-room school, church, hotel, general store, livery stable, blacksmith shop, jail, lumberyard, barbershop, cooper (barrel maker), and saloon. The students are advised that they will become "experts" on these places by examining a Website Mr. Schuck had accessed for each group.

After the students completed their research, they were directed to write a short descriptive paragraph about what they uncovered on index cards. The One-Room School team, for example, wrote:

> Schools in early prairie towns were one-room school houses. There you could find students ages 6 to 16 and eight grades in one room. There was one teacher for all of them, and older students often helped the younger ones.

The index cards were displayed on a large chart to summarize the kinds of buildings that would be found in small prairie towns of the 1860s. Next, students were instructed to use any of a number of cardboard boxes that Mr. Schuck gathered beforehand, along with construction paper, paint, crayons, and other art materials, to construct a model of their buildings. First, Mr. Schuck reviewed with the students the materials that the pioneers used to construct these buildings and advised them that they were free to use the encyclopedia or any library references to gather any additional information. The students went right to work and, after the buildings were constructed, they were arranged on a tabletop display (with the index card information chart behind them). The groups were eager to function as tour guides, telling other classroom "sightseers" about their pioneer town when they came to visit throughout that day.

Visits to "Living Museums"

The world outside the classroom is rich in historical learning experiences. By taking trips outside of school, students experience things firsthand that are impossible in the classroom. Visiting museums and "living history" sites to observe the clothing, houses, furnishing, tools, and other artifacts both enlighten and enrich understandings of any historical period.

A list of places to visit can range from A to Z—from antique shops to the zoo. Regardless, the challenge is to effectively draw out student understandings from wherever you choose to go. A fifth-grade teacher, Kara Boate, who was teaching a unit titled *Colonial Life in America*, makes the value of such trips crystal-clear. She realized that her students' failure to comprehend time and place concepts might stifle their curiosity and interest in studying colonial America, so she arranged a trip to an authentic living museum in Cooperstown, New York.

Words cannot duplicate the effectiveness of watching this metalsmith at work in a Colonial-era living museum.

Classroom Vignette

The children's spirits were instantly aroused as the class walked up the path connecting the parking lot to the restored colonial village called the "Farmer's Museum." Authentic in every detail, the village was an actual working farm in which people dressed in period garb and used authentic implements to perform the duties of colonial farmers. About halfway up the path, the class fixed their eyes in horror as a farmer led a huge ox hauling a cart directly toward them. None of the children had ever before seen a real ox; few knew exactly what an ox was. "Let me outa here!" shrieked Frank guardedly as the snorting animal, oozing slobber, ambled up and stopped next to him. The farmer invited Frank to pat the ox, but Frank was too frightened to try.

"I never knew an ox was so big," Lois marveled as she reached up and patted its wet nose. Following Lois's lead, several classmates approached the ox, some patting it and others commenting on its size, smell, and drawing power for flies. Some children were satisfied to simply look at the ox, while others ran away from it when it made the slightest movement; one or two even made faces at it—they did all sorts of things. The farmer told the class about the importance of oxen to colonial life and they were enthralled with his story. He thanked them for stopping and told them that his tired animal needed to go to the barn for a rest. From that day on the children all knew exactly what an ox was. Pictures of oxen, stories of oxen, or a video of oxen would not approach the sounds, smells, sights, and tactile sensations of the real animal.

The ox experience seemed to transform the students from audience to actors. "That was awesome," said Luke. "What else is there?" The first building the students came to was a school, where a "school marm" taught them about the ABCs exactly as a teacher would have done in colonial times. Spirits soared as the children went to the barn, where workers involved them in the entire process of making linen from flax. By now, it was getting late and Ms. Boate tried to get the class to move on to the tanner, wigmaker, blacksmith, gunsmith, cooper (barrel maker), and glass blower, but the children insisted on staying at the barn for the corn husking bee and gunnysack race. At every stop, the children had an opportunity to touch, handle, and use; Ms. Boate had an equally tough time pulling them away from each so they could visit the next exhibit. It was very apparent to Ms. Boate that her students were quite new to this world of the past, not old hands who have "been there" and "done that."

The next day, when the students returned to the classroom, they were primed and ready to go to work. Ms. Boate involved the class in several activities that focused on the events at the Farmer's Museum. To conclude the day, she read aloud a book that seemed to be a perfect connection to one of the most fascinating adventures from their trip: *Ox-Cart Man,* a Caldecott Medal-winning book by Donald Hall. The story is about a father who loads up his oxcart with the many things his family has been making and growing all year and begins a 10-day trek to market in Portsmouth during early autumn. Naturally, Ms. Boate related their previous experiences to the story. It was amazing to see how much the field trip helped students understand and make sense out of the story (as well as all subsequent learning activities).

The world outside the classroom is a stimulating place for young historians. Explore your community for these "hidden" resources and you, too, may discover new understandings beyond "the here and now."

Oral History

Bringing in someone to tell stories of personal experiences related to particular places or times is an idea with exciting possibilities for any classroom. Known as *oral history,* these stories need not be major projects; all that is needed is someone interesting to spin tales of times gone by. Take the time Jim Mosteller, 93 years old, visited my classroom and captivated my children with firsthand accounts of turn-of-the-century life. "When we got automobiles around here, you couldn't use them in the winter," Mr. Mosteller said. "My father had one of the first cars in town. It was one of those open cars with leather seats and brass lamps. I'll never forget one Sunday; we had eleven flat tires!"

Mr. Mosteller had the children's undivided attention when he told what a dollar would buy in 1939: one dozen eggs, a loaf of bread, a pound of butter, and a half-pound of bacon. He also told the children about a whistle-stop campaign during which Teddy Roosevelt visited town in 1912 ("I can see him to this day") and the transfer of the Liberty Bell on a flatbed car from Philadelphia to San Francisco for safekeeping during World War I.

You should use such valuable sources of historical information both for the children's enjoyment and as a source for researching and recording details from the past. These experiences give students a clearer understanding of and appreciation for people and events of the past.

Interviews and Oral History

Conducting interviews is a superb way to introduce children to the process of collecting historical data. Most children are familiar with interviewing; they see people interviewed on television nearly every day. Help beginning classroom interviewers understand how to conduct an interview by videotaping a television interview and then discussing what an interviewer does. Emphasize the types of questions an interviewer asks; some seek facts while others are designed to find out about personal feelings and opinions. Since children seem to ask many questions that elicit *yes* or *no* answers, point out that interviewers ask few, if any, of those questions because they do not help obtain much information. *Why, how,* and *what* questions produce much more. Offer the children an opportunity to be interviewers by bringing to school an interesting object and encouraging the children to ask questions in an attempt to get at the story behind the object. Play the "interviewee" as the children assume the roles of "interviewers." For example, one day Ms. Lee walked into her classroom with a Native American dream catcher dangling from her left hand. "I have something interesting here," she announced. "Ask some questions to find out as much information about it as you can. I will give only the information you ask for." Twenty excited pairs of eyes took in every feature of the dream catcher: "What is it? What is it used for? Why does it have feathers on the side? Who made it? Why is it important for Native Americans?"

A special interview had begun! Notice that the youngsters started asking *what* questions and eventually moved to other types that gave them a more complete collection of information. Children soon learn how different questions can give them different types of information, but most of all they learn that people possess information, and that they can extract interesting knowledge with a series of strategic queries.

When children are comfortable asking good questions, have them plan an interview with a person who might contribute new insights into an historical topic.

Classroom Vignette

Douglas Chilton, for example, asked his students to brainstorm a series of questions they might ask their grandparents to unlock what life was like when they were in elementary school. Each suggestion was written on a notecard. The students examined the collection of notecards, discarding those requiring simple *yes* or *no* answers and keeping those which stood the best chance of drawing out the information they were looking for.

Once the cards were arranged in a useful questioning sequence, the students were set to conduct their interviews. Some grandparents lived in the same community, but others lived far away so they had to be interviewed by phone or e-mail. Students were free either to use a tape recorder or to jot down notes to help them remember what was said during the interview, but most found that the tape recorder worked best. The children listened carefully to their grandparents' responses to determine whether the desired information was being gathered. If not, they asked probing follow-up questions to clarify points or elicit additional information.

Next, Mr. Chilton's young historians wrote individual accounts of their interviews. Each was read in class, and one set of copies was bound into a class book. A sample is shown in Figure 2-4.

Yet another teacher, Yvonne Perry, used Bonnie Pryor's book, *The House on Maple Street* (Morrow), to set the scene for an oral history project in her classroom. In the story, the past is linked to the present as lost objects from early times are unearthed in a contemporary child's yard.

FIGURE 2–4
A Student's
Interview With
His Grandfather

10/6/92

Interview

My interview is with my grandfather whom I call "Grampy." Grampy was born in 1912 in Rathmel, PA. Grampy is 80 years old and is my father's father. He is the oldest member in my family. Counting him he has 8 brothers and sisters. Three of them are still alive. He now lives in Sayre, PA, alone, with his fat cat named "Mama Cat." Grampy loves to tell jokes and tells them all the time. My dad usually calls him at night and Grampy always has a joke

Question: What kind of hobbies did you have when you were 12?

Answer: He liked to go fishing and hunting He couldn't get a hunting license untill he was 12. Most of his time he worked on the slackpile. His job was helping the family. Once and a while he went swimming. He paid $.10 for a movie. His brother would take them in his Model-T Ford. He still fishes but does not hunt.

Question: What was your first car? How much did it cost?

Answer: Grampy's first car was a 1929 Essex. It cost him $500 to get it. He was 23 years old when he got it.

65

Classroom Vignette

A discussion of dramatic changes in the American family through the years followed the book reading. Students talked about several eras, but became particularly focused on how hard it must have been for people to grow up without television to watch. "What did kids do for entertainment back then?" asked these youth living in the cable television–VCR era. Seizing the moment, Ms. Perry suggested inviting her grandfather to school to talk about what it was like for him as a youngster. Ms. Perry's grandfather was an antique radio buff with an extensive collection of radio tapes from the 1930s and 1940s (the "Golden Age" of radio). Jointly, Ms. Perry and the students developed a set of questions to ask about radio programs and the reasons they were popular.

Ms. Perry's grandfather not only visited class to talk, but he staged an Old-Fashioned Radio Night. Using a replica of an old console-style radio, he recreated an atmosphere of family and friends gathered around the radio. He played excerpts from sports and news events, the "Hit Parade," and even a popular comedy show of the era, "The Jack Benny Program." He dressed in the style of the 1930s and 1940s, and enriched his stories with a snack—small "Baby Ruth" candies, a popular snack of the time.

The children watched, listened, commented, and asked questions as they became immersed in radio's days of glory. After her grandfather left the room, Ms. Perry gathered the children together and invited them to share their thoughts. Sensing their interest was still high, Ms. Perry asked the students to bring in family photos from the 1930s and 1940s, looking especially for pictures showing the family radio. The photos were examined for clothing and hair styles, furniture, and other characteristics.

Simulated Oral History

There are often times when a teacher can stimulate strong interest in an area of historical investigation by simulating the narrative of someone who has lived in the past. The teacher-as-historical figure creates mental images for the students through spoken words and gestures, reacting and responding to audience needs. Based on real events, teachers help children reach beyond the boundaries of immediate time and into the world of the past. Schreifels (1983), for example, brought life to one of her history topics by taking on the role of a famous explorer:

> The day I discovered my fifth grade class had no idea who Vasco da Gama was—and cared less to find out—was the day I vowed to come up with some way to provoke interest in historical personalities. If da Gama and the rest of the early explorers were to become more than hard words to be

stumbled over in a textbook, I realized, something drastic—and dramatic—had to be done.

The next morning during social studies class, I slipped into the hall, plunked an old beehive hat on my head, swept a wraparound skirt over my shoulders and reappeared as an unreasonable facsimile of Vasco da Gama, fifteenth century sea captain. I introduced myself with my best Portuguese accent and invited questions.

At first there were merely giggles, until I threatened to make every student walk the plank unless I got some proper, respectful questions. The first was about how I got there (via a time machine that just happened to look like a filing cabinet). Eventually someone wanted to know just who I was.

"I'm Vasco da Gama, and I'm very famous."

"For what?" they all demanded.

I then proceeded to regale them with stories of my sailing prowess. Ever since that time, I've found I need only lean on the filing cabinet to get everyone's undivided attention. "Is the time machine going to bring us another mysterious person?" students plead. Quite often the answer is yes. And although these time machine visitors may have fuzzy historical memories, they serve to stimulate real interest in people of the past. (p. 84)

As Schreifels advises, you need not "go overboard" to involve students personally and motivate them to learn. Some of the most effective techniques require very little extra teacher preparation time to organize.

Primary Sources

Historians make sense of the past by asking questions. The answers they find appear in two kinds of primary sources historians use to study the past—written and non-written sources.

Written Evidence

Examining original written sources from the past helps students form a personal attachment to history. Anything written down can give historians clues about the people and events of the past—official documents, diaries, letters, songs, and the like. Consider the advertisement for a runaway apprentice from the early 1800s in Figure 2–5.

Among the advertisements in old newspapers, you will often find in such reward notices interesting clues to 19th-century life. Young boys were commonly hired out to artisans to learn a craft. They were housed and fed at the craftperson's expense in return for training, working for free during their apprenticeship. Such advertisements usually include such clues as the clothing the runaway wore and comments on his appearance and temperament. With their "striped row trousers," "gingham roundabouts," "brown flannel jackets," "old straw hats," or "dark fustian pantaloons," these "remarkably ugly fellows" may have "lost two of their fore teeth" or "had three fingers cut off at the first joint on the right hand." These descriptions

FIGURE 2–5
Ad for Runaway Apprentice

RUN AWAY

on the 1st of March instant,
an INDENTURED SERVANT
boy to the
Boot and Shoemaking Business,
named MARBLE LAPLANT:
he is between 16 and 17 years of age,
dark complexion
has a scar on his right cheek,
and is a REMARKABLY UGLY LOOKING FELLOW.

provide remarkable portraits of working-class people of 150 years ago; when children read original accounts of their escapades, they emerge as real individuals rather than as statistics or textbook portrayals. In this case, help your students analyze the evidence by asking questions:

- Why did the apprentices run away?
- What hardships did they face?
- Why did boys want to be apprentices?
- Why were apprentices mostly boys?
- How valuable was an apprentice to his master?

A section of the front page from Benjamin Franklin's *Pennsylvania Gazette* would give students insight into life in our country over 200 years ago. Published in Philadelphia, our country's biggest city in 1787, a typical front page would contain ads such as the one shown in Figure 2–6.

Students can be guided to examine some very interesting aspects of this evidence:

- The front pages at that time were mostly ads.
- The letters *f* and *s* looked alike.

FIGURE 2–6
Ad from Colonial Newspaper

FOR SALE
A large and general affortment of
SCYTHES and SICKLES.
Made by the fufcriber at his
SCYTHE AND SICKLE MANUFACTORY
in Market ftreet,
between Fourth and Fifth ftreet,
next door to the fign of the Black Horfe.
Thomas Goucher.

- Advertisers referred to themselves as "subscribers."
- There were no building numbers. People were guided by signs that had pictures, not words. Many people could not read.

As a follow-up to examining the sample ad from the front page, you could have students research other occupations that could have been advertised in the *Pennsylvania Gazette* during Franklin's time—blacksmiths, milliners, tinsmiths, gunsmiths, wig makers, sail makers, shipbuilders, bootmakers, weavers, and so on. The students could then publish their own replica front page, trying to duplicate the language and other conventions consistent with the early history of our country.

Letters from the past reveal interesting information about the famous as well as the infamous. For example, we have learned a lot about Benjamin Franklin through his letters. They have informed us that he was one of the most vocal opponents of selecting the bald eagle as our national symbol. It is now a well-known story that Franklin felt the turkey would have been far more representative of the newly formed United States. But, how did anyone find that out?

Through calendars and date books, it is known that Franklin was in France when Congress chose the national symbol in 1782. Angry, he wrote a letter complaining of the decision. Of the eagle, Franklin wrote, "He is a bird of bad moral character; he does not get his living honestly; you may have seen him perched on some dead tree near the river, where, too lazy to fish for himself, he watches the labour of the fishing-hawk; and when that diligent bird has at length taken a fish and is bearing it to his nest for the support of his mate and young ones, the bald eagle pursues him and takes it from him. Besides, he is a rank coward; the little kingbird, not bigger than a sparrow, attacks him boldly and drives him out of the district."

Franklin continued, "The turkey is in comparison a much more respectable bird, and withal a true original native of America. . . . He is . . . a bird of courage, and would not hesitate to attack a grenadier of the British Guards who would presume to invade his farmyard with a red coat on."

All of us have heard stories of Franklin's displeasure at selecting the bald eagle as our national symbol, but few have ever read his words of expression. Think about how much actual written accounts, or accurate replicas of them, would enliven the investigative efforts of your young historians.

Artifacts

Artifacts include all the nonwritten objects—photos, paintings, coins, furniture, tools, and the like—that help historians reconstruct the story of human life. I like to use the term *accidental survivors* to describe artifacts because very few of them were intentionally preserved for the future in hopes that others would have them as evidence of life in the past. Photographs, for example, give us interesting clues about the past. The picture on the next page tells us a great deal about school life during the early part of the 20th century—what the students wore, what their hairstyles were like, what their classrooms were like, the kind of school supplies they used, how they were taught, and even what they studied. In an effort to help her students appreciate the value of historical artifacts, Karen Crossman used a photo similar to the one you just examined to help her fourth graders better understand school life of the early 1900s. To extend the experience to their own lives, Ms. Crossman asked her students to pretend that they were a group responsible for informing historians of the future about what the culture of *their* present-day school was like. They were directed to select artifacts to put in a time capsule that would be opened in 50 years. The students were required to explain the significance of each item and describe what historians of the future would be likely to understand about their school culture. The time capsule was sealed and placed in a secure site for a class to open in a half century.

Special collections of artifacts related to historical topics or themes are often assembled by social studies teachers to motivate young historians and give them the opportunity to get a firsthand look at the past. Creating a collection for a particular historical period authenticates the experience for children and helps them learn abstract historical concepts. Teachers report that these artifact collections increase students' interest, enhance their understanding, and make classroom discussions more meaningful and interesting.

Photographs provide historians with useful evidence for unraveling the past.

One teacher, for example, compiled an artifact kit for a unit on ancient Egypt. It contained a model mummy coffin, necklace with a dangling ankh, panel from a false door of a tomb, a tablet with hieroglyphs, as well as photographs and documents. These items not only motivated the students to learn more about Egypt from various resources, but also helped them relate background information to new learning experiences throughout the unit of study.

A major part of studying history is using the tactics employed by historians. Students must examine oral and written records as well as artifacts and books for evidence to explain the past. This is the way you will want to teach history—in a way that allows children to investigate as historians. Provide situations where they can examine historical materials firsthand. Bring artifacts to the classroom, invite guest speakers to demonstrate items from the past, and visit living museums and historical sites to examine original sources. Young historians should use the historian's methods of investigation to study a variety of local, regional, state, national, or international topics.

Classroom Vignette

Patrick Garrett, a fifth-grade teacher, put his young historians to work one day when a question for historical investigation burst forth from a quite unlikely source. A group of students was reading a magazine article describing fads of youth today—their clothing styles, favorite television shows, most popular entertainers, and so on. An extended conversation about whether the poll accurately reflected "kids of today" got nearly everyone involved. As the students talked about and sometimes questioned the article's accuracy, one student looked at Mr. Garrett and abruptly shifted the direction of the conversation: "Mr. Garrett," he asked, "what fads were popular when *you* were our age?"

Realizing that the spark for historical inquiry often comes from meaningful personal questions, Mr. Garrett shared a few of his recollections from life about 20 years ago and was surprised to see that the class was quite interested. Without previously planning a historical inquiry experience, Mr. Garrett found himself in the midst of one. He started by writing the students' central question on the board in an *IWW* format: "*I wonder what* fads were popular among fifth graders in 1985."

The historical problem, stated as a question, is the initial spark for further investigation by young historians. Once his young historians understood the problem and maintained a sincere interest in it, they were ready for action; it was now time to determine the kind of resources that might help them answer their question. The data collection process could have taken several forms, depending on the nature of the investigation but, when Mr. Garrett's young historians delved into conditions of "long ago," they found that the three sources of data we have just discussed were the best options: oral history, written sources, and artifacts. After considering the advantages and disadvantages of each, the students chose to examine the fads of 1985 by gathering an oral record of the past—interviewing family members and neighbors, many of whom were elementary school students in that era. The students designed a written list of questions they would ask their family and friends.

Mr. Garrett's students summarized and rank ordered the responses to each interview question on a large chart. The results are shown in Figure 2–7. Mr. Garrett helped his students make descriptive statements of the data by asking such questions as, "What did you discover about fads of 1985? How does this list compare to present fads? Why do you think this has happened? How might this list change in the future? What makes you think so?"

The class concluded their study by planning an "'80s Day" at school, coming to school dressed in the clothing style of the day, listening and dancing to period music, and examining the popular toys of the time. Parents prepared some special snacks for the "time travelers," including fondue and powdered juice drinks, which were all consumed during a videotape showing of an episode from TV's *Family Ties.*

FIGURE 2–7
Summary of Interviews

Sayings	Toys	Clothing	Entertainers	Sports Stars	TV Shows
Awesome	Cabbage	Camouflage	Bill Cosby	Mary Lou Retton	Cosby Show
Rad	Patch	clothes	Michael J. Fox	Walter Payton	He-Man: Masters
Cool	Kids	Fingerless	Ralph Macchio	Greg Louganis	of the Universe
	He-Man	gloves	Madonna	Larry Bird	Punky Brewster
	Transformers	Legwarmers	Wham!	Evelyn Ashford	The A-Team
	My Little	Jams	Cyndi Lauper	Pete Rose	Webster
	Pony	Parachute	Emmanuel Lewis	Kareem Abdul-Jabbar	Who's the Boss?
	G.I. Joe	pants	Ricky Schroder	Carl Lewis	Silver Spoons
	Stickers	Cutoff	Gary Coleman		Knight Rider
		shirts	Mr. T		Family Ties
		Collars	Michael Jackson		
		turned up	Sylvester Stallone		
		Checkered			
		shoes			
		Jeans			
		Belts below			
		the waist			

Chronology

Up to this point, we have examined important classroom opportunities capable of developing historical thinking in elementary school students—listening to and reading historical stories; examining historical documents and artifacts; visiting historical sites; listening to and interviewing eyewitnesses to history; and constructing historical narrratives. There is, in addition, one area of historical thinking that we have not yet addressed—*chronological thinking.* Chronological thinking is the process of acquiring a sense of historical time—past, present, and future—in order to identify the sequence in which events occurred. Young historians must understand that human events take place over time, one after the other. It is important to establish this linkage in order to reconstruct events and tell the stories that link them. Young historians must learn the measures of time that all historians use—day, week, month, year, decade, and century. Using the calendar is one way to do this, but content related to that chronological thinking skill is beyond the scope of this text. We will look at *event chains* and *timelines,* however—specialized tools that allow young historians to place historical events in meaningful sequence.

Event Chains

At times, a social studies learning experience may deal with a sequence of important happenings (major events of the Revolutionary War) or the significant milestones in the life of a historical figure (the rise and fall of Alexander the Great). An event chain is useful in describing the sequence of these important events (and how one event led to the other) by presenting a series of frames that separate the key parts of the whole. Figure 2–8 illustrates a sample event chain.

Timelines

Timelines are graphic representations of a succession of historical events, constructed by dividing a unit of time into proportional segments. As children study the past, timelines help them put events into perspective by seeing a picture of when important things happened. For the very youngest children, construct timelines on topics of immediate experience. Illustrate routines of the daily schedule, for example. Have the children talk about what they do in school each day and pick out an illustration (or photo) that shows it. The children can arrange themselves in sequence according to which activity takes place first, second, and so on. It is best to begin an abstract concept such as a timeline by relating it directly to what is happening in the children's lives at that time.

Extend their ability to sequence major daily events to something else that has great meaning in children's lives—holidays. Cut out a symbol for each major holiday. These will vary, depending on the culture of your school. The children can use clothespins to clip the symbols in temporal sequence (see Figure 2–9). The children must decide which symbol comes first, second, and so on as they place the symbols in proper sequence.

You can gradually apply the sequencing strategy to specific historical topics being studied. For example, to introduce the general topic of U.S. history, put up a large section of butcher paper (1 yard by 4 yards) and mark off sections to represent centuries. Tell the children that you will be creating a timeline of U.S. history (see Figure 2–10). For most elementary school purposes, the history of the United States begins at 1492, so you can have one segment for each century beginning in the 1400s. As you read about people such as Columbus, Washington, Lincoln, or King, place symbols within the blocks designating separate centuries to help the children see when these people lived.

Another wise way to develop chronological thinking skills with timelines is to use the children's own lives as events to sequence. You can prepare large cards (and add appropriate photos, if possible) with the labels "Birth," "Learn to Walk," "Go to School," "Enter Grade 4," "Go to Middle School," "Enter High School," and "Graduate from High School." Take the children to the playground and assign one card to each child. The child holding the "Birth" card is the starting point for the timeline. The children can next suggest the age at which they began to walk. Have the child holding the "Learn to Walk" card pace from the "Birth" point the number of steps to represent the years from birth to walking. Use the same process for each of the other cards. Discuss the relationships between the distances.

This procedure can also be used to show the sequential development of historical events. The children can first be presented with a random number of cards listing events in history. They can go to the playground, organize the cards in sequence,

FIGURE 2–8
Event Chain

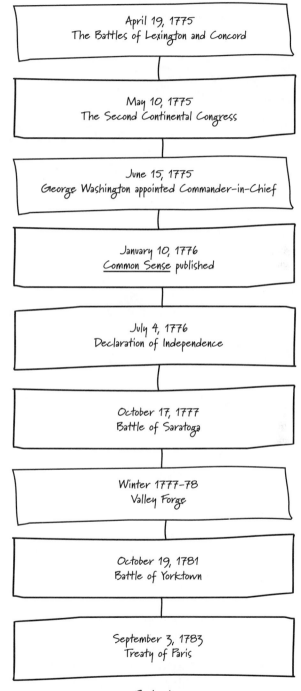

The Revolutionary War
Initiating Events

April 19, 1775
The Battles of Lexington and Concord

May 10, 1775
The Second Continental Congress

June 15, 1775
George Washington appointed Commander-in-Chief

January 10, 1776
Common Sense published

July 4, 1776
Declaration of Independence

October 17, 1777
Battle of Saratoga

Winter 1777-78
Valley Forge

October 19, 1781
Battle of Yorktown

September 3, 1783
Treaty of Paris

Final outcome

FIGURE 2–9
Sequence Skills

FIGURE 2–10
Beginning Timeline

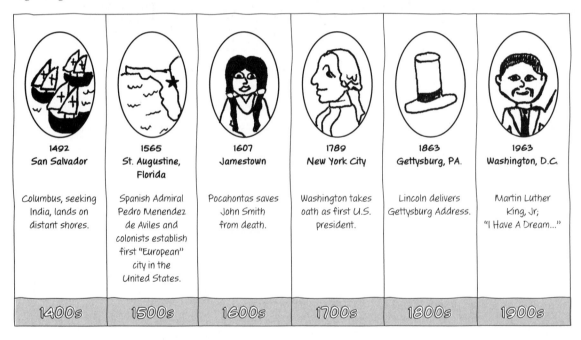

1492 San Salvador	1565 St. Augustine, Florida	1607 Jamestown	1789 New York City	1863 Gettysburg, PA.	1963 Washington, D.C.
Columbus, seeking India, lands on distant shores.	Spanish Admiral Pedro Menendez de Aviles and colonists establish first "European" city in the United States.	Pocahontas saves John Smith from death.	Washington takes oath as first U.S. president.	Lincoln delivers Gettysburg Address.	Martin Luther King, Jr; "I Have A Dream…"
1400s	1500s	1600s	1700s	1800s	1900s

and walk off one step for each year between events. This practice is especially appropriate in the upper grades, since history facts are difficult for most primary children to understand. You would use a similar procedure when transferring this practice into the classroom, but the children will soon realize that if they take one step for each year between events, they will soon run out of room and not be able to

Timelines are excellent tools for helping children develop chronological thinking skills.

complete their task. You can then direct them toward discovering that a smaller unit of measure will be needed, perhaps 1 inch to represent a year.

When working on smaller-scale timelines in the classroom, emphasize exactness and consistency. An inexact or inconsistent scale distorts time relationships and interferes with accurate chronological thinking.

For standard classroom timelines, such as the one in Figure 2–11, give groups of children a set of cards identifying events from a current topic of study. Ask the children to look up information about their assigned event and draw an illustration about it. From there you can direct the children to find the date for each event and place their illustrations in chronological order, with an appropriate scale on the timeline. Following this, have each group share pertinent information about the different events with the rest of the class. By completing this project, your class will have constructed a master guide of the major events under study and will have begun to develop an historical perspective of particular events.

FIGURE 2–11
Timeline

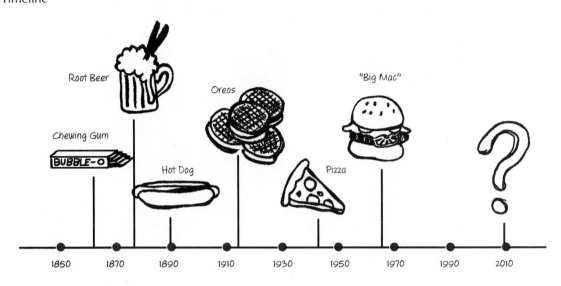

AFTERWORD

History has long been a valued part of schooling in America; it continues to exert a major influence on what and how social studies is taught in our nation's elementary schools. Many have praised its value for producing good citizens over the years, but none has done so more eloquently than Winston Churchill, who once proclaimed, "The further backward you look, the further forward you are likely to see." Such statements underscore the importance of developing the skills and sensitivities of historical consciousness in our schools. In a society steeped in triumphs and tragedies, knowledge of our past helps us to develop pride in our successes and discontent with our errors.

We cannot, however, expect children to become interested in the study of history when all we ask them to do is memorize facts from a textbook. Surely, content is an important part of history, but we must also be aware of the processes of history. Young historians must have regular opportunities to *explore* history rather than simply be *exposed* to it. We must lead students to perceive the nature of history itself. Those techniques will help students acquire a more balanced sense of history— it is not only something one *knows* but also something one *does*.

References

Beatty, P. (1987). *Charley Skedaddle*. New York: Morrow.

Darigan, D. L., Tunnell, M. O., & Jacobs, J. S. (2002). *Children's literature: Engaging teachers and children in good books*. Upper Saddle River, NJ: Merrill/Prentice Hall.

Eisenberg, L. (1991). *The story of Sitting Bull, great Sioux chief.* New York: Dell.

Freedman, R. (1989). *Lincoln: A photobiography.* New York: Clarion.

Fritz, J. (1982). *Homesick: My own story.* New York: Putnam.

Milton, J. (1987). *Marching to freedom: The story of Martin Luther King, Jr.* New York: Dell. 4–5.

National Center for History in the Schools. (1996). *National standards for history.* Los Angeles, CA: Author.

Schreifels, B. (1983). Breathe life into a dead subject. *Learning,* March 1983–84.

3 Young Political Scientists

The Clarksville Senior Center now has a peaceful new flower garden, thanks to Debra Wood's fifth graders. While visiting the center in early autumn, the students noticed that an area between the main building and a shed looked desolate and depressing. After returning to the classroom, the students asked Ms. Wood what they might be able to do as a class project to improve the area. After much discussion, they eventually decided on a flower garden. "A garden is so restful and peaceful," explained Jana. "Yeah," added Vance, "a garden could even bring back memories of World War II victory gardens."

Ms. Wood is an avid gardener and the president of the largest local garden club. She invited several members of the club to visit the class to recommend plants that would need a minimum amount of upkeep in their hot and dry climate. The garden club donated the flowers for the children's project and offered advice and help to the children as they cleaned up the area and planted the flowers. A parent who owned a landscaping company wanted to help, so the students dressed the flower area with his donated mulch. The class then solicited other local businesses for help; several chipped in to help as the garden expanded beyond their wildest dreams. A retired carpenter volunteered his time and expertise to make benches. A local lumberyard supplied the gravel and decorative block for a patio where the beautiful wooden benches now rest. An individual donor contributed a birdbath and feeder. In addition, the students took up a collection for a butterfly bush that added to the overall ambiance of the garden.

The garden is wheelchair accessible, a definite plus for seniors with restricted mobility. The senior adults are now responsible for the general upkeep of the garden and have passionately accepted that responsibility. "It gives them something to take pride in," explains Ms. Wood. "Some senior adults like the tranquility of the garden and will sit on the benches reminiscing. Others find the garden a great place to socialize. I'm so glad the students thought of this wonderful idea."

One child's reaction to this project is shown in Figure 3–1. Buoyed by the success of their flower garden project, Ms. Wood's students tackled new opportunities for community volunteerism with boundless enthusiasm and energy. They adopted a local park and pledged to keep it clean, packed breakfasts for the homeless as part of

FIGURE 3–1
A Reaction to the Garden Project

> It was cool seeing all the beautiful flowers everywhere. It was fun when we got to plant and water the flowers.
>
> I learned it can be fun to do stuff for the community. I never thought planting flowers could be fun but it is. It made the center look very pretty. I hope we keep up with the project and plant new flowers next spring.
>
> Billy

a Martin Luther King Day community service project, collected funds for the March of Dimes, made placemats and took them to a local nursing home to be placed on meal trays on Valentine's Day, and raised $200 for a classmate's family whose house was destroyed by a tragic fire. Through their incredible thirst for involvement, Ms. Wood's young political scientists took their first ambitious steps into the realm of real-life learning called *service learning*. Although service learning cuts across all curricular areas, it is normally associated with the discipline called *civics* (sometimes termed *political science* or *government* in high school or college).

WHAT IS CIVICS?

Civics is the study of the workings of governments and of the rights and responsibilities of citizenship. Individuals involved in the specialized study of government and the obligations of citizenship are called *political scientists*. In this chapter, elementary school students will be referred to as *young political scientists* because, in dynamic social studies classrooms, they participate in citizenship education activities throughout the year, applying their civic knowledge to the solution of real problems—just like professional political scientists.

The unquestioned principal goal of public education over the years has been to prepare students for *effective citizenship*. The National Council for the Social Studies (2001) has defined an effective citizen as one "who has the knowledge, skills, and attitudes required to assume the 'office of citizen' in our democratic republic" (p. 319). To ready themselves to occupy this esteemed office, the NCSS (2001) advises that students participate in well-planned and organized citizenship education programs in which activities "expand civic knowledge, develop participation skills, and support the belief that, in a democracy, the actions of a person makes a difference. Throughout the curriculum and at every grade level, students should have opportunities to apply their civic knowledge, skills, and values as they work to solve real problems in their school, the community, our nation, and the world" (p. 319).

WHY IS CIVICS IMPORTANT?

In 1782, few believed that a unified nation could be created out of a collection of "free and independent states" spread out over a vast expanse of land. Each had incredibly diverse economic interests, was fearful of an overly strong federal government, and remained fiercely loyal to established regional, ethnic, and religious ties. The newly independent states often fought each other over land and money, and, to add fuel to an already blazing fire, there was no national political organization with the power to settle their disputes. By 1787, the situation escalated to such a level that many people proposed that the new union of states could not survive without a strong federal government. Therefore, in the hot summer of 1787, 55 delegates from 12 of the 13 states gathered in Philadelphia for "the Grand Convention" which was charged with the daunting task of creating a strong federal government while protecting the rights of the states and individuals. The delegates, known as the *framers*, took 4 months to draw up the plan: the United States Constitution, a document that spelled out the government of a new nation. It was from this unsettled start that the seeds of one of the grandest political experiments of all time sprouted its roots—a republican nation with a representative democracy.

By the late 1700s, after the Constitution was ratified, many of the early disputes among the states began to fade and a new feeling of patriotism emerged. It was especially fitting that education was considered an important factor that would guide the country and its people into the future. Central to the framers' conception of a successful democratic representative government was an informed public citizenry capable of exercising their rights and responsibilities in an informed and meaningful manner. And this capability, according to Thomas Jefferson, grew from civic education. Civic education would provide the knowledge and courage to "enable every man to judge for himself what will secure or endanger his freedom." If citizens did not know the Bill of Rights, for example, how could they weigh individual freedoms against the needs and welfare of the common community?

From the beginning of our nation's history, civic education has been, and continues to be, central to public education and essential to the survival of American democracy: " 'Government of the people, by the people, and for the people,' in Lincoln's phrase, means that the people have the right to control their government. But this right is meaningless unless they have the knowledge and skills to exercise that control and possess the traits of character required to do so responsibly" (The Center for Civic Education, 1994, p. 2). Therefore, dynamic social studies is based on a conviction that civic education is fundamental to the preservation of our constitutional democracy and, therefore, the primary component of all education in the United States.

WHAT SHOULD YOUNG POLITICAL SCIENTISTS KNOW OR BE ABLE TO DO?

The goal of civic education is the development of informed, responsible citizens committed to the principles of American constitutional democracy. Their effective and responsible participation grows from the acquisition of a body of knowledge and a set of participatory skills. Toward this goal, the Center for Civic Education (1994) established a set of standards specifying what students should know and be able to do in civics. The K–4 standards are summarized in Table 3–1. The standards are not intended to be used as a basis for a national curriculum in civics, but simply

TABLE 3–1
K–4 Content Standards

I. WHAT IS GOVERNMENT AND WHAT SHOULD IT DO?
  A. What is government?
  B. Where do people in government get the authority to make, apply, and enforce rules and laws and manage disputes about them?
  C. Why is government necessary?
  D. What are some of the most important things governments do?
  E. What are the purposes of rules and laws?
  F. How can you evaluate rules and laws?
  G. What are the differences between limited and unlimited governments?
  H. Why is it important to limit the power of government?

II. WHAT ARE THE BASIC VALUES AND PRINCIPLES OF AMERICAN DEMOCRACY?
  A. What are the most important values and principles of American democracy?
  B. What are some important beliefs Americans have about themselves and their government?
  C. Why is it important for Americans to share certain values, principles, and beliefs?
  D. What are the benefits of diversity in the United States?
  E. How should conflicts about diversity be prevented and managed?
  F. How can people work together to promote the values and principles of American democracy?

III. HOW DOES THE GOVERNMENT ESTABLISHED BY THE CONSTITUTION EMBODY THE PURPOSES, VALUES, AND PRINCIPLES OF AMERICAN DEMOCRACY?
  A. What is the United States Constitution and why is it important?
  B. What does the national government do and how does it protect individual rights and promote the common good?
  C. What are the major responsibilities of state governments?
  D. What are the major responsibilities of local governments?
  E. Who represents you in the legislative and executive branches of your local, state, and national governments?

IV. WHAT IS THE RELATIONSHIP OF THE UNITED STATES TO OTHER NATIONS AND TO WORLD AFFAIRS?
  A. How is the world divided into nations?
  B. How do nations interact with one another?

V. WHAT ARE THE ROLES OF THE CITIZEN IN AMERICAN DEMOCRACY?
  A. What does it mean to be a citizen of the United States?
  B. How does a person become a citizen?
  C. What are important rights in the United States?
  D. What are important responsibilities of Americans?
  E. What dispositions or traits of character are important to the preservation and improvement of American democracy?
  F. How can Americans participate in their government?
  G. What is the importance of political leadership and public service?
  H. How should Americans select leaders?

to offer a guide to teachers so they know what they should teach their students and as a framework for curriculum developers on which they might build high-quality civics programs. After examining the standards to see what they might contribute to her social studies program, Claire Boyer, a second-grade teacher at Media Elementary School, decided to take the study of local government beyond the textbook to where the action was. Therefore, she turned her students loose on a civics adventure that involved them in the process of gaining knowledge about the responsibilities involved in establishing the well-being of communities as well as in meaningful opportunities to simulate civic life in the community. Guiding her efforts were the following standards: Standards I.A. ("What is government?"), I.C. ("Why is government necessary?"), I.D. ("What are some of the most important things governments do?"), I.E. ("What are the purposes of rules and laws?"), and III.D. ("What are the major responsibilities of local governments?").

Classroom Vignette

For a special civics project, Ms. Boyer challenged her young political scientists to build a town for themselves and run it. To begin the project, Ms. Boyer asked her students to pretend to be adult citizens of Media 30 years in the future. To draw them into the simulation, she read them a "letter from the Environmental Protection Agency" demanding that all families move from Media as soon as possible due to severe and irreversible pollution problems. The families, traveling together in search of a new place to settle, were led to a large, empty room in the school basement made available just for their project. To keep track of the victims of this unfortunate plight, the "citizens" were required to fill out Official Community Census Forms. After they completed the census, the citizens of this new town (named "New Media" by vote) built homes for themselves from large packing boxes put aside from a shipment of new furniture the school received that summer. The citizens painted and pasted until they were satisfied the boxes looked like "real houses" and brought a sense of reality to the empty room. Streets were laid out and named: a street sign bearing the name "Dunlap Street" was a tribute to Mr. Richard Dunlap, the principal of Media Elementary School, but no one was quite sure of the inspiration for Grape Road or Ice Road. A town newspaper was launched to chronicle the daily progress of New Media's citizens and to keep its populace informed. "Pollution Sends Townspeople to New Land" blared the headlines on January 7, the first day of the project. A subsequent story read, "Townspeople Paint the Town," in reference to the construction of new homes.

The day after the families completed their homes, they held a town meeting to discuss potential community problems, with Ms. Boyer presiding for the first meeting only. As is often the case with youngsters this age, they could foresee no problems in particular. However, Ms. Boyer was quick to suggest some—fires, crime, and problems that might arise if she could not lead future town meetings. Discussion led to the establishment of police and fire departments and an election for mayor. The children quickly set up minimum qualifications for voter registration and went about soliciting candidates for the mayoral position. Seven candidates immediately announced their intent to run, but three dropped out of the race the following day—they were too busy. Campaigning and debating began as students forged their platforms: Alex promised low taxes, and Curtis vowed gun control. Candidates then planned campaign strategies, showing that political make-believe mirrors political reality. There was, to be specific, the "great cookie caper," involving Alex and her closest opponent, Curtis. On the last day of campaigning, Alex distributed "Vote for Alex" pamphlets decorated with paper hands grasping real chocolate chip cookies. Curtis's followers quickly cried "Bribery!" and complained that Alex was trying to buy votes. The matter went to the election board, which found that "No influence was obtained through the distribution of the cookies."

Following her landslide 14–2 victory, Alex immediately appointed Curtis as chief of police and presented him with his first book of tickets. Using his tickets to control the breaking of laws such as speeding (running in the halls) and loitering (daydreaming), Curtis eventually learned the powers of his position. Through it all, Alex made new friends, was subject to the pressures of old ones, and generally learned that a position of authority has its rewards as well as its pitfalls. "I learned I'm never gonna be the real mayor," she reflected. "Even just pretending to be the mayor is a tough job."

Leading New Media through its hectic early days, Alex and her council members provided crucial leadership as the town began to grow. Other classrooms acquired a sense of civic participation and offered to contribute to the growth of New Media. The first graders, studying the topic "Needs of People," contributed a food store and displayed the products themselves (Bob's California watermelon: $20 a pound). The third grade, studying "What Towns Need," built an electric power station, stringing yarn lines from one paper light pole to another all around town. The fourth grade, not involved in a relevant social studies topic at the time, demonstrated the interrelatedness of the physical and social sciences. They wired up streetlights by connecting batteries to light bulbs, thus applying their knowledge of energy to making lives better for people. The fifth grade, anxious to contribute with the rest, made a cardboard trash truck (complete with oatmeal-box "trash cans" for the customers) and a bus from cardboard boxes. Finally, the kindergarten class spruced up the entire town with pink, white, and red paper flowers.

The entire village of New Media grew through the remainder of the school year as the children added new features to show what they were studying. The students served as perfect hosts as visitors from area elementary schools came to Media Elementary School to witness the expansion of New Media.

In this productive classroom experience, Ms. Boyer set in motion a well-chosen project intended not only to address the targeted standards, but also an important educational aim recommended by the NCSS (2001) in its position statement, *Creating Effective Citizens:* "Students are provided with opportunities to participate in simulations . . . and other activities that encourage the application of civic knowledge, skills, and values" (p. 319). Simulations and activities that directly involve the students in experiencing and making decisions provide a rich context for civics instruction.

In General, How Should Civics Be Taught?

As we have learned, our nation's public schools bear a significant and historic duty to enhance the acquisition of civic knowledge and civic responsibility in our nation's youth. Schools fulfill this duty through both *informal* and *formal* experiences that begin during the earliest years of schooling and extend through the upper grades. Informal experiences include all the strategies teachers use to establish a cohesive, supportive, democratic classroom community characterized by warm, friendly interactions among all members. The informal curriculum is centered on the governance of the school community and the relationships among those within it. Adults who govern in accordance with constitutional values and principles and who display traits of character worth emulating are who we need to manage the classrooms and schools of our democratic nation.

Informal Methods of Civics Instruction

Informal methods of civics instruction refer to the classroom management techniques teachers employ to create a spirit of democratic community. Democratic teachers strive to create a strong, cohesive team feeling among the members of the class, fostering an attitude of "we're all in this together." Citizenship education happens, then, when teachers accept students as partners in developing a mini-democratic society.

Elementary school classrooms are matchless settings for helping students undergo early and meaningful experiences in responsible citizenship. These experiences begin on the first day of school, for at no other time during the school year does anticipation and hope soar so high. Returning from a summer's respite, children are excited about returning for a fresh, new start. Sadly, one of the first things many teachers do with these eager young scholars is to squash their enthusiasm by informing them right away who's the boss: "Read them the 'riot act' as soon as they step inside the door. Jump on them for every little thing; don't let anything go by. If you do, it only gets worse!" They go on to suggest, "Don't smile until Thanksgiving!

You've got to be like the Gestapo or they'll run all over you." Not surprisingly, class-room management experts roundly reject this point of view. "That's ridiculous advice," they counter. "Children should see their teacher as a positive influence, not as someone to keep clear of. Teachers need to smile, not scowl. If you don't smile until Thanksgiving, neither will the children!" All too often the only memories the students go home with from a "don't smile" teacher's classroom on the first day of school is how stern the rules lecture was; nothing else of consequence "happened that day."

At the beginning of the school year, teachers need to establish trusting relationships with their students, making them feel that the teacher is on their side and that they are accepted. Students who sense that they are accepted are ready to take on the challenges of a new school year. In other words, the job of a teacher in a democratic classroom starts with acceptance.

Establishing Rules

Establishing rules with children is an important part of the first days of school. When children have a voice in the rule-making process, they develop a sense of ownership and pride that sets the tone for the rest of the school year. Teachers find that

Important lessons in civics take place informally in cohesive, democratic classrooms.

children work best when they have a hand in making the rules and are more inclined to remember the rules, respect the rules, follow the rules, and take a role in group problem solving should classmates have trouble following the rules. Most importantly, the process of collaboratively defining classroom rules builds skills in and respect for democratic processes.

Beginning teachers readily accept the idea of collaboration, but often are uncertain about how to carry out the process of jointly determining rules early in the year. The process is often overwhelming to those who sense a need for establishing classroom control during this beginning phase of their professional careers. There is nothing wrong with starting off your first year of teaching with a set of your own rules and consequences. However, once you acquire greater confidence in your management skills, you will want to give your students a sense of control, too. The following scenario offers an example of the rule-making process as carried out by an experienced teacher in a fourth-grade classroom. The script can be easily followed in its entirety or adapted for different grade levels or individual situations.

Classroom Vignette

Lester Pentek prefers to tap the rich storehouse of children's literature whenever he initiates a new topic of instruction in his classroom. At the start of the school year, he enjoys using Marie Winn's book *Shiver, Gobble, and Snore* (Simon & Schuster), a particularly suitable story for helping students grasp the importance of rules. It focuses on a funny king who made silly rules. In his kingdom lived three unhappy subjects: Shiver, who was always cold; Gobble, who was always hungry; and Snore (guess what he liked to do). Many of the king's rules severely limited the cravings of these one-of-a-kind characters, so they decided to move away to a place where there would be no rules. Alas, the three friends discovered that disputes could not be resolved in their new land—because they had no rules! They finally decided that to live peacefully, they must make rules. After reading the story, Mr. Pentek guides his children through a discussion: "What are rules? Who made the rules in the kingdom? Did the rules make sense or were they foolish? Did all the people want to obey the rules? Why or why not? What did they decide to do? What else could they have done? What have you learned about rules?" The purpose of these questions is to encourage children to realize that rules are necessary to protect people's rights and to keep them safe from others' unacceptable behavior.

Following the general story conversation, Mr. Pentek divides his class into small discussion groups. Based on the story, he directs each group to suggest characteristics that would make a rule a good rule. Here are some of their recommendations:

- The rule should be clear enough so everyone understands what it means.
- The rule must be fair, so everyone can do what is expected.
- The rule must have consequences; everyone must know the outcome of not following the rule.
- Rules should be posted in the classroom for everyone to read.

The next day, Mr. Pentek opens a discussion about rules: "Tell me some rules, or laws, that citizens of our community are required to follow. Why are these rules important? Who made these rules?" He then made a connection to their classroom community: "Are rules important for our classroom? Why are rules important? Who should make these rules?" The discussion always ends with agreement that rule setting should be a collaborative process.

The process of collaborative rule setting starts with a direct question: "What are some important rules that can help us live together harmoniously in our classroom community?"

"You mean like no chewin' gum?" Paul mumbled through a wad of gum pocketed inside one cheek.

"Like no pushing and hitting?" wondered Lisa.

"Yes," answered Mr. Pentek. "Any of these would be fine."

For this phase of the process, Mr. Pentek divides his class into small groups and allows the groups to brainstorm possible classroom rules for one-half hour or more. He then randomly selects one group to write its first two choices on the chalkboard as the other groups cross off similar items from their lists (so they won't be repeated). The groups, in turn, repeat this process until a whole-class list is on the board. This list, and the related discussion, will most certainly be long, as groups will want to bring up specific incidents to go along with each suggested rule. Mr. Pentek usually ends up with a list of over 40 rules, most of which are stated negatively: "Don't fight," "Don't talk out loud unless you raise your hand," "Don't take things without permission." Groups tend to make a negatively stated rule for every miniscule situation they can imagine.

Next, Mr. Pentek and the students review their list. He asks the question, "Do any of these rules look like they might go together?" He offers the students an example from their list: "Here's what I mean—'Don't talk out of turn,' for example, might be combined with 'Raise your hand when you want to talk.'" The students use colored chalk to underline those they want to group together. Mr. Pentek now wants the students to defend their actions, so he asks the question, "Why did you group these rules together?" The question helps students see how one general rule can supplant several specific ones. To further develop this idea, Mr. Pentek challenges the students to create one big rule that can take the place of all the little rules in each group. Four general rules remain. Mr. Pentek says that rules beginning with "no" often seem harsh to him, so he asks the students to rephrase any negatively stated rules as positive ones.

The final task involves constructing a permanent chart listing the classroom rules, having each child sign it, and displaying it prominently in the classroom. Here is the list Mr. Pentek's students constructed:

- Be polite and kind to others.
- Respect the property of others.
- Follow all classroom procedures (they listed several, such as using the pencil sharpener, using the restrooms, getting a drink of water, fire drills, returning assignments, talk among students, and so on).
- Try to solve disagreements ourselves. If we can't, we will ask the teacher for help.

Although the example above was from an upper-grade classroom, the process can be quite easily adjusted for younger children. Marie Clouser, a teacher in the same building as Mr. Pentek, shortened the process of rule making with her second graders. She operated on the belief that, for younger children, the teacher should display four important rules at the beginning of the school year: *be on time; be pre-pared for class; respect other people;* and *respect other people's property.* The students then help with setting guidelines by suggesting specific behaviors associated with each rule—raising hands, putting away supplies, using the restroom or water fountain, headings for papers, and so on.

Using either approach, teachers retain the right to add rules or veto suggestions. As in our federal system of checks and balances, teachers must use their professional judgment to protect the rights and responsibilities of the group as a whole.

Class Meetings

Democratic classroom teachers do not take time away from group learning to focus on individual behavior or the behavior of a few students. Those problems are best addressed by dealing with the specific individuals involved. When conflicts do involve large numbers of students, however, class meetings (like democratic town meetings) are used to deal openly with the issue before it negatively affects the spirit of the classroom. Class meetings will vary according to grade level; the idea presented next is most appropriate for the early grades. It can be easily adjusted to meet the needs of older children, too. Carol Longwell, a student teacher eager to apply some of the strategies she learned in her classroom management course, seized this opportunity to use a classroom meeting when a tense situation arose—children complained about an excessive amount of name-calling that had seemed to grow overnight.

Classroom Vignette

Ms. Longwell: It has come to my attention that some children are using words to make others feel bad or make them feel unhappy. Different words make different people feel bad. For example, I feel unhappy when someone calls me "skinny." Did someone ever use words that made you feel bad or upset?

Patrick: My sister calls me "blockhead."

Ms. Longwell: How does that make you feel? (No response.) Do you like to be called "blockhead"?

Patrick: No. It makes me feel bad.

Ms. Longwell: Would someone else like to share a word?

Lillian: It makes me feel bad when someone says, "Dummy."

Ms. Longwell: I know what you mean; it makes me feel angry, too.

Albert: The big boys say, "Runt, get outa here!" I don't like it.

Patricia: Yeah, the big girls say, "You can't play with us, you peewees." I don't like when they say that.

Ms. Longwell: I can tell by your voice that you feel hurt when someone calls you names. Alice, you look like you have something to say.

Alice: The big kids call me "half-pint." I don't like it.

Ms. Longwell: It isn't a nice feeling inside when others call you names. Did anyone ever say some words that made you feel happy? (Two children start to smile but don't respond verbally to the questions.) I can tell that you're thinking of something that makes you happy because you're smiling.

Leroy: Yes. One time Mr. Kirk told me I have nice eyes.

Ms. Longwell: How did that make you feel?

Leroy: Awesome. I said, "Thank you," and I told my mom and dad.

Jane: (Blurts out) "I like you!"

Ms. Longwell: How does that make you feel when someone says, "I like you"?

Jane: It makes me feel all sunny inside.

Ms. Longwell: I like the way you said that, Jane. Peter, what did someone say to make you feel good?

Peter: Someone said, "You're smart."

Ms. Longwell: How did that make you feel?

Peter: I liked it.

Ms. Longwell: Thanks for telling me about the name-calling. We need to think of a plan for what you can do when someone calls you bad names again.

Everett: I could tell him to stop. If he doesn't, I can run away.

Ms. Longwell: That's a good idea to get away. But that might be hard to do sometimes. Can you think of another idea in case you find it hard to run away?

Brenda: We could tell somebody.

Ms. Longwell: I like that idea, too. Who could you tell?

Brenda: You. But if you're not there maybe we can tell another teacher.

Ms. Longwell: These ideas sound like a good plan. What have we suggested so far?

Gwen: Well, first I would ask the boy or girl to stop.

Cynthia: If that doesn't work, then I can walk away or tell you.

Shayne: If you're not there, we can tell another teacher.

Ms. Longwell: Do you all agree to the plan? (Class agrees.) Thank you for your wonderful suggestions.

Class meetings are best carried out in an environment where children can share freely and where they realize that what each has to contribute is worthwhile. A comfortable seating arrangement in a circle, semicircle, or oval provides the greatest attention and maximum participation, as children are able to see one another as they talk. A democratic group leader, the teacher first serving as a capable model, helps the children to grasp the difficult procedure of discussion:

- Sit in an unassertive position so that the leader is not perceived as the most important person with all the "right answers."
- Clearly present the topic of discussion. Make sure everyone understands what the problem is.
- State the rules of good discussion—raise your hand when you want to speak, listen thoughtfully to the ideas of others, and be respectful of the ideas of others.
- Use *active* listening strategies (e.g., paraphrase a student's comments or ask clarifying questions).
- Summarize the ideas shared. All of the ideas should be concisely repeated: "Let's go back and recall what was said."

Classroom Symbols

One of the factors contributing to the bonding together of states into a new nation was the introduction of patriotic symbols. The children have a spirit for bonding together as a classroom community at the beginning of a new school year, too, and teachers can take advantage of their excitement by planning activities that have to do with building class spirit. Study either national or state symbols and ask the children to develop a list of symbols they would like for their classroom community: flag, song, motto, great seal, flower, animal, tree, cheer, colors, and so on. All suggestions should be welcomed and the class should vote to determine the preferences in each category. For those that can be designed and illustrated, the students should do so and display them proudly throughout the year. Others should be used when appropriate: the classroom song may be sung on Monday mornings or Friday afternoons; the class cheer can be used to reward special academic accomplishments; and the great seal may be used on classroom stationery. Sometimes, students even decide to select a classroom name that they will be known by throughout the year—two that I remember well are "Ms. Beam's Dreams" and "Mr. Carr's Superstars." It is surprising how much of a community spirit can be generated by these special symbols.

Classroom Holidays

We all agree that national holidays such as Independence Day or President's Day help build a feeling of national pride and patriotism. The same can be said for classroom spirit; special classroom "holidays" help enhance a sense of group identity. Special days might include hat day when everyone wears a favorite hat to school; color day when students wear their school colors; or "Retro-Day" when students come to school dressed as students did in the 1950s or 1960s. It is fun to immerse the class totally into the day it is celebrating by gearing your curriculum to the theme, reading appropriate books, displaying suitable pictures, and playing fitting music.

Games

Several games in which students must cooperate to reach a goal can be used to help the group function as supportive classroom members and enhance classroom cohesiveness. John Quigley uses the "Rain Gang" with his second graders, for example.

Classroom Vignette

To begin, Mr. Quigley and his children arrange their chairs in a large circle and sit down. He tells the children that he is going to start an action that he will pass on to the child on his right. That child will imitate the action and pass it on to the next child and so on until it goes all the way around the circle and back to Mr. Quigley. He starts by rubbing his hands together. The child to the right joins in, then the next child, the next, and so on until everyone is doing it. The sound should remind the children of a soft mist that is beginning to fall. When all are rubbing their hands, Mr. Quigley starts snapping his fingers. Each child keeps rubbing hands until it is his or her turn to snap fingers. This signals the beginning of a heavier sprinkle. After the finger snapping has made its way around the circle, Mr. Quigley starts the next sound—hands slapping thighs. Now the rain is getting much heavier. Finally, at the peak of the storm, Mr. Quigley slaps his thighs and stomps his feet in unison. The rainstorm ends as each of the actions is reversed and, one by one, the hand rubbing stops.

Kathryn Burke uses group-cohesiveness activities in her fifth-grade classroom, too, because she feels that they help create the kinds of positive peer relationships that form a pleasant and productive learning community. One of her favorites is a game called "Pipe Line."

Classroom Vignette

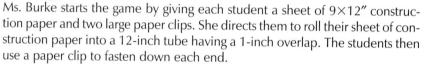

Ms. Burke starts the game by giving each student a sheet of 9×12" construction paper and two large paper clips. She directs them to roll their sheet of construction paper into a 12-inch tube having a 1-inch overlap. The students then use a paper clip to fasten down each end.

Next, Ms. Burke separates the group into two teams and has them face each other in two parallel lines. She directs each team to hold their paper tubes end-to-end to make a long pipe. She then crushes two sheets of tablet paper into balls small enough to fit into the tubes and places one in an end tube of each team. She then gives the order, "Go!"

The students must pass the paper balls from one tube to the next. If a ball falls, the last student to have it in his or her tube picks it up and starts it going again. Passes can be made only from tube-to-tube in sequence. When the paper ball reaches the end of the line, the teams reverse the direction. The first team to get the ball back to the beginning point is the winner.

Although the children learn a great deal from the games themselves, follow-up discussions help solidify the targeted cooperative behaviors. First and foremost, always let the students know that you value their cooperative efforts. Then talk with them about the behaviors that contributed to each group's success (or lack of it). You might want to offer constructive feedback: "While I was watching, I saw that Gary was a bit unsure about how he might get the paper to the next person. But Bryce jumped in right away and showed him how to solve the problem." Or, you might want to ask the students to identify something they did to contribute to their group's efforts: "How did your group make sure that everyone knew how to coordinate their movements?" Ask the students what they might be able to do better in a similar situation the next time they try it.

Don't expect your students to become a cohesive, caring community after only a week or two of these activities, however. A close community takes time to forge—don't rush or become impatient. Use a variety of informal group activities throughout the year and your students will soon exemplify and demonstrate core democratic values and skills.

Formal Methods of Civics Instruction

As children become involved in meaningful democratic classroom life and are given significant opportunities for participating in classroom governance, they will develop a greater insight and appreciation for civic life in the community, nation, and world. The Center for Civic Education, in its influential *National Standards for Civics and Government* (1994), maintains that such knowledge is communicated to students through formal instruction that provides students with "a basic understanding of civic life, politics, and government. It should help them understand the workings of their own and other political systems as well as the relationship of American politics and government to world affairs. Formal instruction provides a basis for understanding the rights and responsibilities as citizens in American constitutional democracy and a framework for competent and responsible participation" (p. 1).

Effective citizenship education, then, has a content base, but not superficial coverage of facts—for example, "The federal system of government divides power among the executive, legislative, and judiciary branches"—without regard for what the facts mean. Students often find such instruction trivial and uninteresting because their teachers fail to establish the relevance of civic knowledge to their lives. Therefore, as young political scientists, students must not only know the facts of civics, but also how to ask questions and answer them. Young political scientists must know what political scientists do. Some of the common civics experiences commonly found in elementary school classrooms follow.

The Constitution of the United States

Normally a part of the fifth-grade social studies program, the children learn that the Constitution is the cornerstone of our government—the basic and supreme law of the United States. In 4,543 words, the Constitution outlines the structure of the United States government and the rights of all American citizens. The Constitution is known as a "living document" because it is flexible enough to be changed (amended) when necessary. The Constitution is divided into three parts:

- *Preamble*—This explains the purpose of the document.
- *Articles*—These describe the structure of the government (legislative, executive, and judicial branches) as well as the process for amending the document. There are seven articles.
- *Amendments*—The changes made to the Constitution. The first 10 are called the Bill of Rights; there are a total of 27 amendments approved since 1791.

Karla Griffin realized that the original United States Constitution places very difficult reading demands on most fifth graders, so she chose to familiarize her students with it by first showing them a replica of the original document and explaining its importance as the basic and supreme law of our nation. Then she asked the students to quickly skim the three main parts.

Classroom Vignette

After each part was skimmed, she asked the students to give a very short description of what it was about. Then, to deepen their understanding of the Constitution, Ms. Griffin passed out copies of *Shh! We're Writing the Constitution* by Jean Fritz (New York: Putnam). In a very engaging style, Fritz transports the reader back to the Constitutional Convention in Philadelphia to tell the story of the birth of our Constitution. Following both of these experiences, Ms. Griffin directed her students to plan and present a skit about the Constitutional Convention. Some students were assigned to role-play specific delegates to the Constitutional Convention (Benjamin Franklin, James Madison, and so on) while others were to portray imaginary television news reporters. The reporters were to interview the delegates about their positions on the real issues. Ms. Griffin recorded the interviews on videotape so everyone could later view the entire dramatic scene.

Following this introduction to the Constitution, Ms. Griffin planned an extended sequence of activities during which she used an easy-to-read version of the Constitution. The students first read the Preamble and Bill of Rights, trying to summarize each. Ms. Griffin did not become impatient with their error-filled efforts, for she realized even college students and adults have difficulty summarizing the Constitution. She did, however, try to capture what was important in those parts. One day, for example, a news article told of the malicious neglect of a dozen dogs that were terribly mistreated by a local owner. A photo accompanying the story depicted several of the incredibly undernourished, scrawny, sickly dogs. The students spent some time discussing the ways these dogs in particular, and animals in general, are sometimes badly treated. "It's like they need a Constitution!" blurted Charlie. "Not a bad idea," chimed in Jaung. Seizing the opportunity, Ms. Griffin divided the students into three groups. Each group was to use the simplified United States Constitution as a model and work together to write a "Preamble" and "Bill of Rights" for an "Animal Constitution." When they finished, the groups compared their "Preambles" and "Bills of Rights," eventually merging them into a single document designed to protect the rights of all animals.

Teachers serious about helping students understand the Constitution look for any opportunity to relate its principles to their lives. Rather than limiting instruction to the memorization of First Amendment rights, for example, Yvonne Jordan encouraged her students to compile a "First Amendment Diary" to personalize what they learned about the rights guaranteed under the First Amendment.

Classroom Vignette

First, Ms. Jordan's students read that when the Constitution was ratified in 1789, some states refused to go along with it because they were concerned that it did not protect certain freedoms. They thought the Constitution should be amended, or changed, to protect these freedoms, so they introduced a Bill of Rights. The Bill of Rights is made up of 10 amendments; the First Amendment is written below:

Congress shall make no law respecting an establishment of religion or prohibiting the free exercise thereof; or abridging the freedom of speech, or of the press; or the right of the people peaceably to assemble, and to petition the government for a redress of grievances.

Building a respect for our nation's special symbols helps promote a sense of patriotism and pride.

FIGURE 3–2
First Amendment Diary

	Religion	Speech	Press	Assembly
Read the morning newspaper			✓	
Went to a basketball game with my friends				✓
Read an article in Sports Illustrated			✓	
Phone my friend to make plans to go to the mall		✓		✓
Searched the Internet for help with my homework		✓	✓	
Went to church with my family	✓			
Attended Sunday School class	✓			

So that the students operated from a common understanding of the First Amendment, Ms. Jordan helped the class rewrite the Bill of Rights in their own words:

> People are free to practice any religion they want. The government cannot stop citizens from saying or writing what they want. The people have the right to gather together to discuss problems they have with the government and to inform the government of their problems.

The students then recalled the freedoms guaranteed by the First Amendment: *freedom of religion, freedom of speech, freedom of the press,* and *freedom of assembly.* Each of these freedoms was listed horizontally along the top of a grid titled, "First Amendment Diary." Along the left side of the grid, students were asked to list the protected activities they were engaged in over a weekend. They were to place a check in a grid indicating which of the guaranteed rights protected them as they engaged in those activities. Tyler's diary is illustrated in Figure 3–2.

On Monday, Ms. Jordan helped her young political scientists compile a class list of all the entries, and led a discussion of the data with questions like these:

- How many times did you rely on your First Amendment rights?
- How might your weekend have changed without the First Amendment rights?
- Which rights were exercised most? Least?

Through this activity, Ms. Jordan not only demonstrated that civic knowledge is indeed important, but that it also must have relevance to the students' lives. Civic education is something more than studying flowcharts of "How a Bill Becomes a Law" or reading a textbook chapter on the "Separation of Powers." Students more clearly understand civics by participating in the processes they are learning about.

As children experience the dynamics of a democratic classroom and learn about the United States Constitution, they should become involved in activities that help them apply their understandings to the classroom community, in general. In any democratic society, including the classroom, everyone is bound by the rules and laws fairly established by its members. To help make this idea very clear, one teacher extended a discussion of the Constitution by encouraging the students to create their own classroom constitution (see Figure 3–3).

National Symbols

Among the events that helped forge a sense of national identity and spirit following the Revolutionary War was the introduction of special national symbols: the eagle as the symbol for America, the Great Seal of the United States, the *Star-Spangled Banner* as our national anthem, and the American Flag. These and more helped Americans become proud of their new nation. National symbols evoke a sense of unity and patriotism. Think back to the tragic terrorist attack on the United States on September 11, 2001, for example. Many Americans turned to the American flag as a symbol of strength and unity and hoped that such patriotism would not end with the events of September 11. As a result, an understanding of and respect for national symbols have undergone renewed emphases in dynamic social studies classrooms.

Classroom Vignette

Students in Nick Bonelli's fifth-grade classroom, for example, gathered around a yellow rug they called their "Conversation Station." The Conversation Station was a meeting place where the class would come together whenever there were special ideas to talk about. Mr. Bonelli designed this unique sharing center because he believes that classroom communication can be most effectively facilitated when the children are able to see one another in a distinctive, functional setting.

FIGURE 3–3
Classroom Constitution

Grade 6 Constitution

We the students of Grade 6, Room 14, in order to form a more perfect class, do establish this *Constitution of the Sixth Grade.*

Article I. Officials

1. There will be two branches of our government: the executive branch and the legislative branch.
2. The executive branch is made up of the President, Vice President, Secretary, and Treasurer.
3. The legislative branch is made up of all the rest of the members of the class.
4. Two candidates each for the offices of President, Vice President, Secretary, and Treasurer shall be nominated the Friday before the third Monday of each month.
5. Election of officers shall take place the third Monday of every month by secret ballot.
6. A student may hold a term of office only once.

Article II. Qualifications of Officers

1. Everyone automatically becomes a member of the legislative branch when entering Room 14 as a student.
2. Students must have these qualifications to be an officer:
   a. must be a member of Room 14 for at least two weeks.
   b. must be honest and trustworthy.

Article III. Duties of Executive Branch

1. *President*
   a. The President shall run all class meetings.
   b. The President shall take charge of the class in the teacher's absence.
   c. The President shall help the substitute (show him or her where things are).
   d. The President shall appoint class helpers.
2. *Vice President*
   a. The Vice President shall help the President when necessary.
   b. In the absence of the President, the Vice President shall take over.
3. *Secretary*
   a. The Secretary shall take notes at all class meetings.
   b. The Secretary shall take care of all class mail (letters, thank you notes, and so on).
4. *Treasurer*
   a. The Treasurer shall take care of all class funds.

Article IV. Duties of Legislative Branch

1. To approve, by majority vote, class helper assignments.
2. To approve, by majority vote, any decision for which the class is responsible.
3. To volunteer for class helper assignments:
   a. clean chalkboard
   b. feed fish
   c. water plants
   d. pass out papers
   e. take lunch count
   f. serve as class librarian
   g. greet room visitors
   h. keep art materials orderly
   i. check attendance
   j. run errands
4. To approve, by two-thirds vote, any amendment to this constitution.

Article V. Presidential Vacancy

The Vice President shall take over if the President's office is vacant, followed by the Secretary, and then the Treasurer.

Article VI. Class Meetings

Meetings shall be held each Friday from 2:30–3:00 p.m.

Article VII. Amendments

1. An amendment may be proposed by any member of the class.
2. An amendment must be approved by two-thirds vote of the legislative branch.

Amendments

Amendment 1.
An elected official shall temporarily give up any classroom helper jobs held during his or her term of office. (Approved: February 10)

On this day, Mr. Bonelli wanted to meet together in the Conversation Station to introduce a new topic of study, "Major Flags Throughout American History." Before beginning any new topic of extended study, Mr. Bonelli likes to make sure his students are able to connect their previous knowledge and experiences to the new information, so he unfurled a large, contemporary flag of the United States and discussed the meaning of the colors, stars, and stripes.

Since the major emphasis in this study was going to be research and writing (finding out and informing) Mr. Bonelli felt his students required a model of expository writing, since children tend to write better original pieces after looking firsthand at patterns they find in books. Therefore, Mr. Bonelli brought to class the book *Flag Day* by Dorothy Les Tina and illustrated by Ed Emberley (New York: Crowell). He read the first six pages of the book—a short history of flags. Mr. Bonelli helped his students understand the author's writing style by discussing how the author developed her ideas and directing his students to complete a story map.

Mr. Bonelli discovered that flags in general fascinated his students. Impressed with their enthusiasm, Mr. Bonelli judged that the time was right to extend the study of "Major Flags In American History." Realizing that this topic was too broad for a single written report, Mr. Bonelli broke it down into small, manageable segments for small-group investigation: "British Union Jack," "Grand Union Flag," "Betsy Ross's Flag," "*Star-Spangled Banner,*" "Flag of 1818," "Flag of 1912," and "Today's Flag." The plan was for small groups to investigate each topic, write a short report, share the reports orally, and compile them as separate chapters of a class book, "Flags Throughout American History."

The next stage of the research project involved developing a list of questions that the students needed to ask about their specific flag. Mr. Bonelli's class decided that each small group should answer the same questions as modeled in *Flag Day:* What colors were used? What symbols were used? When was it adopted? Why was it changed? The questions provided a starting point for further research and served as a focus for the students' writing.

After this organizational phase of the research project, the students searched through a variety of reference materials, including the Internet. Mr. Bonelli monitored this portion of the research process carefully, for he feels that children must learn to choose relevant information and avoid word-for-word copying from a book, encyclopedia, or other source of information. "Nothing can be more deadly," growls Mr. Bonelli, "than to allow a child to stand in front of a classroom and read a report that was copied word-for-word. You'll hear frequent pauses, the child's voice will be in monotone, and children will often stammer and struggle while trying to pronounce unfamiliar words."

To help them organize their reports, Mr. Bonelli gave the students a number of index cards. He explained that these would be referred to as *information cards* for two different reasons: each was to be (1) decorated with a large lowercase "i" (the first letter in the word *information*) and (2) used to

FIGURE 3–4
An "I" Notecard

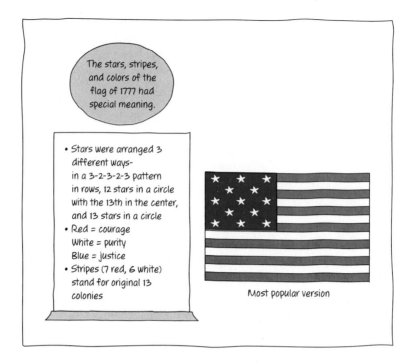

The stars, stripes, and colors of the flag of 1777 had special meaning.

• Stars were arranged 3 different ways-
  in a 3-2-3-2-3 pattern in rows, 12 stars in a circle with the 13th in the center, and 13 stars in a circle
• Red = courage
  White = purity
  Blue = justice
• Stripes (7 red, 6 white) stand for original 13 colonies

Most popular version

summarize important information. Duplicating his model, the students divided a notecard into two major sections: a large circle at the top of the card represented the dot of the letter "i," and the bottom rectangular section represented the stem. The students then printed the main idea (usually the first sentence of a paragraph) on the top section (the dot of the "i"); in the bottom section they listed all the supporting details (see Figure 3–4). This visual aid helped the children summarize the information effectively and guided them as they wrote their reports.

For the sharing stage of flag research, Mr. Bonelli used a tactic he called the "TV News Magazine." In a TV News Magazine format, an "investigative reporter" from each research group sets up an area of the classroom where he or she shares the group's written piece with an audience of four or five classmates (or children from other classrooms). Several investigative reporters function simultaneously and repeat their presentations as their classmates revolve through the areas. Mr. Bonelli's investigative reporters donned simple costumes and used simple props to help the audience get a feel for the historical era. For example, the investigative reporter presenting the "Grand Union Flag" presentation wore a pair of glasses pulled down over his nose like Benjamin Franklin, and the reporter who told of the "Betsy Ross Flag" placed a feather alongside a copy of the Declaration of Independence. As part of its

FIGURE 3–5
Student Report

The Grand Union Flag

The Grand Union Flag served as America's first national flag. It had a British Jack in the top left corner out of loyalty to the crown. It had 13 stripes to show the unity of the 13 colonies. It was raised the first day of January 1776 over George Washington's headquarters. It lasted until 1777.

presentation, each group displayed a drawing of its selected flag and shared a short report describing the flag's historical significance. Figure 3–5 is an example of the report that was read aloud by the first group for our Colonists' earliest flag, the "Grand Union Flag."

After researching several other national symbols in like manner (Uncle Sam, the eagle, the *Star-Spangled Banner*), Mr. Bonelli helped the students connect their new understandings to their own lives. After comparing the community life in their classroom to the community of people who make up our nation, Mr. Bonelli once again divided the class into small groups. This time, he assigned a different symbol to each group: song, flower, flag, animal, motto, and great seal. Each group was challenged to create an original symbol for the classroom. The symbols were displayed proudly in the classroom all year and contributed immeasurably to a feeling of classroom spirit and pride.

Elections and Voting

One of the best ways to educate young people about what it means to be a citizen in a democratic society is to get them involved in voting and other responsibilities of citizenship. Voting offers an excellent opportunity to learn civic responsibility and shared decision making in a meaningful and motivating context. By participating in the entire voting line of action (determining voting issues, suggesting possible choices, casting and tallying votes, and confirming the outcome of the vote) children directly experience how a democracy works.

Initial voting experiences should involve issues of interest for the entire class. It is no secret that unless people are interested in a voting issue, they will not be motivated to participate in the voting process and will likely come away from the experience with a feeling that voting is not important to them. Therefore, initial votes should be taken on issues such as favorite storybook character, a class mascot, foods to serve at a classroom party, which animal makes the best pet, who will win the Super Bowl, and so on.

Children should be encouraged to thoroughly discuss the issue before they take a vote, making arguments in support of the position they are willing to defend and trying to persuade their classmates to vote in their favor. It should be emphasized that opposing points of view should be presented with self-control and easiness. The children must realize that it is normal for people to hold different opinions about things and that these differences should not break down friendly relationships.

The vote itself should be carried out carefully. Try to remember that children do not understand the voting process as an adult does, so some of the procedures we are comfortable with may be confusing to them. Take, for example, the practice of having the children raise their hands and then counting the results. Oftentimes young children will raise their hands as soon as a teacher says, "Raise your hands if . . ." whether they want to vote that way or not. Others will lower their hands before their votes are counted, and some may raise their hands more than once (even though they're directed not to). Instead of raising hands, it might be best to:

> *Poll children.* You can list the children's names on a chart and ask each child how he or she votes. The children can then cast votes by placing a tally mark next to their names. The class can then count the votes for each option.

Construct name graphs. Have the children print their names on 3 × 5" index cards. Ask each child how he or she votes and then place the index card in a line or a stack corresponding to the option. Again, the entire class will be involved in counting the votes.

Use secret ballots. If children become too forceful in trying to influence others to vote for their particular choice, it might be best to give each child one piece of paper, ask him or her to indicate his or her choice, and drop it into a box. Again, the entire class should carry out the vote count.

Voting opportunities should be a part of the older child's classroom, too, for it exemplifies one of the duties and responsibilities of citizenship in a democratic society and demonstrates why voting matters. In the upper grades, however, it is common to extend the students' understandings of the voting process as it applies to the election of officials at the local, state, and national level. Gordon Palmer, for example, centered on voting in national elections by focusing on the question, "So how does one get to be president of the United States?"

Classroom Vignette

"Pat Apatosaurus in 2000!" "We back Rex!" Sound like strange election slogans to you? Not if you were visiting Gordon Palmer's fourth-grade classroom.

Mr. Palmer kicked off the election activities for the 2000 presidential election by having the students search through an easy-to-read version of the United States Constitution for the three requirements a candidate for the presidency must meet: (1) at least 35 years old, (2) a natural born citizen of the United States, and (3) a resident of the United States for 14 years. "But how does one actually get to be president of the United States?" challenged Mr. Palmer. Finding that his children were fairly uninformed about the process, Mr. Palmer put to advantage his students' interest in dinosaurs to help them learn the election process.

To begin, Mr. Palmer reviewed with the children that dinosaurs can be classified in many ways, but the two most popular categories seem to be meat eaters (*carnivores*) and plant eaters (*herbivores*). By randomly picking slips of paper from a box, students were then assigned to the carnivore or herbivore "party." Each party met and selected a party name: the herbivores decided to call their party the *Herbocrats* while the carnivores favored the *Carnublicans*. Mr. Palmer explained that each party would meet to select a candidate for "dinosaur president." However, in order to vote, each student must register.

Mr. Palmer distributed 5×8" index cards to the students and asked them to print their names and addresses on the cards. The students then signed their cards. The Herbocrats and Carnublicans designed symbols for their parties and drew them on their registration cards. To be properly registered, the students went to a prearranged area in the classroom where an instructional aide checked their signatures on the class list, stamped their cards, and crossed their names off the list. They were told to put the cards in a safe place, as they would need to be presented as verification of registration when it came time to vote.

Each party was allowed to select three potential candidates for dinosaur president. This was compared to the primaries during the actual presidential election process. During the primaries, party members vote for the candidate that will represent their party in the upcoming general election. Therefore, the Herbocrats selected Stegosaurus, Triceratops, and Apatosaurus. The Carnublicans went with Tyrannosaurus Rex, Deinonychus, and Raptor. The party members were divided on the issue of which dinosaur would represent them best, so each party mounted a primary campaign, including buttons, stickers, and hats to wear at a rally. Of course, countless speeches, posters, and mock TV commercials extolled the virtues of each potential dinosaur candidate.

While the primary campaign was running its course, Mr. Palmer and his students made a voting booth from a large packing crate the custodians had rescued from the trash pile. They placed an appropriate "United States Polling Place" at the top and decorated it with various patterns of red, white, and blue. It was important that the students have a place to vote in privacy. At last, it was time to complete the first step of the election process, with each party selecting its candidate!

Each party held its election, with officials making sure to check the registration cards and that each registered voter picked up only one ballot. After the election committee tabulated the results, Tyrannosaurus Rex emerged victorious from the Carnublican Party while the Herbocrats selected Apatosaurus. It was then time to focus on the national convention, step two in the election process. During the national convention, the party finalizes its selection for one presidential nominee and each presidential candidate chooses a dinosaur running mate.

After each dinosaur party selected one presidential candidate, step three—the general election process—began. Candidates campaigned in an attempt to win the support of voters. In November, the voters voted for one dinosaur candidate. Each party published an election newsletter containing background information about the candidates. They made campaign buttons, bumper stickers, and posters. As in the primaries, they wrote campaign songs and slogans. Most importantly, each party was directed to write a platform that presented its candidate's view on all issues of importance. The highlight of the whole experience was a spirited "dinosaur debate" with all the protocol observed during regular presidential television debates.

On November 7, 2000, students in Mr. Palmer's class lined up at their packing box voting booth to elect the first dinosaur president. They clutched their registration cards and voted for their choice, again by secret ballot. The votes were counted and the winner was announced later that afternoon. Tyrannosaurus Rex won in a landslide—the students felt that since this was the fiercest meat-eating land animal that ever lived, it would be most suitable to rule the land of the reptiles. Of course, a huge victory celebration took place that afternoon (arranged and hosted by aides and parents). Everyone had a great time. In addition, the voting booth stayed in the classroom and was used for all other important classroom votes that school year.

Mr. Palmer surely realized that when a person casts a vote in the general election, he or she is not voting directly for a candidate. Instead, that person is actually casting a vote for a group of people known as electors. The electors are part of the Electoral College and are supposed to vote for their state's popularly elected candidate. However, the Electoral College system can be confusing even to many adult voters, so Mr. Palmer chose not to include it in his election simulation. Should you try the simulation in the future, it would be interesting to find a way to make the Electoral College a part of the process.

Children should learn that one of the most important ways they can participate in their government is to exercise their right to vote. By voting, people have a voice in their government.

Volunteerism

Today's civics education takes on a much different character than it did in the past. The ultimate goal of civics education has shifted from imparting knowledge to producing citizens who are not only knowledgeable, but also committed to the principle of civic responsibility. A commitment to civic responsibility—voting, letter-writing to public officials, volunteering one's service in various public capacities, and a wide range of other activities—begins in the elementary school classroom and manifests itself in the dispositions of adult citizens who participate fully in our democratic society.

Service learning is a type of authentic learning that involves students in community service, therefore advancing the principles of citizenship and good character. It is a powerful approach to citizenship education that empowers children with an understanding that they can make a difference in their community and develops such citizenship virtues as respect, responsibility, and initiative. Service learning also enhances self-esteem as children learn that their actions can truly make a difference.

Sharon Kletzien, a colleague at West Chester University, has developed a model community service program that has received high national acclaim. Called *Kids Around Town*, the program's goal is to promote among elementary school children an understanding of how active citizenship participation can influence local public policy issues. The *Kids Around Town* model includes seven interdisciplinary steps: *knowing the local government, choosing a local issue to explore, researching the issue, analyzing the issue, solving the problem, taking action,* and *assessing the project.* The children select an issue that affects them locally—for example, cleaning up litter in a community park, abandoned housing, juvenile crime, land-use issues, litter, or bicycle safety. The issue then serves as a springboard for study and action. The students plan strategies and actions to tackle the issue, similar to the way Debra Wood's students did at the beginning of this chapter.

Cooperative learning is particularly recommended for investigating community issues by the program developers because cooperative learning helps create a classroom community that is a microcosm of a democratic society. A specific cooperative structure recommended by *Kids Around Town* is called "Academic Controversies" (Johnson & Johnson, 1999). Useful for discussions of all controversial issues, the basic form for structuring "Academic Controversies" follows:

1. The students choose a topic on which two well-documented positions (pro and con) can be prepared. The students must understand their positions completely and know where to find information to build a rationale underlying their pro or con position.
2. Students are assigned to groups of four. Each group is randomly divided into two pro and con teams.
3. Each pair is assigned its tasks: (a) knowing its position, (b) locating information relevant to its position, and (c) preparing a series of persuasive arguments to defend its position.
4. Each team presents its side of the issue to the other team forcefully and persuasively.
5. Teams reverse perspectives by presenting the opposing position as forcefully as they can.
6. Teams drop their advocacy and attempt to reach a group decision by consensus.
7. The class develops a plan of action to implement its final position.

The concern for encouraging public service and involvement is mushrooming today and, with growing public support, many schools have been quick to adopt community service programs. In a sense, they concur with the sentiments of one astute 11-year-old who observed, "We have freedom in this country but it stinks. Drugs . . . alcohol . . . crime . . . war. There's no good news anymore. We need someone to come in and clean out America!" Education for democratic citizenship means we must practice what we preach in our social studies classrooms. Telling students about the worth of civic participation is one thing; involving them in community service is better. When students roll up their sleeves and get involved, they understand what we really mean by citizenship responsibilities. Students are likely to become highly engaged in community service projects when they see that their actions can make a difference in real life.

Civic Dispositions and Virtues

Up to this point, we have limited our discussion of civics education to helping students develop rich understandings of governmental practices and public policy as well as aiding them to make connections between what they learn in school and

Becoming aware of environmental issues and taking steps to lend a hand helps students to connect what they learn in school to their lives as citizens in a democracy.

their lives as citizens in a democracy. But the vision of contemporary civics education transcends even these two noble goals; an essential part of civics education deals with making a commitment to civic dispositions and virtues—the core values necessary to nurture and strengthen the ideals of American democracy. In its *Curriculum Standards for Social Studies,* the NCSS (1994) described this realm of instruction as going beyond the content of civics to the formation of dispositions required of effective citizenship. Standard X (Civic Ideals & Practices) states: "Social studies programs should include experiences that provide for the study of the ideals, principles, and practices of citizenship in a democratic republic" (p. 30).

In the elementary grades, students are introduced to civic dispositions and virtues throughout the day as they experience life in democratic classrooms. The dispositions and values of our nation are exemplified by the sense of community created in a democratic classroom—everyone operates for the common good of all. How do teachers address the responsibility of teaching civic virtues in ways other than establishing a democratic classroom environment?

In addition to those informal experiences, the Center for Civic Education (1994) established this standard as a guideline for formal instruction in civics: "Students should be able to explain the importance of the fundamental values and principles of American democracy" (p. 22). To achieve this standard, the Center advised that students should be able to explain the importance for themselves, their school, their community, and their nation each of the following fundamental *values* of American democracy:

- individual rights to life, liberty, property, and the pursuit of happiness
- the public or common good
- justice
- equality of opportunity
- diversity
- truth
- patriotism (p. 22)

There has been a resurgence of concern during the past decade that patriotic values like these have fallen out of favor. However, Ryan (1993) comments that this realm of civics education has reemerged as a popular trend "because people are banging on the schoolhouse door. The invitation is coming from the outside. Parents and policymakers are disturbed by a total inability of our culture to pass on its values" (p. 1). Lickona (1993) sees the motivating force as a "growing national sense of moral crisis and what people speak of as a steady moral decline" (p. 1). Society is now turning back to the schools and demanding that certain widely held core values or virtues underlying American democratic society should be at the heart of the school curriculum, the purpose of which is to systematically develop the character of our students.

Literature as a Source for Learning About Character

Maureen Arnold, a fifth grade teacher, read these and similar comments about patriotic values and was tremendously influenced by the well-reasoned commentaries of such respected scholars. She immediately searched for classroom strategies that might help her develop the civic character of her students. Ms. Arnold discovered that one of the most frequently suggested strategies involved personally involving children in stories of the lives of people who have demonstrated exemplary democratic values. Ms. Arnold was already aware that the power of superbly written biographies and other historical narratives helps capture children's attention with their enduring themes of sacrifice and responsibility; power and oppression; failure and achievement. And she knew that these stories vicariously engage students in the lives of others, help them see the world through others' eyes, and make them aware of human potential. But, her major stumbling block was finding a theme that would capture the children's imagination and draw them into the stories of people who made a difference.

Ms. Arnold's quandary was resolved as she watched the opening events of the 2002 Olympic Games on television. She looked on in wonder and respect, thinking how nice it was to watch New York City police officer Daniel Rodriguez sing "God Bless America" with resolute patriotism. She reflected that a year ago—months before the shocking events of September 11—we might have witnessed an entertainment figure standing in Officer Rodriguez's place. However, September 11 refocused our nation's collective vision of heroes. Heroes help us aspire to similar goodness—their character traits and important contributions can be used as models of good citizenship. That was it—heroes—the theme she had been searching for! Ms. Arnold then exuberantly pulled together a series of dynamic learning experiences, including hero biographies that made her civics instruction a great success.

Classroom Vignette

Ms. Arnold assembled her students in the large-group discussion area of their classroom. She wrote the word *hero* on a sheet of chart paper and asked the students to suggest the qualities that they associate with the term. Ms. Arnold listed their suggestions on the chart and asked the students to summarize all of them in one open-ended sentence that began, "A hero is. . . ." The students thought about the sentence carefully and eventually agreed that, "A hero is someone who shows great courage." Ms. Arnold then asked, "From either the past or present, who would you identify as *your* heroes?" Students offered expected names like Abraham Lincoln, Martin Luther King, Jr., and Helen Keller. A few introduced names that were a little less expected from third graders— Eleanor Roosevelt, Neil Armstrong, and Crazy Horse. Others suggested names that were famous, but questionable in terms of attaining hero status: Michael Jordan, Tom Cruise, and Britney Spears. Ms. Arnold felt that in today's society, with messages that come from our media and Madison Avenue, youth seem to be confusing heroes and celebrities. Therefore, she asked, "What's the difference between a hero and a celebrity?" The students debated that question for quite some time until they finally agreed that, despite the fact that both are famous, heroes and celebrities are really quite different. They explained that heroes are endowed with great strength or courage, and are admired for their achievements or beliefs. Celebrities are merely famous; their fame doesn't have anything to do with inner character or the passion, energy, and drive to make the world a better place. The difference, Ms. Arnold's students concluded, is that a genuine hero is more than a celebrity or famous person. A hero expresses values through self-sacrificing acts that benefit others. Heroes inspire others through their selfless contributions to the common good.

Now that the students established a common conception of heroes, Ms. Arnold asked them to examine their list of people and separate the names into "hero" and "celebrity" categories. Individuals like Lincoln, King, Keller, Roosevelt, Armstrong, and Crazy Horse were easily grouped as heroes, while the same was true for grouping Cruise and Spears as celebrities. However, Michael Jordan created quite a stir—everyone had a lot of admiration for him as a person, but some felt he was more of a celebrity while others thought he was more of a hero. Eventually, they reached a compromise that Michael Jordan is a combination of both.

Ms. Arnold scoured the educational literature for classroom activities appropriate for examining the spirit for heroism, for she wanted to explore heroes in greater depth than found in traditional social studies textbooks. Most recommendations centered on the use of various instructional experiences such as inviting "homegrown heroes" to talk to the students formally and informally. Therefore, she kicked off "Hero Week" in her classroom by inviting real-life local heroes to visit the classroom—a firefighter who rescued an elderly woman from a burning building, a retired teacher who was also a Holocaust survivor, a former pilot who flew bombing raids over Germany during World War II, an ex-Negro League baseball player, and a former member of a nonviolent civil rights group who once was jailed for 21 days for her activities. Ms. Arnold and her class viewed a tape on courage from a video program called *Adventures From The Book of Virtues* (available from PBS Video at 1320 Braddock Place, Alexandria, VA 22314-1698; telephone 1-800-645-4727). The tape depicts heroic persons involved in events calling for great courage.

To recall and highlight the major accomplishments of the heroes they had been involved in learning about, Ms. Arnold showed several series of U.S. Postal Service stamps celebrating heroes in American history, and presented a minibiography of each. These stamps were used as models for the children to make their own designs for the heroes they had met in their studies that week. She asked the students to form pairs and directed each committee to design two stamps for the person they wished to honor. See Figure 3–6 for an example.

All of these experiences were well received, but the highlight of Ms. Arnold's program grew from a quote about the importance of using children's literature: "To one degree or another, we all seem to have a taste for learning more about great or important people; who they are and what makes them tick. This predisposition translates into the natural interest children show when they pick up [books] about contemporary heroes [or historical figures]" (Darigan, Tunnell, & Jacobs, 2002, p. 370). Convinced of the power of good books to exemplify wisdom, courage, and inner character, Ms. Arnold pulled together a collection of historical fiction and biographies on a variety of reading levels about people who made a difference—male and female, various cultures, the famous and not-so-famous, adults and children.

Ms. Arnold introduced these books with a short mini-lesson. The mini-lesson consisted of Ms. Arnold reading Barbara Cooney's picture book, *Eleanor* (New York: Viking). The book tells of Eleanor Roosevelt's childhood experiences that led her to become the great person she was. Ms. Arnold announced, "Today, I would like to read to you a wonderful story about one of my heroes, *Eleanor*. From the book cover and title, whom do you think this book will be about? Talk about your ideas with the person sitting next to you." After a few brief moments of sharing, Ms. Arnold continued, "What do you think?" Making these predictions piqued the children's interest and motivated them to hear

FIGURE 3–6
Postage Stamp Design for a Special Hero

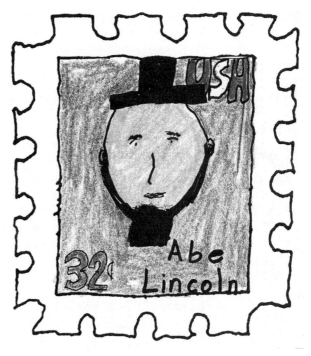

the story. Ms. Arnold then began reading. When she finished, the children talked in great length about how different experiences during Eleanor's childhood helped her grow into a brave and loyal adult.

After the discussion, it was time for the students to go to work and select their own hero books. Ms. Arnold turned them loose to examine all the books she had selected for them to choose from. Her simple directions were that they were to look at the different books, but not select the first one they came to. They were just to sample the different offerings to see which ones they might like to read (she called this a "gallery tour"). As the students examined the books, Ms. Arnold circulated through the area, informally talking with them and making a few timely suggestions for students who needed extra guidance.

After they completed their "gallery tour," Ms. Arnold passed out slips of paper to the children (from a 3×5″ note pad) and asked them to write their names at the top. They were then to list their first through third choices for books. Then, after school that day, Ms. Arnold spread out all the slips and began assigning books according to first choice. Because Ms. Arnold had so many books to select from, most children were to get their first choice. However, there were some alterations that needed to be made, as some children needed to get their second or third choices.

The next day, the children got their books and began reading about their heroes. When they were finished, Ms. Arnold devised a plan that would encourage the students to talk thoughtfully and intelligently about their books. As she considered this mission, Ms. Arnold discovered a list of five traits, called the "five themes of citizenship," that seemed to capture the major attributes of heroic people: *honesty, responsibility, compassion, respect,* and *courage.* She talked about these attributes with the class, making sure they understood and were able to distinguish among the attributes of each. She directed the students to think carefully about their heroes and to decide which of the five themes best described them. Ms. Arnold then exhibited hand-printed signs on which she labeled the five themes; she subsequently placed them at even intervals around the room. The students were directed to go to the area designated by the attribute they selected to form a discussion group. Knowing that open discussions often disintegrate into idle chatter, Ms. Arnold directed their conversations with a printed conversation guide: (1) briefly summarize the actions or values that made your person a hero, (2) explain why you selected this theme to associate with your hero, and (3) make a connection between what you had read and your own lives. Of course, Ms. Arnold modeled these responsibilities by providing examples of what she would say based on the book she had read aloud to the class, *Eleanor.*

To bring together the entire literature-based experience, Ms. Arnold organized the construction of a paper bag timeline. Ms. Arnold modeled the process by showing the students a completed bag she had constructed for Eleanor Roosevelt. Each student was then given a strong paper bag and asked to print the name of his or her hero in bold letters at the top. They were to use this bag to show what they learned about their heroes through writing, drawing, and collage. The students were then given an outline of a coat-of-arms with four blank segments (see Figure 3–7). They were directed to draw an appropriate picture or symbol (or print a word) in the segments that best answered the following questions:

1. What is your greatest personal accomplishment?
2. If you could compare yourself to an animal, what would it be?
3. What is one lesson others can learn from you?
4. What is one word you would most like people to use to describe you?

These coats-of-arms were taped to the front of the bags. The students labeled the bags with dates underneath the hero's name to show when their hero lived and printed two questions at the bottom of the bag. The students were to use questions that could be answered by looking at the coats-of-arms on the front of their bags and by examining illustrations or other clues that they could stuff into the bags.

One student completed Figure 3–7 for Ruby Bridges, who, in 1960 at 6-years-old, was the first African-American student to enter William Frantz Elementary School in New Orleans. The student explained that she drew a school in the first section because it showed that Ruby helped integrate the elementary school by passing through a mob of racist protestors shouting insults and threats. She selected a sheep as an animal to represent Ruby in

FIGURE 3–7
A Coat-of-Arms for Ruby Bridges

section two because she felt Ruby had a wonderful smile and remained calm and peaceful throughout this unusual challenge. In the third section, the student explained that she chose to show an African American child holding hands with a white child because Ruby Bridges would want to teach others that white people and black people should respect and love each other. Finally, the student wrote the word "courage" in the last section because she felt Ruby bravely faced a very difficult situation with very strong character. The story of Ruby Bridges showed that even small children can be heroes for each other.

When all the bags were finished, the students constructed their timelines. They put their bags in chronological order; a timeline based on a unit of measure was unworkable since most students selected heroes from the 20th century and there would be too many bags clumped together in one section. All the students had a chance to explore the fun questions, coats-of-arms, and clues along the interactive timeline.

Ms. Arnold brought her unit on heroes to a close by asking the students to consider three questions. The first was, "If you could bring any real-life hero from the past to the present, whom would you choose?" (A burning class discussion ended up in a close class vote in favor of Abraham Lincoln.) The second question was, "What five questions would you ask the person?" The third question challenged the creativity of small discussion groups: "What kind of hero will our American democratic society need by the year 2050?"

Good citizenship requires a clear understanding of core citizenship dispositions and virtues and a willingness to partake in responsible action directed toward the welfare of one's community. Criticizing the failure of U.S. citizens to become effectively and personally involved in preserving their nation's well-being, Parker (1989) describes the American spirit as being stuck in a state of "individualism." Parker cites de Tocqueville's (1969) 150-year-old description of individualism to clarify: "Individualism is a calm and considered feeling which disposes each citizen to isolate himself from the mass of his fellows and withdraw into the circle of family and friends; with this little society formed to his taste, he gladly leaves the greater society to look after itself" (p. 506).

Parker believes that the demise of effective citizenship today is directly linked to individualism. Where can such an attitude lead? According to Parker (1989), "No one knows to what it may lead. Political chaos, economic collapse, urban warfare, a quiet lapse into tyranny, or something less dramatic—anything is possible (p. 353). Newmann (1989) agrees with this outlook and warns that "[It] threatens the very survival of the human species and the planet" (p. 357).

Critical Thinking

Early in our nation's history, reading aloud passages from the Bible carried out what we now consider civic education. Even into the 1800s, instructional materials maintained a strong religious character, as we see in this selection from *The Boston Primer* (Hersh, 1980):

> *Let children who would fear the LORD,*
> *Hear what their Teachers say,*
> *With rev'rence meet their Parents' word,*
> *And with Delight obey.* (p. 16)

Passages like this were read over and over again until the children memorized them. Therefore, a good education at the time was described as one's ability to distinguish good from evil. Certainly, learning good from evil remains an important goal for everyone, but there is universal agreement that one now needs a far higher degree of learnedness than at any time in the past, and this standard will only continue to grow. Present-day learning demands more highly complex skills than those gained from drill and memorization. Contemporary education must ensure that children not only retain information but that they think deeply about the material—learn from it, reason with it, analyze it, and solve problems with it. Students must learn to think deeply about what they are experiencing in the social studies program— what they experience directly and what they read about in books. In other words, experiencing by itself is not enough; children must be helped to think deeply and critically about what they are experiencing and reading.

What is critical thinking? That's a tough question to answer, for critical thinking is such a complex mental process that psychologists have not yet been able to agree on what it is. Although there are many conflicting opinions, Ennis (1985) offers an explanation that is particularly appropriate for dynamic social studies instruction: "reasonable reflective thinking that is focused on deciding what to believe or do" (p. 54). Critical thinking is *reasonable* and *reflective* when students make earnest efforts to evaluate information and analyze arguments in terms of their accuracy and worth.

More important than a definition of critical thinking itself, however, is the idea that there is a better instructional road to take in social studies education than rote memorization. The NCSS Curriculum Standards for Social Studies (NCSS, 1994) support this vision by addressing citizenship in the context of civic action. The Standards advise that social studies should help students develop the knowledge, intellectual skills, and attitudes necessary to confront, discuss, and consider action on such issues as national security vs. individual freedom or obeying the law vs. the right to dissent: "Social studies educators have an obligation to help students explore a variety of positions in a thorough, fair-minded manner. As each position is studied and discussed to determine the strongest points in favor of it, the strongest points in opposition to it, and the consequences that would follow from selecting it, students become better able to improve the ways in which they deal with persistent issues and dilemmas and participate with others in making decisions about them" (p. 10). The NCSS Task Force on Early Childhood/Elementary Social Studies (1989) takes a similarly strong position on the role of critical thinking and civic action in social studies: "For children to develop citizenship skills appropriate to a democracy, they must be capable of thinking critically about complex societal problems and global problems. . . . Continually accelerating technology has created and will continue to create rapid changes in society. Children need to be equipped with the skills to cope with change" (p. 16).

How do elementary school teachers help build critical thinking skills in dynamic social studies programs? Perhaps because there is so much disagreement about what critical thinking is, and because it encompasses such a variety of sophisticated thinking skills, suggestions for classroom use are sketchy at best. However, if we want our students to be critical thinkers, then the answer lies in discussions about rousing content that stirs intense feelings. Substance is crucial if thinking is to progress higher into the cognitive domain. Children must study the topic in depth, question and challenge the ideas that they read and hear, and, with thinking at its peak, construct evaluative judgments and personal opinions. Critical thinking, then, cannot happen in a vacuum. Before students can think critically, they must understand the underlying facts, concepts, and generalizations and use them as standards for stating and defending opinions and judgments. Therefore, you first teach for understanding. If students don't understand, they have nothing on which to base higher-order thoughts.

Responding to Print

Critical-thinking experiences should begin with oral strategies; students should talk together about literature that speaks to injustice and environmental issues. In talking about literature, students can begin to recognize their own biases and examine alternative ideas. The following is a lesson sequence that occurred in one early grade classroom. It provides a script for designing a similar sequence of critical thinking with other sources of literature. An upper grade example will be shared later.

Classroom Vignette

Today, Jose Arroyo read *The Little Red Hen* by Paul Galdone (Seabury Press) to his eager group of first-grade "political scientists." He knew that the book was useful for many instructional purposes, not the least of which was to offer his students an opportunity to think critically. Therefore, after the story ended, Mr. Arroyo led a group discussion about the ending and the hen's decision not to share her bread with the animals that refused to help her. Did the Little Red Hen do the right thing? Mr. Arroyo encouraged the class to think about why the other animals refused to help and whether they might have had good reasons. The children discussed what they thought about the hen's decision and considered other choices she could have made. Finally, Mr. Arroyo displayed a chart divided into three columns: (1) the child's name, (2) a "yes" column, and (3) a "no" column. Mr. Arroyo explained the chart to the children and asked them to predict whether there would be more *yes* or *no* votes in response to the question, "Did the Little Red Hen do the right thing?" The children were invited to write their names and mark their votes answering the question (see Figure 3–8). Afterward, the class discussed the results.

Young children get most of their social studies content from a variety of sources, but words-in-print seem to predominate: textbooks, children's literature, newspaper items, or information books such as encyclopedias. These resources often contain problems, issues, or ideas that become the lifeblood of meaningful classroom discussions. Whether these discussions involve "meaty" contemporary issues or controversial historical events, they lead students to carefully consider reasons for acts and try to examine issues from all sides. In the upper grades, especially, students should be encouraged to think deeply and critically about what they are learning.

　　An exemplary upper-grade script evolves as Ellen Geiger's fifth graders return to their classroom after recess. The soft, plaintive strains of "Negro spirituals" fill the room. Ms. Geiger will eventually help her students understand that slaves created these songs not only to express their deep longing for freedom, but also as coded messages that passed on secret information for escape, such as where to find escape routes and hideaways: "Steal Away to Jesus," "Go Down, Moses," "The Drinking Gourd," and "Swing Low, Sweet Chariot." But, for now, they helped create an environment for the day's lesson.

FIGURE 3–8
Little Red Hen Graph

Did the Little Red Hen do the right thing?

Name	Yes	No
Connie	X	
Robert		X
MarthA		X
Jud		X
Darrin	X	
Anita	X	
Arriram	X	

Classroom Vignette

Ms. Geiger assembled the class in an area of the room favorable for good listening and read them the book *The Story of Harriet Tubman: Conductor of the Underground Railroad* by Kate McMullan, a fascinating story of one of America's most famous abolitionists. At the story's end, Ms. Geiger asked for the children's opinion, "How would you describe Harriet Tubman? What evidence from the story supports your feelings?" The students contributed many opinions, most of which revolved around the ideas of "toughness," "intelligence," and "courage." Their evidence revolved around Harriet Tubman's hard work in the fields, being whipped by owners, learning to read without her owner's knowledge, being injured while helping a slave escape, going on many Underground Railroad journeys, learning to use special ways to escape (such as navigating by the North Star), and helping slaves whose owners refused to let them go after the Civil War. During this reflective discussion, one student served as a scribe, recording the specifics on a large data-retrieval chart strategically placed near the story reading area.

After spending a good deal of time discussing the story, Ms. Geiger added a new dimension to the experience. "So far," she said, "we have been discussing the many accomplishments that made Harriet Tubman stand out as a famous American. Now I would like you to do some very powerful thinking. Certainly, you have said that each of Harriet Tubman's accomplishments were important, but now I want you to think about which you consider to be her *most important* contribution of all." After much debate, small groups of students agreed that it was her efforts to hide and carry slaves to freedom with the Underground Railroad.

Ms. Geiger then shared a few ideas about the dangers of running away from slavery and of aiding fugitives. She explained how the spirituals the children had listened to as they entered the room helped the slaves find their way to the North. She then read this dilemma faced by Sie, a real slave from Maryland in 1825, when his extremely troubled master came to his cabin with an unusual request:

> One night in the month of January . . . he came into my cabin and waked me up. . . . For awhile he said nothing and sat . . . warming himself at the fire. "Sick, massa?" said I. "Can't I help you in any way, massa?" I spoke tenderly for my heart was full of compassion at his wretched appearance. At last . . . he cried, "Oh, Sie! I'm ruined, ruined, ruined. . . . They've got a judgment against me, and in less than two weeks every [slave] I've got will be . . . sold." I sat silent. . . . "And now, Sie," he continued, "there's only one way I can save anything. You can do it; won't you, won't you?" In his distress he rose and actually threw his arms around me . . . "I want you to run away, Sie, to . . . Kentucky, and take all the servants along with you." . . . My master proposed to fol-

low me in a few months and establish himself in Kentucky. (Henson, 1935, pp. 162–167)

After discussing the situation to make sure the students understood what it was all about, Ms. Geiger said, "Isn't it interesting that after being Sie's master for over 30 years, the plantation owner is so dependent on him? Should Sie help out the owner?" Ms. Geiger divided the children into discussion groups and asked them to decide what Sie should do. She provided a visual aid for them to organize their different viewpoints (see Figure 3–8A). Each group was to consider the statement and discuss reasons to agree or disagree with it. Every time they listed a reason to *agree*, they had to follow it with a reason to *disagree;* They must always have an equal number of reasons in both columns. When the groups finished their lists, they decide to agree or disagree with the statement.

FIGURE 3–8A
Pro and Con Visual Aid

PROS AND CONS

Sometimes an issue is so complicated it's hard to take sides.
Here's a way to help you make a decision.

Sie should help out the owner.	
Agree	**Disagree**

Throughout this lesson, Ms. Geiger employed sophisticated instructional strategies designed to move her students beyond the realm of working with the content itself and into higher-order thinking processes. She did this by sharing rich material that incorporated thought-provoking content. Thought-provoking content sets in motion adventures in critical thinking, for it provides the essential substance of productive discussions dealing with feelings and civic ideals.

Graphic Organizers

As a supplement to good discussion strategies, teachers often help students build critical thinking behaviors with graphic organizers. As an example, let us suppose that your next social studies lesson is supposed to focus on the idea that a stable food supply was one of the prerequisites of an early settled civilization. That is, as early gatherer-hunters learned to domesticate wild animals and to sow seeds, they gradually settled down and became farmers and herders. Then, as farming methods improved, people often found themselves with food surpluses. This unexpected bounty led to trade. A village with an overabundance of grain, for example, exchanged it for tools, pottery, cloth, or other goods from nearby villages. An overabundance of anything is called a *surplus,* and early civilizations had to learn how to face important decisions regarding surpluses—it wasn't always the best move to trade them for other goods. By helping develop these understandings, teachers lay the foundation for higher-order thinking within a meaningful content. Colleen Nemchick believes this can be effectively accomplished by using graphic organizers to help students visualize the evidence they will use in arriving at a judgment. Here is one way she has done that:

Classroom Vignette

Ms. Nemchick has her students play the role of citizens of an ancient village living on the land between the Tigris and Euphrates rivers. Their village is fortunate; unlike others whose farmland has been scorched by the sun, theirs has an irrigation system that helped produce an abundance of grain. While other villages face a devastating famine, theirs is thriving. Ms. Nemchick holds a discussion focusing on the following questions: "How do you feel about having surplus food while your neighbors are starving? Is it fair that one village should have so much while the others face famine? Do wealthy civilizations have a responsibility to help those that have less?"

To delve more deeply into these issues, Ms. Nemchick initiates a cooperative learning experience in which students consider the problem of what a village should do with its surplus grain. She divides the class into base groups of three, and assigns the students a number from one to three. Ms. Nemchick di-

FIGURE 3–9
Graphic Organizer

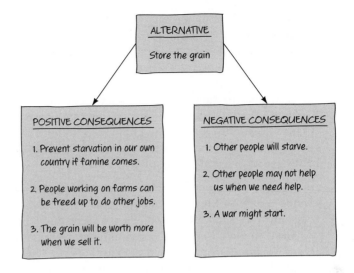

rects all the ones, twos, and threes to move from their base groups into expert groups, with each expert group assigned to examine the positive and negative consequences of one of these alternatives: all the ones will study the "store the grain" alternative, twos will study the "sell the grain" alternative, while the threes study the "give it away" alternative. Completing a graphic organizer helped each group consider the positive and negative consequences of their alternative. The "Store the Grain" group's completed organizer is shown in Figure 3–9.

After completing their graphic organizers, each member of the expert groups returned to his or her base group to share the conclusions. Each base group, after listening to the positive and negative consequences of each alternative, was directed to arrive at a decision.

The possibilities for using graphic organizers are endless. One of my favorites is *the decision tree.*

Climbing a Decision Tree

Richard Remy and his associates (no date) at the Mershon Center of the Ohio State University developed an interesting graphic organizer called *the decision tree.* To begin climbing the decision tree, ask how many students notice that they make decisions every day, such as when they climb a tree or when they decide what to wear to school that day. Explain that decisions are not always as easy as those; in fact, many decisions require deep thought and careful planning. All decisions require choosing from among alternatives. Select a decision such as deciding what to wear to school that day. "What alternatives (options) were available?" In addition to identifying alternatives, decisions also involve studying the consequences (possible results) of each alternative. "What are the possible advantages and disadvantages of choosing either alternative?"

Next, inform the students that they are going to use their ability to use alternatives and consequences in an imaginative situation by helping Sir Lottalance decide what to do about Dingbat the Dimwitted, a fierce dragon causing the villagers severe problems. Tell the students to put themselves in the place of the knight, Sir Lottalance, as they listen to the following story:

> One day very long ago, the country's bravest knight, Sir Lottalance, was riding along on his horse, minding his own business, when he came across some very sad townspeople. They were upset because the nasty dragon, Dingbat the Dimwitted, had lumbered out of his dark cave and carried off the beautiful princess from the king's castle. The grief-stricken king had offered a huge reward for anyone who could destroy the dragon and save his daughter's life. But the first knight to try was barbecued by Dingbat's fiery breath. The second knight to try ran away in panic at the sight of the hulking creature, tripped over his own sword, and became the dragon's shish kabob. Sir Lottalance could hear the princess beating her fists fiercely against the dragon and calling him all the nastiest names you ever heard. He could hear Dingbat's empty tummy rumbling as the dragon waited for another tasty meal of fried knight. Sir Lottalance was the fastest, strongest, and bravest knight in the kingdom. What could he do?

Now point to a bulletin board display showing Dingbat, Lottalance, and a large construction-paper decision tree bedecked with the sign "Occasion for Decision" (as in Figure 3–10). The "Alternative" and "Consequences" areas on the tree are initially blank. For the children to climb the tree, they will have to think of Sir Lottalance's alternatives. Ask them for alternatives. When they have described fighting or fleeing, write the responses on the alternative branches of the tree and congratulate the children for starting their climb up the decision tree.

Help the students climb higher into the branches of the tree by looking at the consequences of Sir Lottalance's decision. Ask, "What would be a good (or positive) consequence of getting out of there fast? What would be a bad (or negative) consequence of getting out of there fast?"

When the students have suggested ideas corresponding to "stay alive," "enjoy dragon burgers for life," or "be called Lottalance the Sissy," add them to the blank areas above the "getting out of here fast" alternative. Again, reinforce your students for doing a good job of climbing and remind them they still have an alternative branch to explore. Ask, "What would be some bad (or negative) consequences of fighting Dingbat the Dimwitted? What would be some good (or positive) consequences of fighting Dingbat the Dimwitted?"

FIGURE 3–10
The Decision Tree

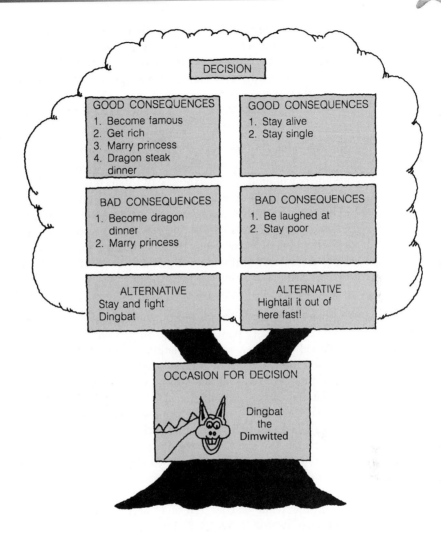

Again, list each contribution as it is offered. Examine the whole tree and look for the students' sense of accomplishment. Then, weighing the consequences, ask the class to vote, deciding whether they should run away or fight. Finally, place their decision high in the top of the decision tree.

To summarize, the decision-tree strategy involves the following steps:

1. Decide what question to examine and label it at the base of the tree.
2. Abbreviate the decision in the "Occasion for Decision" sign.

3. Encourage children to think up alternatives and write them in the boxes on the branches of the decision tree.
4. Discuss positive and negative consequences of each alternative, one at a time.
5. Write in the consequences, ask the children to weigh each, and write in their goal.
6. Congratulate the children as successful decision makers.

Initial experiences with the decision tree should center on relatively uncomplicated problems, such as where to go on a field trip or what to do with friends on a free afternoon. As the children become more competent, increasingly complex issues could be considered, such as the use of tobacco products, great historical decisions, voting for political candidates, or drug prevention.

Myles Spencer, for example, used the decision tree as a graphic organizer as his fifth-grade students read about one of the most significant decisions made in the history of the United States. Rosa Parks made it in Montgomery, Alabama. Mr. Spencer asked the students to think about what they would do if they were faced with Rosa Parks' decision. Then he supplied them with an account of the incident in which Rosa Parks' decision was made.

Rosa Parks' feet hurt, for good cause. Her job as a seamstress in the Montgomery Fair kept her on the run. All that day of December 1, 1955, she had pinned up hems, raised waistlines, and carried dresses back and forth. When the closing time buzzer sounded, she hurried out of the store. Then she boarded a Cleveland Avenue bus, dropped her dime in the box—and hesitated. Where should she sit?

By law and custom, the front rows of seats were reserved for white people. [Black people] sat in the back. Halfway up the aisle there was a no man's land where [black people] might sit until the space was needed for white passengers. There were no signs announcing these rules. In Montgomery, the capital city of Alabama, everybody simply knew them.

But this bus was half empty, and Mrs. Parks sank into the first seat behind the "white" section. Her feet began to feel as if they were almost ready to stop hurting. It was a feeling that made it impossible for her to think about anything else. She scarcely noticed that the bus was getting fuller from one stop to the next. Soon all the seats were taken. A few minutes later people were standing in the aisle. White people.

The bus driver, with one eye on his rearview mirror, called, "All right, you niggers [sic], move back!"

The woman next to Mrs. Parks and two men across the aisle rose and silently made their way to the back of the bus. Mrs. Parks sat still. A white man stood beside her, waiting, but she didn't budge.

The bus driver left his wheel and strode up the aisle. "Get up!" he ordered.

Rosa Parks took a deep breath. (Sterling, 1968, pp. 1–2)

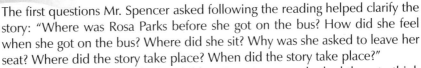

Classroom Vignette

The first questions Mr. Spencer asked following the reading helped clarify the story: "Where was Rosa Parks before she got on the bus? How did she feel when she got on the bus? Where did she sit? Why was she asked to leave her seat? Where did the story take place? When did the story take place?"

Next, Mr. Spencer split the children into groups and asked them to think about what Rosa Parks should do. They were to list her alternatives and the consequences of each alternative on the decision tree. As they worked, Mr. Spencer moved from group to group and offered prompts to guide the children's thinking: "Why do you think some cities had rules establishing separate facilities for blacks and whites? Did Rosa Parks have a right to break those rules? Were the rules fair? What could happen if she did not give up her seat to the white person?" When they finished deliberating, each group shared its decision tree with the class.

The children will naturally be curious about the real outcome of this event, and their questions will flow spontaneously: "What did Rosa Parks really do? What happened to her?" The questions will indicate a perfect starting point for an inquiry session on civil rights issues. You may wish to offer the information yourself, or encourage the children to search for the answers in material you have collected for them. (For your information, Rosa Parks refused to give up her seat and was arrested. Her arrest touched off a Martin Luther King, Jr.-led bus boycott in Montgomery and was one of the major events in launching the campaign for civil rights in the South.) In 1980, Mrs. Parks received the Martin Luther King, Jr. Peace Award for the inspiration she provided to resolve racial differences through nonviolent means.

Distinguishing Fact From Opinion

Critical thinkers must be aware of the existence of both fact and opinion in social studies materials. *Facts* can be defined as statements that are generally accepted as true and can be validated by evidence, such as, "In January of 1863, President Lincoln signed the Emancipation Proclamation." *Opinions* communicate what people feel or believe about something; opinions cannot be proven. For example, the statement, "Harriet Tubman was the most remarkable woman in the history of our country," is, as stated, an opinion. It is true that Harriet Tubman played a major role in the history of our country, but the idea that she was "the most remarkable woman in the history of our country" is an opinion. It cannot be proven as being either true or false; it simply tells how the person felt about her accomplishments. You must help your students to distinguish fact from opinion as they weigh statements for objectivity. Don't, however, create an impression that facts are "good" while opinions are "bad." Learning about people's opinions helps us understand why they act as they do. For example, in 1846, Dred Scott's strong opinions about his status as a slave made him willing to sue his owner for his freedom. Eleven years later, the Supreme Court issued its own opinion—Dred Scott was still a slave because, in the words of Justice Roger Taney, blacks were beings "of an inferior order" and "had no rights which white men were bound to respect." Anti-slavery Northerners reacted to this decision violently. Frederick Douglass called the decision "a most scandalous and devilish perversion of the Constitution." People throughout the country held such strong opinions about slavery since the writing of the Constitution, and the longer the problem went unsolved the stronger the opinions on either side grew. By 1861, feelings had grown to a boiling point and seven southern states seceded from the Union. The Civil War had begun.

It is important for the students to see that two people, or two groups of people, faced with the same set of facts often develop very different opinions. In this case, the strong contrasting opinions of Northerners and Southerners ignited a Civil War. Help students distinguish facts from opinions in social studies materials. Every fact should be backed up with evidence while clue words should signal an opinion. Students should realize that evaluative clue words often signal statements of opinion, such as *best, worst, brave, excellent, immoral, tightwad,* and *admirable.* Several sources of social studies content appropriate for helping students distinguish fact from opinion include political cartoons or editorials from newspapers and magazines as well as advertisements.

Political Cartoons and Editorials

Political cartoons and editorials use words and illustrations in an attempt to sway one's opinion about a particular issue. Sometimes the message of a cartoon, for example, is just plain fun—an illustrated joke. Other times, however, cartoons carry serious messages intended to influence a reader's opinion about an important issue, even though they use humor or sarcasm to make their point.

Most political cartoons deal with one central idea and are fairly uncomplicated. Cartoonists use few words to express ideas because the illustrations communicate most of the message. In addition, cartoonists will exaggerate certain physical characteristics of people to make them instantly recognizable. George Washington's hair or Abe Lincoln's lanky, tall frame, beard, and stovepipe hat are examples of the kind of distinguishing traits cartoonists select to highlight the central cartoon figure. Cartoonists also use standard, quickly recognized symbols (e.g., Uncle Sam, dollar signs, the Republican elephant and the Democratic donkey, and the hawk and the dove) to quickly communicate an idea or feeling.

While selecting political cartoons for your classroom, choose those that convey the simplest of ideas in as uncomplicated a fashion as possible. Help the children identify the standard symbols and central characters, recognize the activity in which

FIGURE 3–11
Political Cartoon

Source: Daily Local News, West Chester, PA (14 March 1993). Reprinted with permission of Rick Cole.

the characters are engaged, analyze the cartoonist's point of view, determine the cartoonist's purpose, and decide whether they agree or disagree with the cartoonist.

Figure 3–11 is a political cartoon. You could ask children the following questions about this particular cartoon to help them understand the purposes of political cartoons in general:

1. This cartoon contains illustrations of people you may know. What do you see here? Are these real people? Can you recognize any of the people in the cartoon?
2. Who are the people in the dark clothing? Who are the others?
3. What is happening? Why are they looking at each other that way? How do they appear to feel about what is going on?
4. Have you ever had anything like this happen to you?
5. What issue do you think the cartoonist is trying to highlight? What point is he making about the issue?
6. How would you react to this situation if you were a tourist? How would you feel if you were someone in the Amish family?

Newspaper editorials serve the same function as political cartoons, but editorials use words rather than illustrations to express a specific feeling or attitude. Like political cartoons, editorials should be discussed carefully. Help the students interpret the main issue, and then help them distinguish fact from opinion by highlighting the facts in one color and the opinions in another. They should see that editorials contain a combination of both. You might want to have the children re-examine the facts and write an editorial countering the opinions expressed by the original writer.

As students begin to understand the nature of editorials and cartoons as persuasive media, they develop strong interests in creating their own. John Kerrigan, for example, became deeply concerned about his principal's decision to remove the hallway door to the boy's restroom as a move to curb vandalism. To express his feelings about the principal's maneuver, John wrote a disapproving editorial. After he read it aloud to the class, his friend Chris drew an accompanying political cartoon. Figure 3–12 displays their joint effort.

Phony Documents

Vanderhoof and his associates (1992) have developed a technique for encouraging critical thinking by using phony documents. For example, a teacher opens a lesson with the following statement: "I have an opportunity to purchase an original letter purportedly written by Harriet Tubman when she was a leader of the Underground Railroad. I have concerns, however, about the authenticity of this letter. Perhaps Harriet Tubman did not write it. I need your help to determine whether I should invest my hard-earned money in this letter." The teacher then reads the letter, which contains several factual errors, to the class. Part of this letter is excerpted in Figure 3–13.

The teacher then asks, "Do you think I should purchase this letter?" Of course, the letter is not authentic, but it motivates the students to read the textbook and other sources of information for detailed information that might corroborate or refute details of the document. The students support a recommendation to purchase (or refuse) the letter based on information contained in the sources they consult. Phony documents like Harriet Tubman's letter can motivate students to read critically and to judge sources for the accuracy of their content.

The primary reason for educating our youth has been, and continues to be, the development of good citizens, proud citizens who make up a rich nation—a nation not made out of one people, but out of a mixture we call American. Americans have come from everywhere and for every reason—some in chains as slaves; others to search for gold, to find land, to flee famine, or to escape religious or political discrimination. Even today, America continues to be a sanctuary for the oppressed as well as a haven for the ambitious. Forefathers of consummate wisdom created a new kind of government for America—of the people, by the people, and for the people. To maintain this prized inheritance, all elementary school teachers must stand up and accept their responsibility to protect, nurture, and renew our healthy democratic society. They can do this by supporting and enacting the influential declaration of the NCSS (1993): "The primary purpose of social studies is to help young people develop the ability to make informed and reasoned decisions for the public good as citizens of a culturally diverse, democratic society in an interdependent world" (p. 3).

FIGURE 3–12
Persuasion Using Both an Editorial and a Cartoon

"Are bathrooms private anymore

Mr. Towson has a great scence of hummer, his last joke was the funnyest of of all. You better sit down for this Ready? Okay – He took... you sure your ready for this... Well, he took the bathroom door off See! I tould you should sit down. Now you propally think all the resonibillaty has gone to his head. Well for once I think he's absolutely almost right. Heres his side. Someone took three rolls of tolite paper in the toilet and flush it. It flooded the bathroom and the boys locker room. But taking the bathroom door off is to much. I mean you ever try and go in the bathroom with about 50 girls standing in front. But, there is a good part, the vandalism has gone down. Now Mr. Towson has something to worry about that is weather the school board impeaches him and if the health board calls the school a health hazrd.

Har! A litel town with a litel school has there own Watergate. I can see the head of linds now "First Princepal to be Impeached." I thought Mr. Towson is a nice guy (sometimes). But the health hazard is yet a nother thing. But don't worry Mr. Towson will figure out some and we hop bathrooms are still private

                                        Chris

FIGURE 3–13
Phony Document

March 31, 1913

When I crossed the Mason-Dixon Line from Virginia into the state of Pennsylvania I finally reached freedom. Free black families, white Quaker families, and friendly German farmers risked their own safeti to give me food and shelter on my journey. The journey was hard; I slept soundly under the stars each night to restore my spent energy.

     I looked at my hands to see if I was the same person now that I was free. I felt like I was in heaven! Crowds of enthusiastic supporters greeted me when I reached Philadelphia and welcomed me to the land of freedom. In the throng of humanity was the one face I had longed to see at the end of my journey—my dear husband, Clarence Tubman . . . .

AFTERWORD

Some experts have considered the way adults assist children through their youth so important that their methods have been singled out from all other pressing issues of our day—drug abuse, war, environmental concerns, crime—as the major factor influencing social change. DeMause (1975), an authority on the history of childhood, asserts: "The major dynamic in historical change is ultimately neither technology nor economics. More important are the changes in personality that grow from differences between generations in quality of the relationships between [adult] and child" (p. 85). What a powerful thought! And what a mighty challenge for you—to affirm your devotion to children by striving faithfully to help them develop the knowledge, dispositions, and virtues of responsible citizenship. Classrooms for young citizens should have a distinctly democratic flavor. Children should know what our country is now, and envision the best our country can be. As a teacher of young children, you will be an important nurturer of maximum civic growth. You will help make society. You must have a vision of good citizenship—much the same kind of vision as Michelangelo had when he peered intently at a monumental slab of Carrara marble and saw within it the Pieta, waiting to be liberated. Will you work with the fervor of Michelangelo to release responsible democratic citizens? Should society expect anything less?

References

The Boston Primer. (1808). In R. H. Hersh (Ed.), *Models of moral education* (1980). New York: Longman.

The Center for Civic Education. (1994). *National standards for civics and government.* Calabasas, CA: Author.

Darigan, D. L., Tunnell, M. O., & Jacobs, J. S. (2002). *Children's literature: Engaging teachers and children in good books.* Upper Saddle River, NJ: Merrill/Prentice Hall.

DeMause, L. (1975). Our forebears made childhood a nightmare. *Psychology Today, 8,* 85.

de Tocqueville, A. (1969). *Democracy in America.* (G. Lawrence (trans.), J. P. Mayer (ed.).) New York: Doubleday. (Original work published 1835–1839).

Ennis, R. H. (1985). Goals for a critical thinking curriculum. In A. Costa (Ed.), *Developing minds: A resource book for teaching thinking.* Alexandria, VA: Association for Supervision and Curriculum Development.

Henson, J. (1935). A slave's dilemma. In B. Brawley (Ed.), *Early negro American writers* (pp. 162–167). Chapel Hill, NC: University of North Carolina Press.

Johnson, D. W., & Johnson, R. T. (1999). *Learning together and alone.* Boston: Allyn & Bacon.

Kletzien, S. B., & Rappoport, A. L. (1994). *Kids around town.* Harrisburg, PA: League of Women Voters of Pennsylvania.

Lickona, T. (1993). In M. Massey (1993). Interest in character education seen growing. *Update, 35,* 1.

National Council for the Social Studies. (1993, January/February). *The social studies professional.* Washington, DC: Author.

National Council for the Social Studies. (1994). *Curriculum standards for social studies: Expectations of excellence.* (Bulletin 89). Washington, DC: National Council for the Social Studies.

National Council for the Social Studies. (2001). Creating effective citizens: A position statement of National Council for the Social Studies. *Social Education, 65,* 319.

National Council for the Social Studies Task Force on Early Childhood/Elementary Social Studies. (1989). Social studies for early childhood and elementary school children preparing for the 21st century. *Social Education, 53,* 16.

Newmann, F. M. (1989). Reflective civic participation. *Social Education, 53,* 255–258.

Parker, W. C. (1989). Participatory citizenship: Civics in the strong sense. *Social Education, 53,* 353.

Remy, R. C. (no date). *Skills in making political decisions.* Columbus, OH: Mershon Center, The Ohio State University.

Ryan, K. (1993). In M. Massey (1993). Interest in character education seen growing. *Update, 35,* 1.

Sterling, D. (1968). *Tear down the walls!* New York: Doubleday.

Vanderhoof, B., Miller, E., Clegg, L. B., & Patterson, H. J. (1992). Real or fake?: The phony document as a teaching strategy. *Social Education, 56,* 169–171.

4

Young Geographers

At the beginning of November, Helen Earnshaw launched her third graders into one of the most enjoyable and instructive projects they had participated in during the year. She began by reading the book *Flat Stanley* by Jeff Brown (New York: HarperCollins). In the book, Stanley Lambchop is squashed flat as a pancake by a falling bulletin board. Stanley is very flat (otherwise he is just fine), and discovers there are many advantages to being flat. He can slide under doors, go down into sidewalk grates, and even be folded up small enough so that his parents are able to put him in a large envelope and mail him off to California to visit some friends. After reading the book, Ms. Earnshaw kicked off the Flat Stanley project by asking her students to design, make, and send their own Flat Stanleys out into the wide, wide world.

Each young geographer began the project by designing a paper Flat Stanley and printing her or his name, return school address, and classroom e-mail address on the back. A brief statement of what the project was all about was also included for the recipients to read: "I am sending you a Flat Stanley and his journal. Please write down some of the things you do with Flat Stanley. Thank you." The students then made journals and, for 2 days, wrote about the major events in Flat Stanley's life. Their journal entries would serve as models for Flat Stanley's hosts. For example, Jerome wrote: "We took Flat Stanley to the mall. He liked seeing all the stores. Flat Stanley ate two chocolate chip cookies!" On the cover were the words "Passport to America." Inside were printed these instructions from Flat Stanley: "Dear Friend: Please finish my journal. Include places I've been and sights I've seen. A souvenir, or best yet, a photo of you and me together at a special place we visited would be nice! Sincerely, Flat Stanley. P.S. Would it be possible to send my owner a postcard from your town?"

The Flat Stanley, the journal, two little souvenirs (a class photo and an inexpensive item of local interest), and a self-addressed envelope with return postage were inserted into oversized envelopes and sent to participating classes (on a prepared list) in other parts of the United States. A send-off party, which included music and food, was a huge success. Each student shared a moment alone with his or her Flat Stanley before saying good-bye. A small spray of confetti was thrown on the box holding all the envelopes and a loud, "Bon voyage!" was chorused before the students took them to the post office to be mailed.

Maps, atlases, and other reference materials became important tools for Ms. Earnshaw's young geographers as they searched for the wonderful places their Flat Stanleys were visiting. For example, locating Fargo, a small city in eastern North Dakota, was a thrill for Andrew. Margaret had received a postcard from a student in Greenville, Mississippi, and the class was eager to find out just where Greenville was located. A sticker was placed on a wall map to mark each location that had been visited by the Flat Stanleys. Students learned new names and new places as well as the physical features of different geographic regions. Some hosts sent e-mail messages to Flat Stanley's original owners, detailing his new adventures as well as providing information about the weather/climate, local sites of interest, population, industries, seasonal activities, and so on.

By December 15, most of the Flat Stanleys had returned from their journeys. The journals were full of marvelous information about their exciting adventures. The envelopes were often overflowing with little souvenirs—local leaves, stickers, postcards, and photographs. Ms. Earnshaw's social studies class spent the next week of school reading through the journals and recording the path of each Flat Stanley's travels. The class honored the Flat Stanleys who had not yet returned with a Missing in Action poster, complete with pictures of the absent Flat Stanleys and their owners.

The project ended as Ms. Earnshaw's students wrote thank-you letters to Flat Stanley's hosts. This fabulous project helped the young geographers in Ms. Earnshaw's room learn much about the world around them.

Ms. Earnshaw constantly strives to provide the best possible learning experiences for her students. As you saw during this glimpse into her classroom, she stirred the students' curiosity about the world and captured their imagination with a wonderfully inspiring, developmentally appropriate geography project. Ms. Earnshaw mirrored the teaching convictions demonstrated by Ms. Roe in the classroom scene at the beginning of Chapter 2. She is convinced that geography, like history, cannot be thought of as a school subject that is limited to memorizing isolated facts. To the contrary, she is assured that geography has more to do with asking questions and solving problems about the nature of the world and people's place in it. Like Ms. Roe, Ms. Earnshaw wants her students to operate as young social scientists striving to find out about things worth knowing. The major difference is that Ms. Roe focused on the temporal dimension of human experience (*history*) while Ms. Earnshaw concentrated on the spatial dimension (*geography*). Time is the context in which history is carried out while space is the context for geography. In fact, these two social sciences often interact to answer questions about where places are and how and why they got there. For example: "How did New Orleans' location on the Mississippi River help it become the nation's fourth-largest city in 1830?" "Why was the Erie Canal built between Albany and Buffalo during the early 1800s?" "What geographic obstacles did the railroad workers face in building the transcontinental railroad during the 1860s?" Understanding temporal patterns is a vital dimension of comprehending Earth as the home of people.

WHAT IS GEOGRAPHY?

As with history, all informed citizens must know geography. There is no doubt that a strong grasp of geography equips people to make better-informed and wiser decisions about how people use the land and its resources; for example, questions like "What can be done to arrest global warming?" and "Is this the best location for the recycling plant?" Geography studies the interactions of people and places on Earth. For example, most people know that there are five Great Lakes in the United States—Huron, Ontario, Michigan, Erie, and Superior. However, not everyone knows that they appeared only about 10,000 years ago and that they were formed by the actions of receding glaciers. The same actions formed the geographic features of the upper Midwest—hilly land and rocky soil, not fit to grow wheat. It was exactly these geographic features that led pioneers to "substitute the cow for the plow" and become successful dairy farmers during the 1800s.

This example of the land of the upper Midwest helps illuminate the main elements of a formal definition of *geography:* "an integrative discipline that brings together the physical and human dimensions of the world in the study of people, places, and environments. Its subject matter is the Earth's surface and the processes that shape it, the relationships between people and environments, and the connections between people and places" (Geography Education Standards Project, 1994, p. 18). In short, geography can be thought of as the science of space and place.

In dynamic social studies, young geographers learn how to describe places on the Earth's surface and explain how these places came to be by using the same questions professional geographers use in their quest to understand and explain the physical and cultural features of the Earth: Where is it? Why is it there? How did it get there? What is the relationship between the people and the environment?

WHY IS GEOGRAPHY IMPORTANT?

The past decade has witnessed a strong push for more geography in social studies classrooms. A major force behind that push has been a set of alarming statistics indicating that students in the United States are woefully illiterate in geography (Denko, 1992). Some examples of our youth's geographic illiteracy include these shocking statistics:

- Forty-five percent of Baltimore's junior and senior high school students could not locate the United States on a map of the world.
- Twenty-five percent of high school students in Dallas could not name the country directly south of Texas.
- Forty percent of high school seniors in Kansas City, Missouri, could not name three of South America's dozen countries.
- Sixty-five percent of 2,200 college students in North Carolina could not name a single African country south of the Sahara.
- In a geography quiz administered to students in many countries, those from the United States ranked at the bottom.

At no time in our nation's educational history has the level of geographic literacy been so low. Although the data you just examined had to do with recalling specific facts, the goal of geography goes far beyond that. Some facts are certainly necessary, but in today's concept of geographic literacy, knowledge is meant to stimulate students' curiosities about the wonders as well as the problems of the world in which they live. Information is used more as a prelude to discovery than as something to be stored in the brain and pulled out for game-show-type quizzes. Geography is about ocean dumping, droughts, voting patterns, the revitalization of cities, and not only knowing *that* Washington, DC is our nation's capital, but *why.* New geographers are curious: "Why is there so much forest in northeastern Pennsylvania? Why is the Pennsylvania Turnpike located where it is? Why did Pittsburgh develop into a large city?"

Succinctly, the Geography Education Standards Project (1994) advises that students need to study geography for reasons that range from the most profound to the most utilitarian:

The existential reason. Humans want to understand the intrinsic nature of their home—that pale blue dot in the vastness of space. Geography helps them understand where they are.

The ethical reason. Life is fragile; humans are fragile. Geography provides knowledge of Earth's physical and human systems and of the interdependency of living things and the physical environment. This knowledge, in turn, provides a basis for people to cooperate in the best interests of our planet.

The intellectual reason. Geography focuses attention on exciting and interesting things, on fascinating people and places, and on things worth knowing because they are absorbing and because knowing about them lets humans make better-informed and, therefore, wiser decisions.

The practical reason. With a strong grasp of geography, people are better equipped to solve issues at not only the local level, but also the global level.

WHAT SHOULD YOUNG GEOGRAPHERS KNOW OR BE ABLE TO DO?

One of the most intensive efforts to stress geography education was completed by the Geography Education Standards Project (1994). The project addressed the issue of geographic literacy and issued its *National Geography Standards* as a blueprint of the scope of geographic study. The subject matter framework consists of two levels. At the first level, the subject matter is divided into six essential elements: *The World in Spatial Terms, Places and Regions, Physical Systems, Human Systems, Environment and Society,* and *The Uses of Geography.* These are called *essential* elements because each piece is central and necessary; citizens must look at the world in this way. At the second level, each essential element incorporates a number of geography standards containing a set of approaches to the subject matter of geography. A total of 18 standards are distributed among the six essential elements. Table 4–1 displays the framework of the National Geography Standards.

The National Geography Standards serve as a guide for current efforts to improve geography instruction in our elementary schools. The standards are applicable to all school districts in the United States, but districts are encouraged to adjust each to local interests and needs. To illustrate how this adjustment process can work, let us examine Standard 17: *How to apply geography to interpret the past.* It offers classroom teachers a wonderful opportunity to combine history and geography, as we can grasp from Barbara Fleming's original approach to studying the Erie Canal.

TABLE 4–1
The 18 Standards

THE WORLD IN SPATIAL TERMS
Geography studies the relationships between people, places, and environments by mapping information about them into a spatial context.

The geographically informed person knows and understands:

1. How to use maps and other geographical representations, tools, and technologies to acquire, process, and report information from a spatial perspective
2. How to use mental maps to organize information about people, places, and environments in a spatial context
3. How to analyze the spatial organization of people, places, and environments on Earth's surface

PLACES AND REGIONS
The identities and lives of individuals and peoples are rooted in particular places and in those human constructs called regions.

The geographically informed person knows and understands:

4. The physical and human characteristics of places
5. That people create regions to interpret Earth's complexity
6. How culture and experience influence people's perceptions of places and regions

PHYSICAL SYSTEMS
Physical processes shape Earth's surface and interact with plant and animal life to create, sustain, and modify ecosystems.

The geographically informed person knows and understands:

7. The physical processes that shape the patterns of Earth's surface
8. The characteristics and spatial distribution of ecosystems on Earth's surface

HUMAN SYSTEMS
People are central to geography in that human activities help shape Earth's surface, human

settlements and structures are part of the Earth's surface, and humans compete for control of the Earth's surface.

The geographically informed person knows and understands:

9. The characteristics, distribution, and migration of human populations on Earth's surface
10. The characteristics, distribution, and complexity of Earth's cultural mosaics
11. The patterns and networks of economic interdependence on Earth's surface
12. The processes, patterns, and functions of human settlement
13. How the forces of cooperation and conflict among people influence the division and control of Earth's surface

ENVIRONMENT AND SOCIETY
The physical environment is modified by human activities, largely as a consequence of the ways in which human societies value and use Earth's natural resources, and human activities are also influenced by Earth's physical features and processes.

The geographically informed person knows and understands:

14. How human actions modify the physical environment
15. How physical systems affect human systems
16. The changes that occur in the meaning, use, distribution, and importance of resources

THE USES OF GEOGRAPHY
Knowledge of geography enables people to develop an understanding of the relationships between people, places, and environments over time—that is, of Earth as it was, is, and might be.

The geographically informed person knows and understands:

17. How to apply geography to interpret the past
18. How to apply geography to interpret the present and plan for the future

---

Source: Geography Education Standards Project. (1994). *Geography for life: National geography standards 1994.* Washington, DC: National Geographic Research & Exploration, pp. 34–35.

Classroom Vignette

Ms. Fleming felt that a recording of an old folk song and excerpt from a journal entry would be useful artifacts to help her fifth-grade students begin to understand the importance of the Erie Canal in America's growth. The canal, completed in 1825, linked Albany to Buffalo, and eventually New York City, and greatly reduced freight costs for shipping goods between the established eastern cities and the expanding western frontier. Goods that once cost $90–$100 to ship between New York and Buffalo now cost only $4 a ton. Towns along the canal—such as Buffalo, Rochester, and Syracuse—boomed. Because of the Erie Canal, New York City became the largest center of trade in the United States. The success of the Erie Canal set off a burst of canal building throughout the country.

Ms. Fleming brought to class an old recording of "The Erie Canal" (also known as "Low Bridge") she had purchased from the Erie Canal Museum in Syracuse, New York. She asked the students to recall songs like "I've Been Working on the Railroad" that were sung by workers to make tedious tasks easier. Ms. Fleming then played the recording of "The Erie Canal" for her students, and they listened intently. Afterwards, she explained that this was a song the mule drivers sang as they traveled along the canal. The tempo of the song was determined by the slow and steady plodding of the mules as they pulled the barges back and forth along the length of the canal. Ms. Fleming then passed out copies of Peter Spier's book, *The Erie Canal* (in which the words to the song are printed along with fascinating details of the canal system), played the recording again, and invited the students to sing along. She told the students that social scientists often use songs, diaries, and other primary resources to learn about the past. In this case, they were to act as young historians and young geographers while using the song and the book to learn how the construction and operation of the Erie Canal affected people's lives during the early 1800s.

The Erie Canal (Low Bridge)

I've got an old mule and her name is Sal
Fifteen miles on the Erie Canal
She's a good old worker and a good old pal
Fifteen miles on the Erie Canal

We've hauled some barges in our day
Filled with lumber, coal, and hay
And every inch of the way we know
From Albany to Buffalo

Low bridge, everybody down
Low bridge for we're coming to a town
And you'll always know your neighbor

And you'll always know your pal
If you've ever navigated on
The Erie Canal

We'd better get along on our way, old gal
Fifteen miles on the Erie Canal
'Cause you bet your life I'd never part with Sal
Fifteen miles on the Erie Canal

Git up there mule, here comes a lock
We'll make Rome 'bout six o'clock
One more trip and back we'll go
Right back home to Buffalo

Low bridge, everybody down
Low bridge for we're coming to a town
And you'll always know your neighbor
And you'll always know your pal
If you've ever navigated on
The Erie Canal

Ms. Fleming next passed out a map of the Erie Canal and the students traced its route from Lake Erie to the Hudson River, making sure they located the cities mentioned in the song. She then distributed a copy of a diary entry she had fabricated after reading several first-hand accounts of travel on the Erie Canal (see page 145). It helped unlock the meaning of several passages from the song. The students were to use the words in the song as well as the journal excerpt to find evidence to solve the following mysteries:

What source of power moved the boats through the canal?

At what ports did the boats stop?

What did the boats carry?

What was the length of the canal?

What is the meaning of, "Low bridge, everybody down?"

After examining two primary resources (the song and the journal entry) and two secondary resources (Spier's book and the map), students were directed to think about additional data sources that might help them visualize the physical characteristics and patterns of human behavior related to the Erie Canal during that historical period. Ms. Fleming asked them to suggest any other sources they might consult for more information about life on or near the canal. The students recommended newspaper articles, paintings, a visit to the Erie Canal Museum in Syracuse (only about 10 miles from the school), and researching Internet sites (there are a number of good sites, including the Erie Canal OnLine http://www.syracuse.com/features/erie canal/). After conducting additional research, the young historians created a large annotated wall map of the Erie

An Excerpt from a Canal Traveler's Journal

May 5. Arrived in Schenectady at about one o'clock. As soon as the stagecoach stopped, up stepped a muscular gentleman who cried out, "Gentlemen, do you want to go to the West? We start at 2 o'clock, gentlemen. Only 3½ cents a mile with superior accommodations!"

"Don't take passage in that boat," called a second fellow. "I'll take you for half the money in a fine boat." In no time there were at least a half dozen more boat captains, all anxious for our passage and at almost any price we pleased. We wanted to see the boats ourselves before we were to take passage in any, so we sallied forth to the canal and selected a superior boat of the Clinton Line. The Captain agreed to take us to Utica, a distance of 89 miles, for one cent and a quarter per mile!

Climbing up to the roof of the boat, I joined several other travelers who were conversing on deck. Everyone enjoyed the fresh air and scenery until the helmsman blew his horn and bellowed, "Low bridge, everybody down!" The first few times this happened, we merely ducked to avoid the bridges overhead. However, we soon resorted to lying down to avoid several low ones. (Many bridges are no more than 8 feet, while the boat itself is a full seven.) It required great attention and care in passing the bridges, for you may get knocked down and join the company of the canal fishes! A woman next to me told of a child being knocked from the roof and the mad attempt to rescue her. Thank God she was rescued! Such things, however, do not often occur.

Our boat was one of the fastest on the canal, for horses instead of mules pull it. The Captain was a bit adventurous as he regularly exceeded the four miles an hour speed limit. The day seemed to move slower than the boat, but, at last, nightfall came and I was immensely exhausted. The sleeping quarters were quite uncomfortable. Three layers of canvas cots extend from the walls. The canvas was prone to wearing thin so if the person above is a bit plump, the canvas can sag so much that the person may end up lying on top of you.

Canal at its completion, labeling the canal locks, towns, and other waterways. Relating the project to the children's lives today, Ms. Fleming asked her young geographers to think about the sources they used for information about the Erie Canal and how they might relate to any current form of public transportation. What types of resources might be useful to a future historian as he or she attempts to uncover information about transportation in 21st-century America?

IN GENERAL,

HOW SHOULD GEOGRAPHY BE TAUGHT?

To help nurture young geographers as they strive to acquire the *geographic literacy* necessary for effective citizenship in an increasingly pluralistic society, teachers in dynamic social studies classrooms, as with history, must balance a "*what* to teach" and a "*how* to teach" approach. The "what to teach" component specifies the essential knowledge and skills that form the foundation for making decisions important to our geographic well-being; the "how to teach" component identifies the strategies and activities that help children work in school as young geographers.

Teachers want to create programs that produce geographically informed students who (1) acquire the cognitive skills that help students see meaning in the relationship between people and places and (2) read, analyze, and make the primary tools of geographers—maps and globes.

Acquiring the Skills of a Young Geographer

Implementing a geography program in the elementary school that places students in the role of young social scientists requires an understanding of four distinct skills that underlie competent geographic inquiry. These skills are: (1) *observation*, (2) *speculation*, (3) *data gathering and analysis*, and (4) *evaluation*. These skills may be accomplished through guided discovery strategies with their emphasis on building higher-order cognitive processes or through problem-solving experiences that focus on projects and research activities.

Guided Discovery

Teachers who prefer to use guided discovery strategies begin by selecting objectives for instruction and helping the students acquire them by giving them opportunities to make observations and participate in class discussions. One example of teaching the four geographic inquiry skills through guided discovery is described below.

Observation

The process of geographic inquiry begins with observing an actual physical location (such as a pond, building, or road) or representation of a physical location (video or film, photograph, slide, and so on) and seeking answers to the question, "What do you see here?" The process of understanding begins with what the students can directly observe.

Classroom Vignette

Troy Watkins, for example, initiated a thematic study of "Homes Around the World" by inviting his fourth graders to draw models and tell stories about their own homes and families. He wanted to connect the concept of *homes* with their own backgrounds of direct experience. After discussing the likenesses and differences of the homes in the classroom, Mr. Watkins displayed large study prints of homes from around the world and guided the observation and discussion with these questions: "What kinds of homes do you see here? In what ways are these homes similar to your homes? In what ways are they different?"

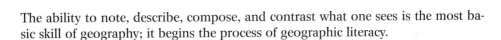

The ability to note, describe, compose, and contrast what one sees is the most basic skill of geography; it begins the process of geographic literacy.

Speculation

Geographic inquiry begins with observation, but it doesn't stop there. It continues to move into a more complex realm of understanding by asking questions about why things are where they are and how they got there: "Where is it located? Why is it there? What do you think has caused it to look this way?" Students must not stop the process of studying a geographic location simply by looking at and describing it; they must speculate about possible answers to "where" and "why there" questions.

Classroom Vignette

Mr. Watkins, continuing the topic of "Homes Around the World," asked his young geographers to look for clues in the pictures so they could think about how climate, natural resources, and other physical features of locations around the world might have influenced the selection of building materials as well as the design of a home. He asked, "How have the people in these pictures adapted their homes to the environment?" This process should not be pure guesswork, but rather a systematic form of deductive thinking, asking students to bring elements from other experiences to the process of deliberation. These speculations guide the search for information.

Data Gathering and Analysis

In this third phase of geographic inquiry, young geographers gather information about locations as well as about the human characteristics of those places. To answer geographic questions, students must be helped to gather information from a variety of sources in a variety of ways. In the earlier elementary grades, literature, videos, resource persons, or any other suitable informational sources might do this. In the middle and upper elementary grades, students can engage in investigative activities such as fieldwork (taking photos, distributing questionnaires, conducting interviews, collecting samples, reading and interpreting maps), or library research. Regardless of the grade level, all pertinent information should be recorded on a chart or study sheet, followed by a probing discussion: "What has happened? Where has it happened? How did it happen? Why did it happen? What took place before it happened? How does it compare (to something we experienced)?"

Classroom Vignette

While examining a special kind of home (adobe, for example), Mr. Watkins asked, "What kind of homes are these? Where are they found? What are they made of? How were they made? Why are they made this way? How do these homes compare with ours? Why are adobe homes so popular in different parts of the world? Do you think that homes in these areas will always be built this way?"

After discussing similar questions about several different types of house-building materials, Mr. Watkins helped the students construct a general statement about all the pictures being studied. When the class is able to synthesize information into a generalized statement, successful geographic inquiry has been demonstrated; in this case, the students said, "People from different places on the Earth build their homes from materials commonly found in their environment."

You may extend and reinforce the explanation by using small-group writing activities, constructing models, making maps, studying tables and graphs, or examining all kinds of literature. Having had such a stimulating data-gathering experience, children will have much to share.

Evaluation

The final phase of geographic inquiry is making personal judgments about the situation: "Have the people been wise in using the environment in such a way? Is this the most productive use of the land or its resources?" All personal opinions must be supported with sound reasons: "If not, why not? If so, why?" Since so many conditions of our world are intimately associated with the wise use of our physical environment, skills in this realm are of primary importance to children, the adult citizens of the future.

It is this physical environment–people linkage that makes geography such a valued component of today's dynamic social studies. Geographic terms and place locations are important; students need to know and use correctly the appropriate geographic terms and concepts. However, to become truly geographically literate, students must apply their basic skills and understandings to discovering relationships among people, places, and other phenomena. For example, generalizations about the influence of weather on homes or elevation of a region on agriculture are fundamental geographic relationships necessary for increasing geographic literacy.

Geography Projects and Research Activities

The project approach to geography is an appropriate teaching strategy that provides students with geography-related problems that serve as a focus for research activities. Projects are basically in-depth studies that engage children in working with a problem. Project activities can involve the whole class or small groups; they can extend over several months or take as little time as a week. Regardless, children take the initiative in completing projects. The teacher sets the scene and falls into the role of a facilitator, supporting the children's efforts. Projects can be initiated from the children's spontaneous interests or they can flow from an interesting experience provided by the teacher. An example of a geography project that started out through interest generated by a fifth-grade teacher, Marquis Lambert, is described below.

Classroom Vignette

Mr. Lambert designated sections of his classroom as "travel agencies," posted a United States map at each, and included the names of selected regions of the United States at each designated section: Middle Atlantic region, Great Lakes region, and so on. Groups of students were assigned to each agency and given the charge to design an advertising campaign for its respective region. They were required to produce a guidebook, travel poster, travel brochure, magazine advertisement, and a radio/television campaign.

Before they began the project, Mr. Lambert asked the students in each group to focus its research activities on what its region was noted for: sandy beaches, scenic mountain trails, historic sites, industry or agriculture, and so on. Some found information in magazines and books, others decided to watch a video. After they finished, each subgroup brought its information back to the region study group where they organized their information into a chart. The children then illustrated the major features on a large poster and used a few well-chosen words to characterize the region (for example, the group investigating America's Southwest focused on the Grand Canyon, labeling its poster, "The Grand Canyon—One of the seven Natural Wonders of the World").

The brochures were eye catching, too. Patterned after commercial brochures, the covers were adorned with scenes associated with each locale, such as "The Central Plains: Our Nation's Breadbasket." The children chose six important features of the region to highlight on the inside, each feature accompanied by text and illustration. The back of each brochure included a small map of the region, directions on how to get there, and sources of further information.

The purpose of the magazine ads and radio/television campaigns was similar—to draw visitors by describing the essential features of the region.

The guidebooks were much more comprehensive. Mr. Lambert showed commercial guidebooks as models, which students used to create their own, with these features:

- A *preface* that briefly promoted the attributes of the region.
- An *introduction* that served as a slightly more comprehensive guide to the region's attractions, climate, geographical features, chief products, and so on.
- A *history* section that included significant dates and events as well as the important people who have lived in the region.
- A *calendar of events* that highlighted fairs, celebrations, and seasonal attractions.
- A *places to see* section that included descriptions of major cities, museums, parks, zoos, recreational activities, historical sites, businesses, and the like.
- A *food and shelter* section that provided information about hotels, restaurants, and campsites, as well as their costs.

To culminate the activity, Mr. Lambert requested that each travel agency share its advertising campaign with the class. All children learned a great deal about the various regions through participating in this enjoyable, productive activity.

The current interest in projects stems from our growing knowledge about the probing, wondering minds of the children we teach. Geography projects are a source for nurturing their natural responsiveness and interest in the world around them. For that reason, geography is one social science that the children can really throw themselves into.

Using and Making Maps—The Tools of Geographers

It is impossible to teach or learn geography without using a map, or even several maps simultaneously. Maps are absolutely necessary for collecting data about places on the earth's surface. Thus, desk and wall maps, globes, and atlases are essential requirements for all dynamic social studies classrooms. The Geography Education Standards Project (1994) added: "Geography has been called 'the art of the mappable.' Making maps should be a common activity for all students. They should read (decode) maps to collect information and analyze geographic patterns and make (encode) maps to organize information. . . . For students, making [and using] maps should become as common, natural, and easy as writing a paragraph" (1988, p. 43). One of your major tasks in developing a solid geography component in your social studies program is to help children acquire the basic skills necessary to construct and interpret maps.

Basic Map Skills During the Early Grades

If there is a single area of controversy in geographic education that stands out above all others, it may be agreeing on the most appropriate time to introduce children to specific map skills. Although social studies textbooks and scope and sequence charts often include various map basics, there is considerable disagreement whether children are cognitively ready to do anything with maps before grade 2. Despite this discord about the best time to begin map skills instruction, there is widespread agreement that when instruction is carefully organized and sequenced, children are capable of learning the basics of map reading and map construction. The first step in organizing and sequencing a helpful map-reading program is becoming aware of the specific skills required to read and make maps:

1. Locating places.
2. Recognizing and expressing relative location.
3. Interpreting map symbols.
4. Developing a basic idea of relative size and scale.
5. Reading directions.
6. Understanding that the globe is the most accurate representation of the Earth's surface.

Three-Dimensional Classroom Maps

Siegel and Schadler (1981) advise that first mapping experiences should be with a location thoroughly familiar to the children. Their research demonstrated that attachment to the mapping location significantly influenced the students' understandings of maps as well as the accuracy of the maps they constructed. Therefore, the most appropriate place to begin map skills instruction appears to be the classroom.

Begin an initial classroom map-making project (it usually works best with second graders) by making available one empty, clean half-pint milk container for each student. Direct the children to cut off the tops so that they have a square, open-top box. Turn the box over so it is standing on the open end and cut away parts along the sides with scissors so that the cartons appear to have legs (see Figure 4–1).

Discuss the cardboard desks with the children, focusing on how they are models of their real desks. Encourage the children to paint the desks with tempera paint,

FIGURE 4–1
Stages of Constructing Milk
Carton Desks

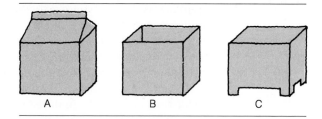

glue construction paper books or pencils on top, and put their name cards on the fronts of the desks.

When the desks are completed, ask one child to put her desk on a large sheet of cardboard (such as a side of a refrigerator packing box) that you have placed on a worktable or on the floor. Explain that the cardboard represents the classroom floor, but on a smaller scale. Let the student observe where her desk is located in the actual classroom and then examine the cardboard to determine where her desk would be located. Once she places her desk on the cardboard, the construction process moves quickly. A child who sits next to the first child places his desk on the proper location. As children take turns placing their desks on the classroom model, they are experiencing and using *three* basic map-reading skills:

1. Recognizing that their milk cartons stand for their real desks (*interpreting map symbols*).
2. Finding where their desks should be placed (*locating places*).
3. Determining the placement of individual desks in relationship to the other desks (*recognizing and expressing relative location*).

Next, ask the children to bring to school the following day empty boxes they might have around the house, from small jewelry boxes to boxes about the size of a toaster. Divide the class into committees, each of which is responsible for constructing a classroom feature such as the piano, teacher's desk, learning center, and so on. Keep a careful eye on the children as they select the boxes most appropriate for their particular feature. Often, the group responsible for the teacher's desk will select the largest box, even though that box is much larger than the relative size of the teacher's desk, because children think of their teacher as an extremely important person in their lives and thus deserving of the largest available box. However, you should encourage the children to look carefully at their own real desks in comparison to yours so they can eventually select a box that closer represents the true size relationship. Sometimes children come up with quite amusing ideas as they work on making relative size decisions. For example, one teacher found that all students had completed their assigned classroom features and were ready to place them on the growing three-dimensional map, but Alice remained at the back of the room busily working on her model wastebasket. When the teacher approached Alice to see what was going on, he discovered that Alice's progress was being delayed by a powerful urge to be as accurate as possible with her job. Since the classroom wastebasket was full of paper scraps, Alice was folding and cutting dozens of tiny pieces of paper for her wastebasket and folding or crunching them in order to precisely reproduce the real thing.

To complete their assigned features, the selected boxes should be decorated and painted so they are ready to join the student desks on the growing classroom map. This phase of map construction is critical as it contributes to what Preston and Herman (1974) describe as acquiring a "bird's-eye-view" concept. They found that a major reason children fail to understand maps in the upper grades is the lack of ability to form a mental picture of something if it were viewed from above, much like the way a bird might view something as it flies over. You can help develop a bird's-eye-view concept by encouraging the children to look at their features from directly above as they place them on their growing classroom map. That way, they will see only the tops of the desks, tables, file cabinets, and so on, and begin to understand that this is how a real map is constructed. Notice that during this phase of construction, the three previous map-reading skills are extended and reinforced, and a new skill is introduced: developing an idea of relative size and scale.

Once the classroom has been properly arranged, you can further extend and reinforce map skills by using their map as a learning tool:

Locating Places

"Place the teacher's desk where it is located in our classroom."

"Where would you place the piano? The file cabinet?"

"Point to the box that shows the puppet stage . . . the worktable . . . the teacher's desk."

"James, can you find Michelle's desk? Put your finger on it."

"Put your finger on the aquarium. Now trace the path you would take to answer the door."

Recognizing and Expressing Relative Location

"Whose desk is closest to the coat rack?"

"Trace the shortest path from the reading corner to the door."

"Which is closer to the door, the science center or the teacher's desk?"

Interpreting Map Symbols

"Pick up the box that represents the puppet stage."

"What does the red box stand for?"

"How can we show the coat rack on our map?"

Developing an Idea of Relative Size and Scale

"Which is larger, the file cabinet or the piano?"

"Which box should be smaller, the teacher's desk or the worktable?"

"Point to the smallest (or largest) piece of classroom furniture."

Flat Maps

The three-dimensional model of the classroom can be easily transformed into a flat map by following a few simple procedures. First, have the children look at their

Making models of the classroom or nearby school environment is the best way to begin map skills instruction.

three-dimensional map from directly above and discuss what they see. Then ask the students to put a piece of construction paper beneath each feature and trace around the outside of each with a crayon. As the three-dimensional features are removed and the outlines are cut out with scissors, the children should label the remaining outlines, such as "file cabinet" or "Bart's desk," and glue the outlines in their appropriate places. The three-dimensional map gradually becomes a flat map as traced outlines replace the models. For the children to understand how the flat map functions in the same way as the three-dimensional map, you should ask questions like those used for the three-dimensional map. Effective discussion is as important for this flat map phase as for the three-dimensional phase.

Model Communities

Building models of real places should not stop with the three-dimensional classroom. During grades 1 or 2, the process of building a model community is a thrilling and challenging project that can last all year long. It provides a way for students to learn about their community and master beginning map skills as well.

As with the model classroom, most materials for the model community consist of packaging items easily found around the house. Oatmeal boxes and toilet paper or paper towel rolls are excellent building materials for trees, cylindrically-shaped buildings, silos, or structures such as large oil storage tanks; cereal boxes, tissue boxes, candy boxes, and pasta boxes easily represent tall buildings, apartment houses, or stores; and different size milk or juice cartons make nice houses with peaked roofs. All you need to do is provide the proper work materials (construction paper, tempera paint, crayons, marking pens, school glue) and the children will go right to work. To give the children an idea how these materials might be used, show them how you would make a model of your own house. For example, cover a milk carton with red construction paper, glue on yellow windows and green doors, paint on a blue roof, and a house begins to take shape. A cardboard cylinder painted brown and covered with plumes of green adds a tree to the landscape.

I like to start the actual community model with something the children can directly observe—the school building. Walk around the outside of the school and carefully note its shape and size. Then, encourage the students to look over the boxes and decide on one that might be best to represent their school. Discuss the colors of construction paper that might be used to cover the box as well as any possibilities to highlight special features (trees, flagpole, parking lot) so it would look as much like the real school as possible. Toy cars, school buses, bicycles, and other vehicles also add to the realism.

The model community can expand outwards from the school as students visit the immediate neighborhood, a shopping center, the zoo, community services buildings, and so on. You might limit the project to the immediate school neighborhood, or extend it throughout the year as the children learn about all the elements of communities or neighborhoods.

Story Maps

Good children's books offer superb opportunities for exposure to additional informal mapping activities. Some books are perfect; *chain stories* that the children (or teacher) can illustrate easily are excellent selections. For example, in Eric Hill's popular "lift-the-flap" story *Spot's First Walk* (Putnam), a curious puppy meets all kinds of new animal friends as he wanders behind fences, by a chicken coop, and near a pond on his first venture away from home. As you read the story, invite the children to predict whom they might meet under each flap. When the story is finished, ask the children to draw pictures of the snail, fish, bees, hen, and other friends Spot met along the way. Their simple illustrations can be arranged as a floor display, creating a sequence map of Spot's travels. The children can retell the story by telling what happened at each point as they walk along their "map." Other stories appropriate for story maps include *Katie and the Big Snow* by Virginia Lee Burton (Houghton Mifflin), *Rosie's Walk* by Pat Hutchins (Macmillan), and *Harry the Dirty Dog* by Gene Zion (Harper). A sample story map for *The Three Billy Goats Gruff* is shown in Figure 4–2.

Mental Maps

Geography Standard 2 states that, to be geographically informed, students must be able to understand "*how to use mental maps to organize information about people, places, and environments in a spatial context.*" A mental map is a very effective way of keeping in mind a lot of information about some aspects of the Earth's surface.

FIGURE 4–2
Billy Goats Gruff Story Map

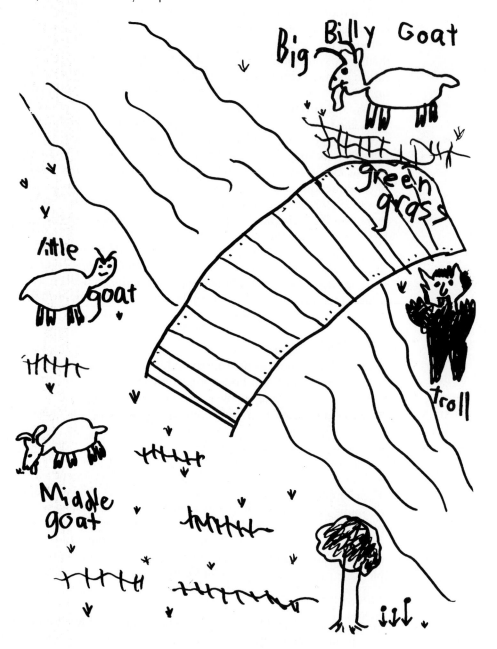

FIGURE 4–3
Mental Map of the United States

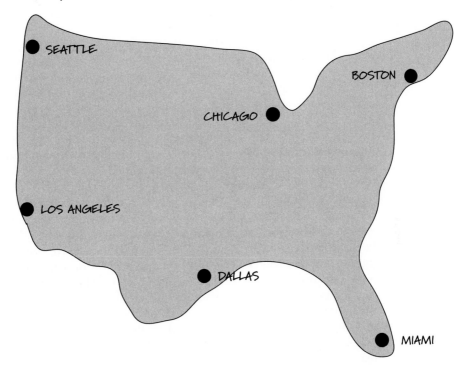

Basically, it is a drawn representation of what the person knows about the locations and characteristics of a variety of places on the Earth. For example, look at Figure 4–3, a mental map drawn by one of my students. It is a quickly sketched, simple outline, but still has a fairly good recognizable identity. You've probably figured out that it is the United States because its sizes are roughly proportionate, the bulges and indentations are roughly where they should be, and important cities serve as helpful location markers. All of us have used mental maps throughout our lives. Did you ever make a map of your bedroom as a youngster? What about sketching out a map to help someone find your house? Mental maps can be of continents, countries, cities, and oceans, or they could be the layout of your campus. Whatever the representation, people develop and refine their mental maps through personal experience and from their social studies program. The development and refinement of mental mapping is an important part of geography because they reflect students' skill in observing and thinking about the world in spatial terms. Young geographers will add detail and structure to their mental maps as they learn more about the world around them. Therefore, mental maps for young children include their schoolroom, school, bedroom, home, the neighborhood, or community. As they read, observe, and experience more about the world around them, students can draw mental maps of geographic features such as oceans, continents, countries, or mountain ranges.

Map Games

As the children gain experience with maps in appropriate contexts, it is fun to challenge them with game-type activities. For example, clear a large area of your classroom or take the children to a spacious area such as outdoors or to a multipurpose room. Make a large diamond with a long piece of yarn or masking tape. Have the children stand above the diamond shape and look at it carefully from above. Trace a rough outline of the diamond shape on a large sheet of drawing paper. Discuss how your outline represents the diamond shape they are looking at, only that it is much smaller. Next, ask the children to draw their own diamonds with crayon on sheets of drawing paper (large, simple diamonds are fine). Place a ball on the floor near one corner of the large diamond and invite the children to stand over it and look at it carefully. Then ask them to draw a picture of the ball on the corresponding corner inside the diamond they drew on their drawing paper. Add several more objects to this activity, each time following the same procedure. Use objects such as books, chairs, blocks, toys, and so on. Compare the relative size and distance of each item so children will see how position and proportion affect the accuracy of maps. Now split the class into groups of three or four. Have each group draw a special "map" illustrating how they would arrange four objects inside the large diamond on the floor. The groups then exchange maps and challenge one another to arrange the real objects inside the large diamond on the floor as they were represented on the maps. Now invite the groups to further challenge one another by having one group rearrange the objects on the floor and asking the others to make the map of the resulting configuration.

Refining Map Skills

Because map construction projects or story experiences cannot guarantee complete mapping competency, you will need to offer various opportunities to reinforce the specific map and globe skills. With appropriate experiences spread throughout the early grades, most new skills can be developed and strengthened in a rewarding and satisfying manner.

Map Symbols. To reinforce the idea of map symbols through activities that go beyond classroom maps, present the children with a few easily recognized symbols they might see every day, such as those in Figure 4–4. Ask the children what each symbol or sign means to them. What do they stand for? Emphasize that symbols represent real people, places, things, or ideas. Following the discussion, ask the children to pair up and have each child draw a secret symbol without allowing his or her partner to see. Then have each pair try to guess what each other's picture stands for. If they have difficulty getting started, offer suggestions such as road signs, punctuation marks, math symbols, and the like. Prepare a bulletin board display of their efforts after they have shared the symbols with each other.

FIGURE 4–4
Recognizable Symbols

Explain that a *symbol* is a sign that stands for something. Illustrate this idea by sharing a familiar object, such as a toy airplane. Ask the children to draw an airplane, say the word, and write the word on the chalkboard. Help children understand that some sounds are symbols, some pictures are symbols, and some printed words are symbols. The children should now be ready to understand that the special set of symbols groups of people commonly use to communicate ideas is called a *language*. In the United States, most citizens use the English language to communicate orally and in print. In school, we learn English as our main language, but we also learn other languages, such as the language of maps.

To move from this introductory lesson on symbols, write the word *tree* on the chalkboard and discuss what a tree looks like. After talking about their ideas, emphasize that the written word *tree* stands for a real thing the same way a map symbol does. Ask the children to suggest what the illustrated symbol for *tree* would be in the special language of maps. Follow this procedure while helping the children make up their own symbols for houses, factories, stores, libraries, lakes, roads, mountains, and so on. Don't be overly concerned if their symbols are not the same as standard map symbols; at this point you are more concerned with the overall concept of symbolization rather than the accuracy of representation.

FIGURE 4–5

Introducing Map Symbols

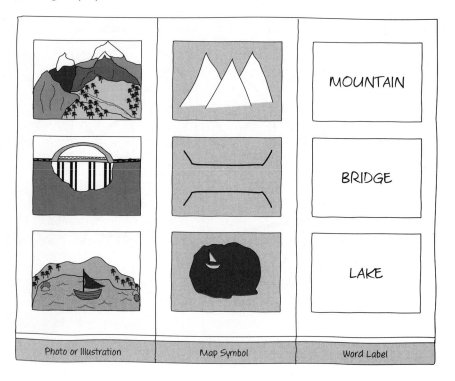

You will now want to show your students real maps and introduce standard map symbols once the children are familiar with the concept. Remind the children that maps are symbols on paper, and that the symbols on maps represent real things. Show them photographs or slides of a railroad, a bridge, and a building. Then display the three corresponding map symbols. Ask the students to match the symbols to their corresponding pictures and add the word labels. Follow this progression (photo or picture—symbol—word label) whenever new map symbols are introduced in your classroom (see Figure 4–5).

Direction. The best method of introducing young children to the skill of finding directions on a map is through a participatory learning experience. Primary-grade children enjoy going outdoors with simple compasses to find the cardinal directions (north, south, east, and west). After they locate north on the compass, the children will soon learn that south is behind them, east is to the right, and west is to the left. If the children are outside at noon on a sunny day, they will find a new clue for determining direction—in our northern hemisphere, at noon, their shadows will point in a northerly direction. Once they determine north this way, the other directions will be easy to find. To help them remember the other directions, ask the children

As children acquire map concepts, they enjoy creating their own maps of real places on the earth's surface.

to search for outstanding physical landmarks. Have one child face in a northerly direction and select the first obvious feature, such as a large building. Give the child a card labeled *North* and ask him or her to stand facing in that northerly direction with the card in his or her hands. Then select a second child to stand back-to-back to the northerly child. Ask the children what direction this child is facing. If no one says "south," tell them. Ask this second child to find an outstanding physical feature to the south (such as a large tree) and give him or her a labeled card to hold. Repeat this procedure when explaining the east and west directions. By associating landmarks with directions, the children begin to understand that directions help us locate places in our environment. You may ask, for example, "In what direction must I walk if I want to go to that large hill?" To help reinforce these directional skills, provide a number of follow-up activities; for example, "Simon Says" can be adapted to a directional format—"Simon says, 'Take three steps west.' Simon says, 'Turn to the south.' "

After the children have had this fundamental introduction to direction, extend the understandings to their map constructions. Ask the children to place their direction labels on the appropriate walls in the map construction (for example, after completing the outdoor experience). On the classroom maps, do not label the front of the room *north* and the back of the room *south* if these are not the true directions. After checking the classroom directions by using the compass or checking with the previously established reference points, teach the children to orient maps in the proper direction whenever they use them. This may involve turning chairs or sitting on the floor; however, by always turning themselves and their maps in the direction of true north, children avoid the common misconception that "north" is the direction toward the front of the room.

Scale. The idea of scale should be introduced in a relative way rather than in a mathematical sense in the primary grades. Children must realize that maps need to be small enough to be easily carried and readily used. Give the children sheets of drawing paper in shapes that approximate their three-dimensional constructions and tell them to construct their own maps using the map construction as a model. Some children will immediately reduce the size of the classroom features proportionately. Others will have greater difficulty trying to reproduce the large classroom features on their smaller papers.

The Globe. Since the early primary-grade child's concept of Earth is fairly restricted, planned instruction in globe-reading skills is often not successful. Nonetheless, you should not totally omit globe-related activities from the primary classroom. With simplified 12-inch globes, children can understand that the globe is a representation of the Earth, much like their constructed maps were representations of a real place. The globe should include a minimum amount of detail and preferably should show the landmasses in no more than three colors and the bodies of water in a consistent shade of blue. Only the names of the continents, countries, largest cities, and largest bodies of water should be shown. Globes that show more detail easily confuse the very young child.

   With young children, globes can become valuable informal teaching tools. When reading stories, children may wish to find where their favorite characters live; you can show them the geographic location. For example, if you are reading the children a story about Los Posados in Los Angeles, you may want to show them where Los Angeles is in relation to your community. However, even this would be a meaningless activity unless you relate it to the children's own experiences. You may say, "A globe tells you where places are. Pictures tell you what the place is like." Then, point out where Los Angeles is located and show a large picture with familiar landmarks. Children who hear about the North Pole at Christmastime may want to find out where it is located. Television stories or newspaper articles may suggest other places the children are interested in. The teacher can use times like these to familiarize young children with characteristics of the globe and with the fact that they can use the globe to locate special places. The basic globe concepts for development in the primary grades are: (1) to understand the basic roundness of the Earth, (2) to understand the differences between land and water areas, and (3) to begin to locate the poles, major cities, and the United States. Teaching suggestions follow:

• Have the children distinguish large land areas from bodies of water. Primary-grade children can learn that Earth is mostly made up of water and seven large land areas known as continents. They should know that our country is called the United States of America and that it is located on the continent of North

America. North America is made up of several countries: north of the United States is Canada; south is Mexico; the Atlantic Ocean is east of the United States, and the Pacific Ocean is west. The United States is made up of 50 states; most touch each other. Two states that don't are Alaska and Hawaii. By the end of the primary grades, children should locate their home state on a map of the United States (see Figure 4–6.)

• Talk about how it would feel to be an astronaut and to be able to look at the Earth from a satellite. Have the children describe how the landmasses and bodies of water would look. Show a satellite photo and map of the Earth.

• When studying about families around the world, tape small pictures of people in typical dress to their corresponding countries on a large papier-mâché globe constructed by the children. Discussion of the need for different types of clothing can lead to an awareness of warm and cold regions of the Earth.

FIGURE 4–6
Primary Grade Globe Concepts

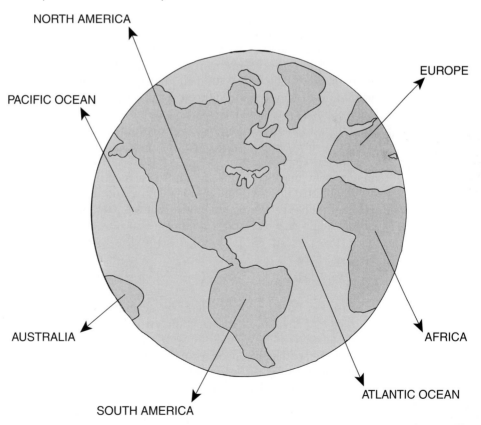

Map Skills in the Middle and Upper Grades

The identical six map and globe skills introduced and reinforced during the early grades are again the ones used in the later grades, but in more highly sophisticated contexts and for different purposes. Children in the early grades were developing concepts of what a map is and learning the basics of how to read maps by constructing their own maps and being helped to acquire the six basic skills through related teacher-initiated questions and prompts. They see maps in their textbooks that (1) are simple in nature, usually depicting familiar places such as a zoo, park, or neighborhood or imaginary places such as Candy Land or Space Land, (2) contain mostly pictorial or semi-pictorial symbols, (3) often represent an environment such as a neighborhood or community that can be directly experienced by all the children, and (4) gradually expand outward to show, with minimal detail, the continents, the United States, or the route a letter would take if mailed from their home town to Mexico City, for example. In later grades, children put these skills to use by employing maps to gather information and to solve problems. In short, the map reading is transformed from *learning to read maps* to *reading maps to learn.*

Place Location and Direction

As children move through the elementary grades, they continuously use directions and examine a map to find places. The major difference, however, is the level of sophistication of the map itself. Comparing this process to reading a book, you might say that children are taught to read stories in first grade but only at a level appropriate for that stage of development. Later, as the children's skills mature, they are introduced to newer, more difficult stories. Likewise, the maps that early primary-grade children read gradually evolve into more specialized maps, many of which require them to understand grids, a concept that combines knowledge of *place, location,* and *direction.* Middle- and upper-elementary-grade children move toward functionally employing maps with grids, such as road maps, to acquire information for a specific purpose. Dawn Irwin used a road map to help her fourth-grade students reinforce their symbol, place location, and direction skills while, at the same time, learn about the ever-expanding highway system in the United States.

Classroom Vignette

First, Ms. Irwin and her students reviewed the concept that various signs and symbols communicate information on road maps, helping travelers get to their destination without trouble. However, as is the case with all maps, the students were aware that they need to know what the symbols mean if they were to be of assistance. For this lesson, Ms. Irwin asked the students to work in pairs and distributed a road map of the United States showing two major types of roads: interstate highways and United States highways. Today, the students were going to learn some interesting facts about interstate highways.

FIGURE 4–7
Interstate and U.S.
Highway Route Signs

INTERSTATE   U.S.

First, Ms. Irwin displayed a sample highway sign for each of the two major highways (see Figure 4–7). Treating the map experience as a guided discovery activity, she asked the students to examine their road maps and decide which of the two sample signs represented an interstate highway. After a period of discussion during which they called on past experiences and word knowledge skills, the students decided that the first sample sign represented interstate highways because they recalled seeing such signs as they traveled with their families, because their road maps showed these signs were on the roads that connected two or more states, and that "interstate" means traveling between two or more states. Ms. Irwin then asked her students to carefully examine the interstate highway system and see if they could see any pattern by which the highways were numbered. Quickly, Jose detected that interstate highways going east and west are assigned even numbers while those going north and south are given odd numbers. Ms. Irwin passed out a yellow highlighter marker to each group and asked the students to trace each north-south highway with yellow. After each group did so, Ms. Irwin asked the students to identify the states through which the highways passed. She then challenged them to see if they could find a definite pattern with only the highlighted north-south interstate highways. It took a little time, but the students eventually discovered that the higher the number, the farther east the highway: for example, Route 95 runs north and south along the east coast while Route 15 does the same in the west.

Next, Ms. Irwin passed out a pink highlighter to each group and asked the students to highlight the east-west interstate highways. In like manner, she directed the students to identify the states through which each passed and challenged them to uncover a numbering pattern. They found that the highest even-numbered interstate highways were in the north and gradually got smaller as one traveled south: for example, Route 90 was northernmost on their maps while Route 40 was southernmost.

Ms. Irwin then assigned one interstate highway to each group and asked them to use a marking pen to trace their highways on a large political map of the United States that was previously taped to the wall. They then made properly shaped interstate highway signs to label their routes.

To culminate the lesson, Ms. Irwin passed out to each group a smaller political map of the United States and asked each to recommend a route for a new interstate highway, number it according to the pattern discovered previously, and justify why the highway would be beneficial to the people of our nation.

Latitude and Longitude

In the upper elementary grades, children extend their knowledge of grids as place location devices to the system of latitude and longitude. This system consists of east-west lines called *parallels of latitude* and north-south lines called *meridians of longitude* (see Figure 4–8).

The parallels of latitude, imaginary lines encircling the Earth, measure distances in degrees north and south of the equator (designated as zero degrees latitude). The parallels grow smaller in circumference as they approach both poles. The meridians of longitude, also imaginary lines encircling the Earth, converge at the poles and measure distances in degrees east and west of the prime meridian (designated as zero degrees longitude).

The importance of grids as a means of locating places can be illustrated with a large, unmarked ball. Lead a discussion comparing the similarities of the large ball and the Earth as represented by the classroom globe. Glue a small plastic ship to the ball and ask the children to describe its exact location, imagining themselves shipwrecked and needing to radio their location to be rescued (the ship marks their wreck). They will discover that this is nearly impossible, since there is no point of reference from which to describe an exact location. For example, if the children say the ship is located on the front side of the ball, you can turn the ball and the statement will be incorrect. If they say the ship is on the top of the ball, turn it back again to the original position. Gradually, the students will experience the frustration of locating places on a globe without agreed-on reference points. After some deliberation, they will most likely suggest the addition of parallel east-west lines and instruct the rescue squad to search an area "three lines down from the middle line."

FIGURE 4–8
The Earth's Grid System

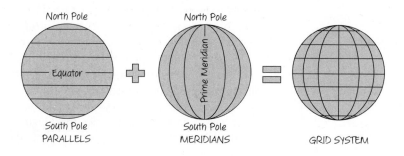

On closer examination of this arrangement, and after prodding from the teacher, the children will discover that the rescuers need to travel all around the world along the "third line down from the middle line" to find them unless given even more precise locations by devising *meridians,* or north-south lines. The rescue squad then only needs to find where the two points meet. Eventually, the children can be led to locate many well-known places in the world using latitude and longitude. Determining precise locations by using actual *degrees* of latitude and longitude may be beyond the capabilities of most fourth- and fifth-grade children. Guide them, however, in using latitude and longitude for locating general areas, such as the low latitudes (23 1/2° north and south of the equator), the middle latitudes (between 23 1/2° and 66 1/2° north and south of the equator), and the high latitudes

A number of specialized skills are necessary before students are able to read globes with meaning.

FIGURE 4–9

Yearly Rainfall and Vegetation Maps of Africa

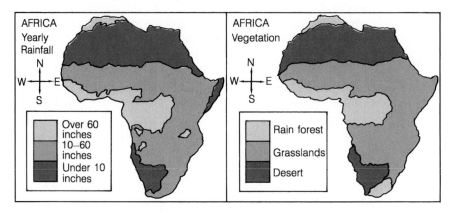

(between 66 1/2° north and the North Pole and 66 1/2° south and the South Pole). Children can generalize about the climatic similarities within these areas. In which latitudes are most cities located? Where is the weather warm (or cold) throughout most of the year? Show them how to find places east or west or north or south of their location by using meridians. After careful observation, they may find many surprising facts. For example, Rome, Italy, is *nearer the North Pole* than New York; Detroit is *north* of Windsor, Ontario; Reno, Nevada, is *farther west* than Los Angeles; the Gulf of California does not touch California at any point; and the Pacific Ocean is *east* of the Atlantic Ocean at Panama. All early grid instruction should be general and avoid as much as possible the use of degrees in place location.

Relative Location

Simply locating places on maps does not give students a true understanding of geography. Children must understand the influence of place location on people's lives and how one physical feature may influence another. Recognizing and expressing relative location is a sophisticated map skill because it involves not only locating places but also understanding interrelationships among geographical features, such as the influence of rainfall on the vegetation of a region. For example, upon examining a rainfall map of Africa, shown in Figure 4–9, children should be able to determine the type of vegetation that might grow in each region and how this influences the ways people live.

In the early primary grades, children develop concepts of relative location by explaining why different places in the classroom are located where they are ("Why do you think the science center is next to the sink?"). The concept is expanded in the intermediate grades as children develop the ability to interpret the influence of physical factors, such as mountains, deserts, valleys, and oceans, on human life. The effects of these geographic features should be discussed, along with humans' attempts to change conditions for their own benefit, as Sharon Filetti does in this example.

Classroom Vignette

Sharon Filetti found that a simulation activity was a good way of helping her students understand how people make decisions to adapt physical conditions of a location to meet their wants and needs. First, Ms. Filetti distributed a map of Newtown, a hypothetical community. It has a busy downtown section through which a new highway must be constructed. Ms. Filetti assigned each student a building through which the highway must pass (Mark was assigned the bakery, Lifeng the florist shop, Gustav the hospital; others were given a place of worship, historic house, apartment building, department store, pizza parlor, movie theater, YMCA, auto dealership, supermarket, and so on). Ms. Filetti displayed a large wall map of the downtown area and asked the students to write their names next to their assigned locations. She informed them that the highway had to be completed with minimal delay, so the entire business community needed to meet to decide where to put the highway. Ms. Filetti decided that tunnels or bridges were acceptable, but she didn't suggest them before the students had a chance to think of those ideas by themselves. Similarly, Ms. Filetti's students took part in simulations where decisions about other interactions between humans and nature needed to be made, such as the effect of a large shopping mall on an undeveloped rural area, the effect of a toxic chemical plant on local streams and lakes, and the effect of a huge dam on what had been pristine farmland.

Often, teachers have their students construct special-purpose maps that highlight the people–place connection of geography. Product maps are good examples. See Figure 4–10 for a sample.

Map Symbols

Maps and globes use symbols to represent a region's characteristics. In the primary grades, pictorial or semi-pictorial symbols are recommended; as a rule, the younger the child, the less abstract the symbols should be. However, as with younger children, always present a picture along with the pictorial symbol. Emphasize the importance of looking at the legend before using a map. David Barnes expanded his students' understanding of map symbols by offering them the following experience.

FIGURE 4–10
Special Subject Maps

Classroom Vignette

To start, Mr. Barnes played a CD of Woody Guthrie's song, "This Land is Your Land." The folk song describes many of the landforms found in the United States. Mr. Barnes made a transparency of the lyrics so the students could follow the words as they listened. After the song finished, Mr. Barnes asked the students to discuss the images that formed in their minds.

Next, Mr. Barnes invited the class to join him on a simulated airplane trip across the United States from the west coast to the east coast. "Hello, this is your captain speaking," announced Mr. Barnes. "Please fasten your seatbelts and prepare for takeoff." After turning up his player full blast, Mr. Barnes played a sound effect CD of an airliner taking off. As the sound diminished and the captain announced the plane had reached cruising altitude, the students were advised they could unfasten their seatbelts. During the imaginary flight, Mr. Barnes called the students' attention to the many landforms that spread out beneath them: "Notice that the brown and yellow Mojave Desert is disappearing behind us. If you look carefully out of your window, you will see the green-forested peaks of an enormous range of mountains stretching out below. Of course, these mountains have a much cooler and wetter climate than the desert."

As Mr. Barnes' imaginary flight continued eastward, his monologue continued to include descriptions of the major landforms: Great Plains, Central Plains, Mississippi River, Appalachian Highlands, the Atlantic Coastal Plain. Each time a new feature was introduced, Mr. Barnes showed large pictures of the landform as well as a map of the major landforms that cover the United States.

After the plane "landed," the students reviewed the map and discussed the colors and symbols located on the legend. The students were then organized into groups, each responsible for researching more information about what they would observe and experience if they were able to visit the landforms they were introduced to on their transcontinental flight—deserts, mountains, rivers, plains, and highlands.

Lastly, Mr. Barnes decided to reinforce the idea that maps and globes use color or shading as a symbol to show elevation of land from sea level; his students should understand that color is a special kind of symbol. Therefore, Mr. Barnes mounted a large sheet of paper to the wall on which he drew a blank map of the United States. Each research group used crayons to color the landforms according to the way they are conventionally represented on maps.

Scale

To portray geographic features of the Earth on a globe or flat map, you must use the concept of scale to ensure accurate size and space relationships among the features. This is accomplished by reducing the size of every real feature in an equal percentage. Introduce children to the concept of scale by comparing a class picture to the actual size of class members. Lead them to realize that the picture represents a real group of children, but in a much smaller way.

Perhaps the most appropriate formal map scale to use at the elementary school level is the graphic scale. Place a scale of miles at the bottom of the child's map. Children can place a cardboard marker between any two points (Los Angeles and San Francisco) on their maps, place a dot for each city, and then lay the edge of the marker along the scale. The segments of the scale on Figure 4–11 are of equal length and represent miles on the map. Comparing the marks on their cardboard marker to the scale, children will see that the distance between Los Angeles and San Francisco is approximately 350 miles.

Reading a Globe

Recall the basic globe-reading skills we discussed for the early grades: informal instruction aimed mainly at helping the children realize that the globe is a model that represents the Earth. Their major formal map-reading experiences up to this time

FIGURE 4–11
A Graphic Map Scale

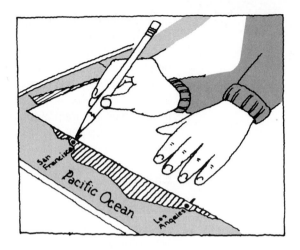

dealt with flat maps on which they located cities and other places of interest. They learned how to tell direction and how to compute the distance between one place and another. Now they must learn that a globe is the only accurate map of the Earth, and is an even better tool for studying locations, directions, or distances than a flat map. To emphasize this, you may want to show a satellite photograph of the Earth and compare it to a classroom globe. It is fairly easy to find satellite photographs; one option is to request them through the United States Weather Service.

After you compare satellite photographs to a classroom globe, illustrate just why the globe is more accurate than flat maps. Using a large, thin rubber ball or a globe made from papier-mâché, cut the ball in half and draw an outline of North America (or any random shape) on the ball. Have the children apply hand pressure to flatten the ball and discuss the resulting distortions. Then use scissors to cut through the ball along lines that represent longitude lines. Have the children try to flatten the ball again. Although the ball flattens more easily, the drawn outline still becomes distorted. Help the children discover that this is a major problem faced by mapmakers (*cartographers*) when they attempt to make flat maps of places on the Earth (see Figure 4–12).

Globes help to show shapes of areas exactly as they would appear on the Earth's surface. Unfortunately, maps are not able to do this. Representing a curved surface precisely on a flat map has confounded cartographers for years. The resulting distortion has been responsible for such honest comments as, "I didn't know Greenland was such a large country!" In fact, it isn't. Although it appears huge on some map projections, Greenland is actually one-third the size of Australia. A classroom globe shows shapes and areas more accurately than maps. Therefore, globes and maps should be used reciprocally while developing the skills outlined in this chapter.

As the children explore the globe, they should understand that the Earth can be divided into *hemispheres* (*hemi-* is a prefix meaning "half of"; thus, "half of a sphere"). If we live in the United States, we live in the northern half of the globe, or the North-

FIGURE 4–12
Globe as a Flattened Ball

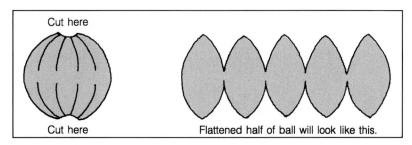

ern Hemisphere. At the same time, we live in the western half of the globe, or the Western Hemisphere. The *equator* and the *prime meridian* split the Earth in half in each direction to form the hemispheres. Other significant lines that encircle the Earth and run parallel to the equator are the parallels of latitude, discussed previously. Two important latitudes are the Tropic of Cancer and the Tropic of Capricorn. The region between these two lines, including the equator, is called the *tropics.* The Tropic of Cancer is north of the equator; the Tropic of Capricorn is south of the equator.

Map Selection for the Classroom

It is important to provide a wide variety of maps for the young geographers in your classroom, both as sources of information and as models for their own map construction:

Political maps show the boundaries of countries, states, cities, and other political entities that were created by people as opposed to naturally occurring boundaries such as rivers or mountain ranges.

Physical maps highlight the physical features of a place, such as rivers, lakes, mountains, and so on.

Topographic maps are representations of three-dimensional surfaces on flat pieces of paper. They typically use shading to show landscape features and elevation.

Raised relief maps, a version of topographic maps, use raised surfaces rather than shading so you can actually "feel" the mountains.

Weather maps are perhaps the type of map most frequently seen by most people. Weather maps show the temperature, precipitation, sunshine, storms, and other weather-related features of a region.

Special-purpose maps are maps that depict special subjects or features of an area: the world's language families, average annual rainfall in the state, natural resources, goods produced in various parts of the state, how states voted in the national election, or the soil types of the Northeast.

Historical maps show events and patterns that occurred in the past. They may show the routes of the explorers to North America, for example, or how something such as a nation's borders had changed over the years.

Road maps are maps that help us figure out a good route between two places. If you want to travel from Boston to Atlanta, a good road map will tell you how far and in which direction you must travel.

Classroom Vignette

Irina Carlson helped her students understand that space and place analysis is the cornerstone of geography, and that maps are the principal tools in performing that analysis. Therefore, she asked her students to list all the different maps they have seen or used, which she placed on the chalkboard. Ms. Carlson then grouped her students into teams of four and directed each team to look through its textbooks for 5 minutes in an effort to find as many different kinds of maps as they could. The groups each made a list of the different maps and identified the page numbers on which they were found. They compared their new list to the one developed at the beginning of the lesson. The students were surprised to find that there were so many different kinds of maps in their textbook.

Next, Ms. Carlson asked the students to discuss why there are so many different kinds of maps. They concluded that it is impossible to display everything about a place on a single piece of paper. Therefore, for a map to communicate information clearly, it must be restricted in the number of things it shows.

Ms. Carlson then showed a large, clear aerial photograph of Pittsburgh and its environs. The students were asked to discuss what they observed, focusing specifically on the variety of geographic features—highways, mountains, rivers, bridges, center city, residential areas, shopping malls, buildings, and so on. Each group was then randomly assigned a different type of map—political, physical, topographic, raised relief, weather, historical, road, and thematic map of choice. The groups used the large photograph as well as any needed reference materials to produce a specialized map of Pittsburgh, their home community.

Students must be helped to understand that they can learn many things about a region from different types of maps, and also that they are able to communicate to others what they know about a place by constructing many kinds of maps of their own. If elementary school students are going to acquire geographic competency, they must acquire the necessary tools and techniques to think geographically. Geographic information is compiled, organized, and stored in many ways but, to think geographically, maps are central to understanding and analysis.

AFTERWORD

A lot has changed in the field of geography education in the past 10 years. Geography is once again recognized as a core curriculum subject in our elementary schools. The standards for teaching geography identified what American children should learn in kindergarten through grade 12. Many states now use these standards as guidelines for instruction and for curriculum development in the public schools. However, the most dramatic changes affecting geography education in the past decade are changes in the world itself. Many countries do not have the same names they once had, borders have changed, and some countries have dissolved while others have burst into existence. Advanced communications make it possible to become instantaneously connected to people all over the world. The global economy has become increasingly competitive, and precious natural resources are becoming scarcer. The environment becomes more fragile with each passing day.

How can geography help us successfully face these challenges? Studying geography helps develop the skills and understandings that enable us to become geographically informed. That is, geography's tools and techniques help us to better understand where things are and how and why they got there. Geography helps us better understand our relationships to other cultures and environments. Such vision and understanding forms the basis for making reasoned political decisions on all levels, from global trade to the best location for a new community high school.

From creating community maps to deciphering road maps, this chapter offered simple, enjoyable suggestions to teach young geographers the fundamentals of geography. As the power and beauty of geography are unlocked in dynamic social studies classrooms, children can much more clearly see, understand, and appreciate the web of relationships among the people and places on Earth.

References

Denko, G. J. (1992). Where is that place and why is it there? *American Educator, 16,* 20.

Geography Education Standards Project. (1994). *Geography for life: National geography standards 1994.* Washington, DC: National Geographic Research & Exploration.

Natoli, S. J. (1988). Implementing a geography program. In S. J. Natoli (Ed.), *Strengthening geography in the social studies, NCSS Bulletin 81.* Washington, DC: National Council for the Social Studies.

Preston, R. C., & Herman, W. L., Jr. (1974). *Teaching social studies in the elementary school.* New York: Holt, Rinehart and Winston.

Siegel, A. W., & Schadler, M. (1981). The development of young children's spatial representations of their classrooms. In E. M. Hetherington & R. D. Parke (Eds.), *Contemporary readings in child psychology.* New York: McGraw-Hill.

5

Direct Instruction

Nicole Maier enjoyed doing everything she could to enliven social studies by connecting her students' lives to other times and places. She found that primary documents were especially useful, as they had the power to draw in students to the rational and emotional experiences of real people. One topic that lent itself perfectly to Ms. Maier's approach was the misery of slavery in America.

Ms. Maier understands that she cannot simply pass out copies of a document and ask the students to read them; many students find the documents too difficult to read. Therefore, Ms. Maier begins by developing a context for understanding; that is, she helps students associate their previous knowledge of the topic to the new learning experience. Directing the students to pair up, Ms. Maier asks each team to make a list of facts about slavery they "know" as true. Then, for each fact, Ms. Maier asks them to describe a mental image that comes to mind when they think of that aspect of slavery. An intense discussion brought out heated feelings about the atrocities of slavery. Ms. Maier then admonishes the students that, as young historians, they have just reached conclusions without examining all the evidence. She agreed with the students that slavery was an immoral, disgusting part of our history, but asked if their feelings were representative of the way all slaves were treated.

Ms. Maier passed out the two excerpts shown in Figure 5–1. She directed the students to examine each and judge which they felt most closely represented the true conditions of slavery.

FIGURE 5–1
Controversial Issue

Statement A

Rules on the Estate of P. C. Weston

"The Proprietor wishes the Overseer to understand that his first object is to be, the care and well being of the negroes. The Proprietor never can or will excuse any cruelty, or want of care towards the negroes [but] it is absolutely necessary to maintain obedience, order, and discipline; to see that the tasks are punctually, and carefully performed." (Metcalf, 1992, p. 6)

Statement B

Recollections of His Experiences in Bondage: Frederick Douglass

Mr. Severe was rightly named: he was a cruel man. I have seen him whip a woman, causing the blood to run half an hour at the time; and this, too, in the midst of her crying children, pleading for their mother's release. He seemed to take pleasure in [showing] his [cruelty.] From the rising till the going down of the sun, he was cursing, raving, cutting, and slashing among the slaves of the field in the most frightful manner. (Douglass, 1845, p. 9)

The narratives were read aloud dramatically so the students could get a feel for the emotions of each writer. When Statement A was finished, Ms. Maier asked the students to write a short reaction to someone who might claim that, "The master was good to his slaves." Two samples are illustrated in Figure 5–2. The same process was used as the students listened to a reading of Statement B, but the students were to respond to the claim that, "All slave owners were brutal."

After both reactions were recorded, Ms. Maier asked the students to decide whether or not this new evidence could support their initial impressions of slavery. All agreed that slavery was dreadful, but were uncertain whether most slave owners were like sadistic Mr. Severe, who enjoyed beating up on slaves; fatherly slave masters who were kindly and civil; or someone in the middle who was just looking to profit from an investment. The young historians weren't sure they had quite enough evidence to support any of the characterizations, so they insisted on tapping into more direct evidence before coming to a conclusion. Ms. Maier congratulated the students on their excellent historical thinking and directed them to Internet sites and other resources containing these documents.

FIGURE 5–2
Student Responses

I think the owner was honest. I Think he was honest because he seemed to really care about the way slaves were treated. Not all slave owners were mean, or were they?

By,
Chris Smith

I think the slave owner was mean. When he said that it was absolutely necessary to maintain obedience, he probably use a lot of cruel techniques to keep the slaves in line.

By,
Becky Reilly

The primary role of social studies teachers is to promote learning. They have at their disposal a number of general strategies to help achieve this goal. You have seen illustrated examples of many of these strategies throughout Chapters 2, 3, and 4. Now, in Chapters 5 and 6, we will examine the professional skills you will need to develop in order to carry out the general strategies most often utilized in dynamic social studies classrooms—*teacher-directed, inquiry,* and *creative problem solving.* Although each is explained separately in this text for ease of understanding, similarities exist among them and teachers will often incorporate features of more than one while finding their unique teaching style. As Ms. Maier gained teaching experience, for example, she made adjustments in these strategies until she was able to settle on one that best fit her personality and helped meet her students' needs. Ms. Maier eventually settled on a variation of *direct instruction* because she felt it had two primary advantages over the others: (1) it was more time-efficient and allowed her to cover more content in less time, and (2) it offered her greater security in terms of establishing a lesson focus, leading student discussions, and establishing classroom control.

Direct instruction is characterized as a *teacher-centered approach,* meaning that the teacher accepts primary responsibility for precisely stating the instructional objectives, structuring the content or skills to be taught, carefully presenting lessons in a logical sequence to the whole class, providing immediate feedback, and monitoring student performance closely. The teacher is in control of instruction at all times. Direct instruction should not be totally unfamiliar to you since the strategy has historically been at the forefront of teaching practices. It is likely that your elementary school teachers employed varieties of the teaching model when they taught you social studies concepts and skills, carefully sequencing and structuring what you learned. Although teachers assume primary responsibility for organizing and presenting the content, however, they are not the only active individuals in the classroom. Teachers who use direct instruction properly reach out and captivate students with a logical presentation of important ideas while strengthening the educational experience with creative activities, appropriate resources, and challenging instructional materials. It is the teacher's direct presentation and control of the instructional episode, not the perceived lack of involvement on the part of students, which defines the approach. The teacher directs and models, but learner involvement must be high in order for this approach to be effective.

Teachers using direct instruction in dynamic social studies classrooms view students as active travelers on a journey toward understanding. They eagerly take the role of "tour guide" on this journey, bringing in as much of the world as they can to school, creating a climate in which children are actively involved in their learning, and escorting the students as they search for answers and construct increasingly mature understandings. The direct instruction model is particularly attractive to beginning teachers, and one that you will most likely put to use as a student teacher or beginning teacher, because it allows you to assume more direct control of all the children (whole-class instruction is usually employed), requires all the children to use the same learning materials (management concerns are minimized), and pro-

Direct instruction is a teacher-centered strategy, but student involvement must be high for this approach to work well.

vides specific objectives and clear direction for teaching/learning behaviors (it helps maintain lesson focus).

To succeed, lessons utilizing the direct instruction approach must be carefully and logically structured. Although there are numerous variations of the direct instruction model, the format chosen for dynamic social studies occurs in five distinct but interrelated phases:

1. *Task and concept analysis.* Because all new concepts and skills spring forth from what the students already know or can do, teachers must carefully select, learn about, and sequence both the prerequisite knowledge to which new information can be linked (*concept analysis*) and the subskills that help establish the foundation for a new skill (*task analysis*).

2. *Introduction.* Teachers must help students become aware of the relationship between the new content or skill that they will learn and what they already know or can do. Teachers also motivate the children and establish a clear purpose for the instructional activity that is to come.

3. *Presentation.* Teachers select learning materials appropriate for the students' abilities and deliberately sequence and structure the way they are

used. Essentially, teachers facilitate learning by describing, clarifying, or explaining a new concept or modeling a targeted skill.

4. *Learner response.* Teachers help students actively organize and make sense out of the information rather than simply absorb it and recall it upon demand (*concept mapping*). They also provide an opportunity for students to practice the skill so that it can be performed routinely with little conscious effort (*guided practice*).

5. *Functional application.* The students independently put to use their newly acquired knowledge or skill.

TASK AND CONCEPT ANALYSIS

Thoughtful *concept and task analysis* is critical because it helps teachers and students focus their attention on important and interesting information rather than trivial or uninteresting details. A social studies lesson needs a good *introduction* because it arouses curiosity, stimulates the imagination, relates new content to past experiences, and creates enthusiasm for what is to follow. It needs a good *presentation phase* because it is during this phase that major concepts and skills are delivered to the students. It needs a good *learner response phase* because the students organize the information they received. It needs a good *functional application phase* because the students are able to use the new skill on their own or to apply new understandings in new contexts.

Task Analysis

One major goal of social studies instruction is helping children acquire specific skills such as constructing and interpreting charts and graphs, using the computer, composing a meaningful written report, making a timeline, reading a map, creating a model, outlining information from reference books, planning an interview, making a mural, learning the steps of an ethnic dance, reading maps, comprehending textbook material, or testing hypotheses. *Skills* are mental operations having a specific set of procedures that are developed through practice. Those who favor the direct instruction model claim that students cannot learn skills without a teacher's help; these important skills are best taught and reinforced as separate lessons with explicit assistance from the teacher followed by many opportunities for practice. This process begins by carefully breaking down the skill into a number of separate components, each of which provides the foundation for the next. The process through which the component parts of the skill are identified and sequenced is referred to as *task analysis.* In learning to read maps, for example, learning what a map *is* would certainly come before a lesson where students are required to locate the state capital. In other words, instruction is sequenced so that more complex procedures grow from fundamental abilities.

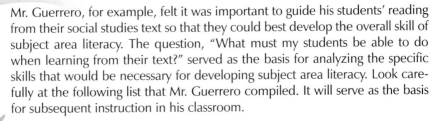

Classroom Vignette

Mr. Guerrero, for example, felt it was important to guide his students' reading from their social studies text so that they could best develop the overall skill of subject area literacy. The question, "What must my students be able to do when learning from their text?" served as the basis for analyzing the specific skills that would be necessary for developing subject area literacy. Look carefully at the following list that Mr. Guerrero compiled. It will serve as the basis for subsequent instruction in his classroom.

Students must be able to:
1. identify what they already know about the topic
2. raise questions about what they do not know
3. predict what the text will be about
4. predict what information will be found in the passage
5. relate new information to previous knowledge
6. focus their attention on the reading task

Mr. Guerrero wanted his fourth graders to relate their personal experiences to the textbook topic, so he began the social studies lesson by displaying a photograph of a small rural community and the rich, lush farmland that surrounds it. He used the photograph as a springboard for discussion of the ways people interact with and adapt to their environment, focusing especially on the need to protect natural resources: "How do people and animals rely on the fertile topsoil? Trees? Water?" Mr. Guerrero discussed wasteful practices such as over-watering lawns and using paper unwisely with his students. He then asked the class how they might protect the Earth's topsoil, trees, and water supply. Following this short discussion, Mr. Guerrero directed his students to turn to a specific page in their social studies text. He pointed to and read aloud the chapter title, "Saving Our Land." He explained to the class that when he sees a new chapter title, he always thinks for a moment about what it might mean. "It seems to me that the chapter could be about ways of protecting our natural resources like water and topsoil and trees," he suggested. "What do you suppose gave me that idea?" After they discussed his idea for a short while, Mr. Guerrero asked the class to talk with a partner and come up with an idea of their own about the title. Then, as the pairs shared their ideas, Mr. Guerrero wrote their suggestions on the chalkboard. Next, Mr. Guerrero asked the students to look at the text photos and the first major heading, "The Need to Protect Our Land." He said to the class, "I always look at the illustrations and headings before reading because they raise questions in my mind about what might be ahead. For example, one question I had about the photo at the top of the page is, 'Why is conserving topsoil so important?' What are some other questions I might have asked myself about the heading and the photos?" Mr. Guerrero went through the next section, "Conservation Efforts," the same way and continued with the succeeding sections, writing each set of comments and questions on the chalkboard.

Modeling is a process of motivating through example; if the students see you are excited about performing certain reading skills, they will pick up your enthusiasm and become stimulated to follow your lead.

Concept Analysis

The mental structures that help us sort out, order, connect, and make sense of new knowledge are called *concepts.* Concepts can be thought of as idea categories into which we sort all that is characteristically associated with or suggested by a term. Concepts are much like mental file folders into which we sort out and store new information. For example, the concept category *riverbank* is a mental structure into which we place all information we have accumulated about *the border of any stream.* *Riverbank* is considered a concept label and the descriptive information we place into it is called its *defining features.* Concepts are designated by a label, a word that helps us communicate any idea to someone else. In the direct instruction model, learning depends a great deal on the teacher's ability to identify the defining features of the concept being taught. For instance, before she taught her lesson on riverbanks, Mrs. Yearsley identified the following defining features:

Riverbank
- the rising ground bordering a river or stream
- the rising ground forms the shore
- a riverbank encloses the water of rivers and streams

The concept analysis process provides focus for the upcoming lesson and ensures that the information will be accurate. Many concepts, such as *riverbank, latitude, coin, flag, iceberg, fjord, merchant,* or *shelter,* have well-defined characteristics and are relatively easy to analyze and teach. They are called *concrete concepts* because they are easily identified by their physical appearance. Others, such as *democracy, peace, freedom, justice, equal opportunity, liberty,* and *war* have characteristics that are much harder to specify and, therefore, they are much more difficult to analyze and teach. These terms, called *abstract concepts,* are much more difficult to conceptualize because we cannot simply look at concrete examples. That is why you will find concepts like *riverbank* or *home* or *police officer* taught at the first-grade level while *democracy* or *freedom* rarely find their way into the instructional program before the upper-elementary school grades or even middle school.

Concept analysis is the process of breaking down a concept by identifying its name, definition, and defining features. To better understand how this relates directly to dynamic social studies, take a minute and think about a concept many fifth graders learn about in their social studies classes—*Buddhism. Buddhism* is the word label for the concept we will be teaching. Once the word label for the concept is determined, it is important to include a brief *definition* of that term; for example, *the teaching ascribed to Siddhartha Gautama holding that one can enter into nirvana by mental and moral self-purification.* Definitions are easily found in dictionaries, encyclopedias, textbooks, and other reference materials. The word label and definition are important, but concept analysis is not complete until we break down its defining characteristics, or subordinate concepts that help us distinguish Buddhism from Christianity, Hinduism, Judaism, Confucianism, Islam, or any other religion. Think about Buddhism for a moment; what thoughts come to your mind when you think of the term? Are you picturing a sculpture of the Buddha? Maybe you see an image of a Buddhist monk. Perhaps words like *Siddhartha Gautama, the Buddha, India, religion, Four Noble Truths,* or *dharma* pop into your head. We store information in our minds in a number of different ways; images and symbols are two of the most common. The images and symbols you associated with Buddhism are a result of your past experiences with Buddhism. The forms they took were determined by the way you sorted out, ordered, connected, and made sense of the details of those past experiences.

Factstorming

The first step in expanding the concept, then, is a process commonly called *factstorming.* Virtually any concept can be factstormed by a team of teachers or even by an individual teacher. All that is required is to create a graphic representation, such as a web, of what you already know about the concept. First, write the word or words describing your theme in the center of a large sheet of paper and draw a circle around it. Then, by asking probing questions (for example, "What are the teachings of Buddhism?"), identify related subordinate concepts. Write them on the paper, too, draw a circle around each, and draw lines to connect them to the central concept. Repeat the process with each subordinate concept. The chart is only the beginning. Most teachers do not know enough at this initial stage of the planning process to identify all the necessary content, but are only brainstorming "starters," or ideas to put the planning process in motion. As you continue your plan to teach about Buddhism, you must add to and reorganize the concept, for it must continue to change and grow. An analysis of the defining characteristics of the concept of *Buddhism* as it was researched and organized by one teacher to serve as the content source for a mixed group of fifth and sixth graders is depicted in Figure 5–3.

FIGURE 5–3
Concept Map for Buddhism

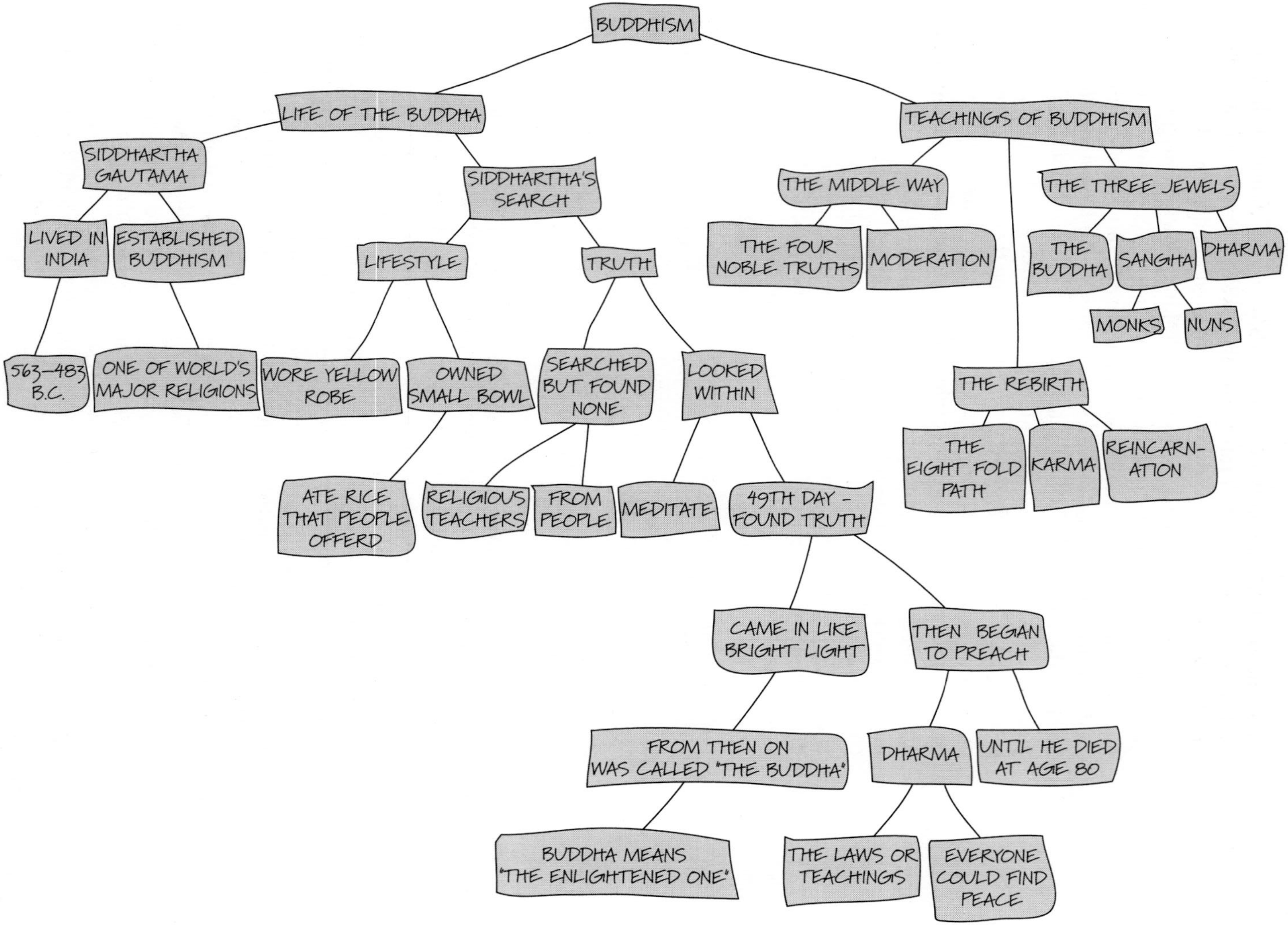

BUDDHISM

TEACHINGS OF BUDDHISM

THE THREE JEWELS
- DHARMA
- SANGHA — MONKS, NUNS
- THE BUDDHA

THE MIDDLE WAY — MODERATION
- THE FOUR NOBLE TRUTHS

THE REBIRTH
- REINCARNATION
- KARMA
- THE EIGHT FOLD PATH

LIFE OF THE BUDDHA

SIDDHARTHA GAUTAMA
- ESTABLISHED BUDDHISM
- LIVED IN INDIA
- ONE OF WORLD'S MAJOR RELIGIONS
- 563–483 B.C.

SIDDHARTHA'S SEARCH

TRUTH
- LOOKED WITHIN
- SEARCHED BUT FOUND NONE
- MEDITATE
- RELIGIOUS TEACHERS FROM PEOPLE

LIFESTYLE
- WORE YELLOW ROBE
- OWNED SMALL BOWL
- ATE RICE THAT PEOPLE OFFERD

49TH DAY – FOUND TRUTH
- THEN BEGAN TO PREACH
  - UNTIL HE DIED AT AGE 80
  - DHARMA
    - EVERYONE COULD FIND PEACE
    - THE LAWS OR TEACHINGS
- CAME IN LIKE BRIGHT LIGHT
- FROM THEN ON WAS CALLED "THE BUDDHA"
- BUDDHA MEANS "THE ENLIGHTENED ONE"

Notice how the most general category (*Buddhism*) is at the top of the hierarchy. It is the concept label under which the teacher diagrammed subordinate concepts. Taken together, these comprise the defining features that will be used to teach about Buddhism. If the teacher were to offer a narrative explanation of the hierarchy, it would include statements such as those that follow:

- Buddhism is one of the world's great religions.
- Buddhism was started in India somewhere between 563 and 483 B.C.
- Siddhartha Gautama established Buddhism.
- Siddhartha was called "the Buddha," or "enlightened one."
- Buddhist laws and teachings are called the dharma.
- The Four Noble Truths explain the Buddha's beliefs about suffering.
- The Buddha preached "The Middle Way"—moderation.
- Buddhists believe that the Eightfold Path leads to enlightenment. The deeds a person performs during his or her lifetime are called karma.
- Buddhists believe in a process called reincarnation.
- The good or bad karma travels with a person to the next life.
- The Buddhist tradition is made up of three parts called the Three Jewels: (1) the Buddha, (2) the dharma, and (3) the sangha (religious community, including nuns and monks).

Although they have been abused by overly excessive rote, drill, or practice exercises, facts are highly essential ingredients of concept development. They help learners distinguish *continents* from *countries, glaciers* from *icebergs,* and *Buddhism* from *Hinduism.* Concepts grow from facts; facts are what give concepts their defining features. Facts serve as building blocks, furnishing the details necessary to develop concepts. Concepts do not materialize magically out of the thin air; students gradually construct them as knowledge accumulates through varied learning experiences. Without a system of organizing the wealth of information about our world, though, each fact becomes isolated and students have very few options other than to memorize it—and to complain that "social studies is boring!"

A word of caution about the defining characteristics of concepts must be given here, however, for although subordinate concepts and facts provide the defining features that make a concept what it is, they must be selected carefully. For example, details of George Washington's $60-a-set dentures (made from ivory, wild animal teeth, or lead covered with gold), as interesting as they may be, would not contribute much to enriching the concept of our *presidency.* However, they could provide engaging facts to support an understanding of *health care* during colonial times. Concepts are superb organizational devices, but they can be formed accurately only when learners gather meaningful information through sound, developmentally appropriate activities. Concept learning is a process of learning what key features (*defining characteristics*) are essential components of a concept and what other features (*irrelevant characteristics*) are nonessential. Learning to differentiate defining features from irrelevant features takes time and experience; when teachers present students with imprecise examples, this process only takes longer and becomes more difficult to master.

Teacher as Scholar

It should be obvious now that, in order to use direct instruction effectively, dynamic social studies teachers must be tuned in to subject matter; teachers need information. Each concept has its own body of subordinate concepts and facts; each has a precise set of data. Teachers must be in command of this knowledge and use it to support and deepen their direct instructional practices. To appreciate how important it is that a teacher truly understand the concept being taught, let us imagine that you are teaching a lesson dealing with the native Arctic people commonly referred to as "Eskimos," but more accurately identified as Inuit (IN-yoo-eet). (The native Arctic prefer to be called *Inuit,* a word from their own language meaning *the people,* not *Eskimo,* a word given to them by Native Americans and anthropologists meaning *raw meat eater.*) We shall assume that you have already helped the children establish a background of information about the Inuit through meaningful classroom experiences: There are many Inuit nations; each has its own language and traditions; most Inuit live in the Arctic (the area around the North Pole) but many live in the northern part of Alaska. The purpose of the current lesson is to help students deepen their understandings of the Alaskan Inuit by learning about their igloos (iglus). Focus for a minute on the image that comes to your mind: Is it a domed snow house sitting on a frigid, treeless, barren blanket of snow and ice? If you are like most, this concept of "igloo" is quite clear and would probably guide your instruction, perhaps to the point of culminating the study of Inuit homes by involving your students in a project of constructing tabletop models from sugar cubes or marshmallows. With such a clear idea of igloos, is there any need to dig up additional information? You bet there is! If you had been thorough enough to do an information search, you would have discovered that Alaskan Inuit generally used snow iglus only as temporary hunting shelters. Even in the past, Alaskan Inuit built their permanent homes from logs, whalebones, and sod rather than the domed snow and ice structures we stereotypically associate with them. Actually, *iglu* is the Inuit word for "house," and Inuit now live in modern houses, or split-level, colonial, condo, or apartment iglus, very similar to those you and I live in.

What would children learn about the Alaskan Inuit if we limited their exposure to what we *think* we know? In this case, the result certainly would be unfair, stereotypical, and incorrect, abusing all the major responsibilities of a contemporary multicultural society. In any lesson, we *must* uncover and verify considerable information so that students acquire a genuine concept of the culture, period of time, or phenomenon being studied. One experienced teacher's recollection highlights the need for a spirit of continuous learning:

Good social studies teachers are tuned into the subject matter; they make sure they thoroughly understand the concepts they will be teaching.

> The biggest surprise of teaching, for me, was that I didn't know my subject matter. That was the one thing I had been most confident about. I had almost an "A" average in my major and felt really on top of my field. When I began teaching and had to explain concepts, I found that I had only a very superficial understanding of them. I knew stuff in kind of a rote way and when I had to explain it to someone else I kind of just fell on my face. I learned more about my subject in my first four months of teaching than I did in my four years of college. (Ryan, Burkholder, & Phillips, 1983, p. 177)

You must continuously strive to deepen the background knowledge you bring to teaching. As a teacher of dynamic elementary school social studies, this means that you must not only have much information at your fingertips, but should also have access to resources that supplement your knowledge. The computer and a wide variety of print sources are the places to start. Check with your school or public librarian for references on a concept you wish to pursue. Constantly search for information to bolster what you already know; you should find yourself saying, "I'll look it up," many times a day. Books and the computer will provide most of the needed information, but you will also want additional resources. Consult specialists either in person or by telephone, visit museums and other sites, view films or filmstrips, listen to audiotapes, and seek other opportunities to broaden your background. You will need to spend a great deal of time uncovering and organizing information: What is the key information necessary for a satisfactory understanding of this concept? Do not be satisfied only with accumulating a mass of information; be sure the information is accurate. You must pass on true understandings, not careless misinformation to our citizens of tomorrow.

INTRODUCTION

The actual classroom presentation phase of a direct instruction lesson begins with what is commonly referred to as its *introduction*. The introduction is an extremely important part of a direct instruction lesson, for it serves three critical functions: (1) it activates the student's prior knowledge related to the concept or skill targeted for instruction, (2) it draws students into the lesson, and (3) it focuses their attention on the instructional task by establishing a clear purpose for learning.

Activating Prior Knowledge

A good direct instruction lesson helps students connect what they already know or can do to the new information or skill to be learned. David Ausubel (1961), a respected advocate of direct instruction, supports this notion most emphatically, stating, "If I had to reduce educational psychology to just one principle, I would say this: The most important single factor influencing learning is what the learner already knows" (p. 16). Ausubel refers to the way teachers encourage students to retrieve relevant knowledge and connect it to the new material as providing *advance organizers*. There are several other terms used to describe this process of prior knowledge activation. For example, Madeline Hunter (1982) calls it an *anticipatory set* and Lev Vygotsky uses the term *external mediator* (Bodrova & Leong, 1996). Regardless of what we call it, each of these respected educators emphasize that the ability to connect the new and old is not a naturally acquired skill, so it must be taught to students with the goal that they can eventually develop the ability to organize and monitor their own thinking. Remy (1990) explained why this ability is especially important for future citizens: "One of the most important attributes of competent citizens in a complex society is the ability to connect things that seem superficially to be discrete. Such an ability is a clear sign of higher order cognition . . . and is a highly prized goal of social studies education" (p. 204). Van Doren (quoted in Boyer, 1982) adds, "[t]he student who can begin early in life to think of things as connected, even if he revises his view with every succeeding year, has begun a life of learning" (p. 384).

Constructivism

This perspective is fundamental to a *constructivist* view of learning. Constructivists believe that learning occurs as students relate what they have learned from their past experiences to the new situation. Because children rarely learn new knowledge in isolation from things they have already learned, early experiences provide an essential foundation for subsequent cognitive development. The paragraph below may help illustrate this point (Freedman, 1987).

> Five doctors worked over the president that night. Now and then he groaned, but it was obvious that he would not regain consciousness. The room filled with members of the cabinet, with congressmen and high government officials. Mary waited in the front parlor. "Bring Tad—he will speak to Tad—he loves him so," she cried. Tad had been attending another play that evening. Sobbing, "They killed my pa, they killed my pa," he was taken back to the White House to wait. (pp. 125–126)

To attach meaning to this paragraph, a student must have already built an understanding of *presidents of the United States.* Associating those understandings with new information in the paragraph, the student is able to tell that the selection is about a president. However, if the student's background does not include information about *assassinated presidents,* the paragraph will have very little or no meaning. However, let us say that the student's background does include knowledge of assassinated presidents. The student will then search through the paragraph, scouring it for clues to determine the president's identity. To identify the president, the student must know something about the *four former presidents* who were assassinated, even though none was specifically identified in the passage. Relying even more heavily on past experiences and the clues in the paragraph, the student recognizes names and events that eventually help pinpoint this passage as a description of the horrific assassination of Abraham Lincoln. In other words, the student *constructed* meaning by associating past personal experiences with the clues contained in the paragraph. Therefore, constructivist views of learning hold that effective learners rely heavily on past experiences to unlock meaning from the new learning experience. Constructivist learning, then, is a strategic process by which learners construct or assign meaning to a learning situation by using their own prior knowledge. The meaning the learner constructs does not come solely from the learning experience; it comes from the learner's own experiences that are triggered or activated by the challenge of the new learning situation.

Whether we call this part of the introduction an *advance organizer, anticipatory set,* or *external mediator,* it is important to help the student access previous experiences to help unlock meaning from the new subject matter. Consider the advance organizer to be like a thumbnail sketch, a general overview of the new material that shows how it relates to the students' previous experiences or to what they have previously learned. Of course, the value of an advance organizer depends on how good it is and how well the teacher uses it.

Teachers have used a variety of techniques over the years to provide advance organizers, but the most effective appear to be (1) *class discussions* that include thought-provoking questions and (2) *external mediators* that include graphic outlines displaying a visual overview of the material to be covered.

Class Discussions

Perhaps the most widely used type of advance organizer is to make an introductory statement about the material to be learned followed by a general discussion that connects the students' existing knowledge to that new material. The best way to understand how this can be done is to visit a social studies classroom where advance organizers are used in direct instruction lessons.

Classroom Vignette

Larry Chavez's sixth graders, for example, had been involved in the study of the oil industry and are about to read a short textbook selection about the different ways oil is pumped from the ground. Mr. Chavez begins the lesson by making an introductory statement about the material to be covered: "Yesterday, we discovered where some of the Earth's major oil fields are located. Today we will find out how the oil is drawn out from under the Earth's surface." To activate the students' prior knowledge of this new material, Mr. Chavez decided to encourage intuitive guesses. Instead of giving a definition of the type of oil pump they will read about, he asked them to recall anything they might have experienced with grasshoppers. Some students talk about the insect's thin, powerful back legs and how far grasshoppers are able to leap; others mention the antennae; the oddly shaped head; the long, thin body; and the wings. Most children volunteer something special, whether their previous involvement with grasshoppers was limited to simple observation or something more concrete such as using them as fishing bait. Mr. Chavez holds up a large illustration of the insect and asks the students if they would agree that this was the creature they were all describing. The students agree that it was and once again point out and describe some of its distinguishing characteristics. Next, Mr. Chavez asks a question that raises a few eyebrows: "Here's a question I really want you to think about. . . . How might grasshoppers be used in the oil industry?" The students glance silently at one another and eventually giggle quietly in disbelief. "Are you serious, Mr. Chavez?" they ask suspiciously. "Of course I'm serious," replies Mr. Chavez, "I know it's hard to believe, but grasshoppers actually are used in the oil industry. Turn to a partner and talk about how you think grasshoppers might be used."

After a few minutes, Mr. Chavez writes down the students' thoughts: "They have special senses that cause them to jump around real fast if oil is underground." "They help find new oil wells." "The brown 'tobacco' that they spit on your hand when you hold them can be collected and used to lubricate the machinery until the well begins to produce oil." After listing each thought, Mr. Chavez makes a connection to their reading: "Your thoughts are very interesting. Please read pages 78 to 81 in your textbooks. On those pages you will discover which, if any, of your ideas explain how grasshoppers are actually used in the oil industry." The children quickly find the pages in their textbooks and begin reading. In a short time they are incredulous: "You tricked us, Mr. Chavez! The oil industry doesn't use the insects, they use grasshopper pumps!" The students discovered that oil industry "grasshoppers" were large low-pressure pumps shaped somewhat like the actual insect. Mr. Chavez displays a large picture of a grasshopper pump as well as a model next to the insect illustration and invites the students to discuss the similarities and differences. Although the oil "grasshoppers" weren't of the insect variety, signs of learning were obvious as the students continued to talk about the oil industry grasshoppers during a lively and informative discussion.

Mr. Chavez pursued this line of instruction because he knew it was important for students to connect their prior knowledge of grasshoppers to this new instructional topic before they were asked to read from their books. Compare his approach with another teacher who began his class discussion with the statement, "Yesterday, we read about where the major oil fields are located throughout the world. Today we're going to continue learning about the oil industry." Then, "Have you ever seen a grasshopper? Today we're going to learn about how oil is pumped from the ground with a special pump called a grasshopper. Read pages 78 to 81 to find out how this special pump works." Notice that the second teacher's background development was vaguely related to the main concept of the book, and the mystery and intrigue associated with Mr. Chavez's approach was missing. Mr. Chavez's strategy was much more productive because it created an appropriate puzzle in the children's minds. Disequilibrium brought the students' existing knowledge to the forefront, offered the teacher some insight into the students' current level of cognitive functioning, and furnished a base from which to reconstruct existing concepts.

The advance organizer is not intended to become an extended discussion. Consider it only a quick review of what the children already know, not something that ends up being a separate lesson in itself. Below are five very useful types of comments and questions that can be used for connecting one's background knowledge to the new learning task. I have written examples of questions or comments for each category that might be used in preparation for a lesson on the Wampanoag Indians:

1. *Existing knowledge.* "Tell me what you now know about the *Wampanoag Indians.*"
2. *Thought association.* "When you hear the names *Massasoit* and *Hobbamock*, what do you think of?"

3. *Rapid recognition.* Display key terms and ask the students to tell what they already know about them—*wigwams, longhouses, deer stew, breeches, loincloths,* and *petticoats.*

4. *Quick lesson review.* Ask questions that help connect the new learning experience to information that was learned in previous lessons—"Yesterday, we learned about a native people called the *Wampanoags.* Who were the *Wampanoags?* Describe daily life in a *Wampanoag village."*

5. *Open discussion.* Sometimes an open-ended question will offer the best connection to past experiences—"We are going to read about the *Wampanoag Indians.* What have you learned about the *Wampanoags?"*

External Mediators

In addition to arousing attachment to the new content through carefully planned discussion strategies, teachers often find it helpful to use attention-grabbing external mediators such as diagrams, charts, drawings, and other visual displays. The term *external mediator* was taken from the works of Vygotsky, and refers to graphic displays of information that help students consciously connect their past experiences to the targeted skills or concepts under study. Using external mediators to assist students' thinking is not a new idea; adults use them all the time. Have you ever heard of someone tying a string around her or his finger as a reminder of something or laying out her or his daily schedule with a pocket calendar? These external mediators regulate our thinking and keep us on task. In the Vygotskian framework, these "tools of the mind" must be used with children so that they can eventually develop the ability to organize and monitor their own thinking. External mediators, then, build the scaffolds that help children make the transition from being regulated by adults to becoming self-regulated learners.

Bubble Trees

One example of an external mediator that helps direct student attention to what is important in the coming material is called a *bubble tree.* Bubble trees work best when relevant information can be categorized beneath a *core understanding.* The top of the structure labels the core understanding; in Figure 5–4, this is "The fall of Rome." Information related to the core understanding should fall neatly into major *idea categories:* "barbarians" and "other causes." The tree begins, therefore, as a top bubble containing the core understanding and branches labeled as major idea categories. Finally, students go through the main instructional activity and use the bubble tree to help organize the *facts* they uncover within each of the major idea categories, eventually constructing a complex structure of branching information.

Bubble trees can be used in a variety of ways in a dynamic social studies classroom: (1) completed by the teacher before a lesson, they can present an overview, or advance organizer, for the students; (2) presented only as blank bubbles, they can tap students' prior knowledge before the lesson—as the lesson progresses, students add to, modify, and refine their suggestions; and (3) after a lesson, they can be used by students to review and organize content.

FIGURE 5–4
Bubble Tree

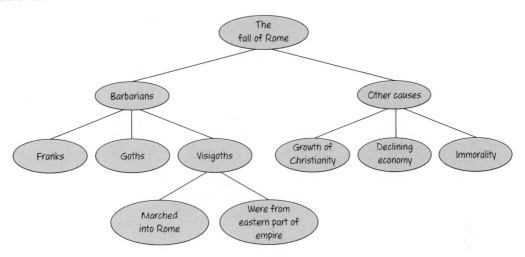

Prediction Charts

Another example of an external mediator is the *prediction chart.* Prediction charts contain a statement or series of statements related to concepts or issues under study. Although one statement can be used, typically three to five statements are found on most prediction charts. Students are asked to individually respond to each statement by indicating their agreement or disagreement. Then a group discussion helps the students clarify their thoughts by eliciting reasons for each prediction.

A sample prediction chart is shown in Figure 5–5. Keep in mind that the purpose of the chart is to draw out students' honest predictions, not "correct" answers. Therefore, you must be nonjudgmental, accepting students' thoughts even if they are contrary to your own. Students are quick to correct their prediction charts once they find new information in their reading. For example, their image of plantation owners living in huge, palace-type homes quickly reversed as they learned that some owners lived like that, but most lived on medium- and small-sized farms.

K–W–L Charts

A final example of a highly useful external mediator that is popularly used to help guide children's thinking before, during, and after the learning experience is the *K–W–L chart.* Each letter represents a different activity that guides learners not only prior to, but also during and after, the learning experience. *K* represents what the students already *know* about the topic. Before the learning experience actually takes place, students discuss and brainstorm all the ideas they can associate with the topic and record their ideas on a chart, as shown in Figure 5–6. The teacher might

FIGURE 5–5
Prediction Chart

	SLAVES	SMALL FARMERS	PLANTATION OWNERS
HOMES	ONE-ROOM CABINS WITH DIRT FLOORS. OFTEN, MORE THAN ONE FAMILY LIVED TOGETHER.	LOG CABINS USUALLY BUILT IN THE MOUNTAINS OR FOOTHILLS.	HUGE HOUSES WITH MANY ROOMS. YOU COULD COMPARE IT TO A PALACE.
WORK			
CROPS			
LIVING CONDITIONS			

elicit suggestions with a question such as, "Before we read this material on *oceans*, let's take a few moments to jot down some of the things we already know about oceans." You may need to model one or two suggestions so that students begin to see what you mean. *W* represents what the students *want* to know. As the students reflect on what they already know about the topic, they form questions related to gaps in their understanding. Again, the teacher might need to model a personal question and write it down on the chart to give the children an idea of what they are expected to do. *L* represents what students *learned* about the topic. After participating in the learning experience, students record what they uncovered. They check their questions to see if each has been answered; if not, you may want to suggest other sources of information.

To illustrate, the following sequence of activities were used by Ms. Anderson to help her students understand how beach sand is created.

FIGURE 5–6
K–W–L Chart

K = What We Know	W = What We Want to Find Out	L = What We Learned

waves
salt water
large bodies of
  water
sandy beaches
vacation spots
sea shells
Atlantic Ocean
Pacific Ocean
	Where does sand
	  come from?
	Why do some beaches
	  have rocks
	  instead of sand?
	What animals live
	  in the ocean?
	What are the names
	  of the oceans?
	How many are there?
	Which ones border
	  the United
	  States?	

Classroom Activity

1. Begin by calling the children's attention to a study print of a beach scene. Discuss what the children see and encourage them to talk about any times they have visited the beach. Encourage them to consider the vastness of the beach. Ask, "Who would care to guess how many grains of sand are on the beach?"
2. Hold up a large plastic jar filled with sand. Ask, "How many grains of sand are in this jar?" Add, "We may never know the answers to these questions, but we do know one thing—how sand is made."
3. Help students generate a list of all they know about how beach sand is made on a "K–W–L" chart. Write everything they know (or think they know) in the *K* column.
4. Have the students list what they want to know about the topic under the *W* column.
5. Read the book *Oceans* by Seymour Simon (Morrow). Invite the students to add questions to the *W* column if they wish.
6. Direct the students to record in the *L* column all they learned about how sand is made. In addition, any inaccurate information listed in the first column should be corrected.
7. To bring closure, ask students to make an entry in their social studies journals summarizing what they have learned about beach sand.

The K–W–L strategy is a favored social studies activity because it helps students actively associate their previous knowledge and experiences while establishing personalized purposes for becoming involved in a new learning experience. This procedure effectively bridges direct instruction and more independent learning in social studies.

Drawing Students Into the Lesson

I am sure you have heard the old adage, "You can't get people to do anything unless they *want* to do it." The drive that propels us to do something is called *motivation,* an internal state that pushes us toward certain actions and keeps us engaged in them. Let's consider what you are doing right now—reading this book. What is provoking you to read it? Is it because you find elementary school social studies a fascinating subject? Do you clearly want to learn the subject matter? Do you have a thirst to become the best possible teacher? Are you primarily interested in good grades? What about outperforming your classmates? Maybe you want to please your professor or parents? Or, perhaps you just want to complete the assignment as quickly and as painlessly as possible? All people do things because they are motivated in one way or another. Basically, there are two types of motivation: (1) *extrinsic motivation,* where the driving force for doing things lies outside the individual—they bring rewards from others; and (2) *intrinsic motivation,* where the driving force lies within the individual; you perform the task because it is enjoyable or worthwhile in and of itself. Looking back at the possible reasons you are reading this text, which are examples of intrinsic motivation and which are examples of extrinsic motivation?

Researchers have found that extrinsic motivation (grades, extra time at recess, a pizza party for achieving a certain class learning goal, praise, a star on a chart) can certainly promote learning in social studies. However, extrinsically motivated students are likely to exert only the minimal effort they need to accomplish a task and often stop an activity as soon as reinforcement disappears. Intrinsically motivated students, however, are more likely to exhibit initiative, persistence, independence, ambition, and enjoyment in their learning. They enjoy taking risks, show creativity in performance, experience pleasure in what they are doing, and learn information in a meaningful rather than rote fashion (Ormrod, 2000).

Ideally, whatever children do in our classrooms would grow from an internal desire to do it—they would meet all assignments with complete absorption, focus, and concentration. Although this is an optimal state in the social studies classroom, it is rarely achieved. A child, like all of us, often needs to be drawn into experiences, and social studies is no exception. Motivational activities arouse enthusiasm, a desire to discover what the social studies activity has to offer. They involve many kinds of experiences, such as *puppets, drama, puzzles, hands-on experiences, relating the activity to the children's own lives,* and so on. Mr. Chavez, for example, used a mental *puzzle* to motivate his students to read about the grasshoppers. Another teacher may use *hands-on experiences* to pique her children's interest. Knowing her students would be much more interested in watching a video about coal mining if a few lumps of coal, a bandanna, and a miner's helmet were placed in a guess box, she challenges her students to look into the box and tell what the items bring to mind. Before reading a biography of Martin Luther King, Jr.'s life, yet another teacher helps students *relate the activity to their own lives* by asking them to talk about times when they were not treated as well as others. Another teacher, a frustrated thespian, may use *drama* to motivate students to learn about the globe:

Classroom Vignette

Entering the classroom looking every bit a space alien in her metallic shirt and pants, Betty Hunter introduces herself as Retem Nomis, a citizen of Zaxton, a planet 25 light years from Earth. "We Zaxtons have been tracking you earthlings for some 300 years now. We have had a very difficult time understanding your language, but even more difficulty learning about you. In the 6 Earth years I've been observing your planet, I've had a chance only to learn a few things." She continued, "My studies of history and geography tell me that in all likelihood you are familiar with such primitive things as trees, dirt, and running water. These are things I've only read about or seen in pictures on my viewing screen. Such tales are difficult to believe because our planet became so crowded that Zaxton is now depleted of all those things. But, my favorite kinds of lessons about Earth are lessons about geography and history."

Ms. Hunter then goes on to what she called Geography 101, *Life on Earth*.

Motivational activities, then, are any kind of experiences that help draw in students to the learning task and help them experience pleasure as they undertake the challenging aspects of the lesson. When we make early learning experiences positive and nonthreatening, we stand a good chance of producing students who will be enthusiastic about, and willing to become actively involved in, learning throughout their lives.

Establishing a Clear Purpose

"Why does my teacher want me to do this?" "What am I supposed to get out of it?" Students have asked questions like these in classrooms all around the country when given vague directions for engaging in a learning experience: "Read pages 43 to 45 in your textbook. Be prepared to answer my questions when you're done." A child who does not sense a clear purpose for this assignment often considers textbook reading (or any other learning experience) impossible, unnecessary, or completely frustrating; he is confused, so he does not read with any purpose and views his efforts as a sign of failure and incompetence. A second child may read for such a vague purpose because she wants to be able to answer the teacher's forthcoming questions. However, motivation to become engaged in the text material is very low and will most likely result simply in rote learning. A teacher informs the third child; he knows why he is about to read the assignment—to find out something he is vitally interested in knowing, to follow certain directions, to get a central idea, even to stimulate some personal thinking on a subject of deep interest. This is the child who will delve into the learning material enthusiastically, knowing what is expected at the end—uncovering new insights or knowledge, making discoveries, solving a challenging problem, experiencing an emotion, or simply laughing.

It is critically important to remember, then, that once students are motivated, the next immediate step is to transfer that interest directly to the purpose for engaging in the main learning experience, since the motivational strategy and purpose are directly connected. Therefore, after showing her children a ship's bell and ringing it several times, primary-grade children are directed to watch a video "to find out why bells ring every half-hour on a ship." After showing her students a large photo print of a mushroom cloud and holding a short class discussion of the horror of the atomic bomb, upper-grade students are directed to read the picture storybook *Hiroshima No Pika* (Translation: *The Flash of Hiroshima*) by Toshi Maruki (a book that vividly portrays in words and pictures the horror of an atomic attack with hopes that it will never happen again): "Read the book to learn about the pain and suffering one family experiences when the flash interrupts their breakfast of sweet potatoes on August 6, 1945 at exactly 8:15 a.m." These purpose statements focus students' attention on what to look for as they proceed through the learning activity, directing them to a particular aspect or several aspects of importance. Sometimes these will simply be oral statements while, at other times, they could be written on a chalkboard, chart, or handout so students can periodically refer to them.

All good learners use sound learning strategies, including the ability to independently establish personal purposes for learning. That is why you will want the children to gradually learn to set their own purposes for learning. Students set their own purpose when they examine the photos in their textbook prior to reading the text and ask the question, "I wonder why banana farmers cut the stalks while the fruit is still green?" They then read to find out. Learners also set their own purpose when they check the heading in a newspaper, "India Arming Pakistan Border," and turn it into a question, "Are these two countries close to war?"

LESSON PRESENTATION

After you have built the framework for learning with a sensible, captivating introductory sequence, students should be ready to enter into the main learning experience itself. During the second, or *presentation phase* of a direct instruction lesson, the teacher either illustrates and explains the concept being taught or models the skill to be acquired. It is during this phase that the teacher uses a variety of learning activities to make the topic or skill meaningful to students.

Selecting Materials for Instruction

Most school districts encourage teachers to employ a variety of materials in their social studies program because not all pupils learn in the same way, and different media appeal to different learning styles. Each medium has particular strengths and limitations in the way it conveys messages; the impact of a message is likely to be stronger if more than one sensory system is involved in receiving it. Different sources can provide different insights on the same subject, while some discrepancies or inaccuracies may go undetected if a single source is used.

Selections of learning materials and activities for direct instruction in social studies classrooms must be consistent with the ways children learn. For example, Jerome Bruner (1966) identified three levels of learning that children move through as they encounter new information—*enactive, iconic,* and *symbolic* (see Figure 5–7). The *enactive level* includes objects, people, places, trips, visitors, and real-life classroom experiences. Within this level children represent and understand the world through direct experience and actions on objects. For example, suppose your goal for today is to help students learn about the emergence of cotton as an important cash crop in the Southeast during the late 1700s. Your enactive level possibilities include bringing in cotton bolls for the children to handle as they examine the soft fibers, observe the tiny seeds, and touch the prickly shell. By handling the real item,

FIGURE 5–7
Three Modes of Knowing

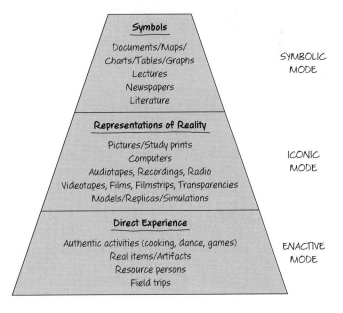

the students can appreciate the discomfort slaves would experience as they were forced to pick out the little seeds from within the boll in order to free the attached fibers.

The *iconic level* involves "imagery," or representations of real objects when the objects or places themselves cannot be directly experienced. For example, there is no way to have your students directly observe the slaves picking cotton in the hot fields all day and emptying their sacks until a huge heap of cotton took shape. And, you cannot have them observe the subsequent 8 to 10 hours when the slaves would clean the fiber out of the cotton bolls. Time and place limitations prevent you from doing this. A well-produced video documentary or possibly a set of study prints with clear explanations would then be a useful substitute for the real experience. Representations help children form concepts when real things are not available. Although pictures and models are not always as motivating as the real thing, these learning experiences are much more effective than trying to connect new learning to words alone. Good pictures, especially large poster-size photographs and prints, will deepen children's concepts of people and places. Dynamic social studies teachers build substantial picture files by searching hard for just the right pictures to furnish experiences that students can connect to new ideas. Models, too, bring valuable stimulation to the classroom. A good social studies classroom contains models that mirror displays found in our best museums; not a hands-off place where students simply stand and gawk, but one that invites handling and touching.

The *symbolic level* involves using abstract ideas, symbols, language, and logic to represent the world. Actions and images should still be used in thinking, but they do not dominate. For example, good children's literature or a social studies textbook can enliven and deepen the children's understandings of how cotton is grown and processed and how the Southeast became an important cotton growing region.

Understanding Bruner's three levels of learning is important for planning and organizing the presentation phase of your social studies lesson. They help you recognize the need for *balance* among the activities you choose, so that there is not too much symbolism (workbooks, practice sheets, talking, reading) and too little enactment (or vice versa) in your program. As a rule of thumb, you should remember that all students thrive on a balance of solid learning experiences, but that younger students need direct contact and real experiences (a visit to an orchard) and visual representations (a videotape of apple-growing procedures). Older students still require concrete experiences but are increasingly able to gain knowledge from abstract sources (a folktale from China or listening to a well-planned lecture).

The Teacher's Role

Certainly, the act of constructing concepts in dynamic social studies classrooms does not happen simply by exposing children to learning experiences. Piaget (1964) writes, "Experience is . . . necessary for intellectual development . . . but I fear that we may fall into the illusion that being submitted to an experience . . . is sufficient. . . . But more than this is required . . ." (p. 4). In emphasizing the need for a child's *construction* of knowledge, Piaget stresses the crucial role of the teacher; teachers must not only be effective *organizers* who are able to select rich materials and intriguing situations that arouse learners' interest in learning, but they must also be thought-provoking *guides* who know how to stimulate and direct learners' thinking to effectively organize the information taken in during the experience.

Because the goal of concept learning is helping students construct big ideas (concepts) from the facts and information associated with the learning experience, the presentation of learning materials and activities must be carefully organized and sequenced. What follows are several key points to keep in mind while teaching concepts through direct instruction in the social studies classroom:

- *Offer a clear definition of the concept or topic of instruction.* You will recall that this was an important responsibility during the concept analysis phase of planning. You defined the main concept early; now is when you will use it. Suppose *plantation* was selected as the central concept for this day's lesson. You will want to define the term for the students so they are clear that *Southern Plantation Society* in the early 1800s consisted of three interdependent elements—the plantation owners, the slaves, and cotton (the crop that made the plantation possible).

- *Present numerous and varied examples of the concept you are teaching.* Students must encounter many illustrative examples of the defining features of the concept. In teaching about *Southern Plantation Society,* a teacher would include such defining features as *cotton, cotton gin, cotton fields, big house, gardens, more than 100 acres, slave quarters, slaves, slave labor, abolitionists,* and *plantation owners.*

- *Present nonexamples to demonstrate what the concept is not.* Students construct concepts more easily when they are able to contrast negative instances of the concept. Nonexamples help students define a concept's limits and what fits and doesn't fit into its boundaries. For example, nonslave owners often made a living as independent small crop farmers who grew corn and other vegetables because their land was not fertile enough to grow cotton. So, nonexamples of *Southern Plantation Society* would include characteristics of small crop farmers such as *corn, vegetables, log cabins, poverty,* and *less than 100 acres.*

- *Ask students to identify, organize, and classify all positive and negative examples accurately.* This process should demonstrate if students truly understand the concept or whether they have simply learned the information through rote memorization. Because this final responsibility is extremely crucial to the success of the direct instruction model, I have chosen to discuss it under a separate heading: *Learner Response.*

LEARNER RESPONSE

Once students complete the actual learning activity, they should take part in exercises that require them to think, to explain, or to organize the information that emerges from the learning experience. Two of the most popular approaches to accomplish this task require the teacher to (1) lead class discussions with good prompts (questions) or (2) involve the students in concept mapping, a relatively new strategy where they "diagram" their understandings of the content.

Leading Class Discussions

Class discussion, a valuable teaching procedure in any subject, is essential to successful use of direct instruction. Over the years, social studies teachers have believed that the most efficient and effective way to learn is by leading discussions with well-designed questions. However, the way these questions have been asked often results in little or no benefit to the learner; this is especially true when teachers use questions as "tools of interrogation" rather than as aids to help students seek and master information, organize it, and apply what they learned about the world around them. "Interrogations" are characterized by the exclusive use of closed questions that are asked simply to string together unrelated bits of information. For example, a closed question such as, "On what date did the Congress vote to accept the Declaration of Independence?" permits only one child to respond. If that child knows the answer, the teacher must then ask a second question to elicit another fact. If the first child fails to come up with the right answer, the teacher keeps on going until someone does. Monotonously, the remaining questions are handled in the same way. A series of closed questions inevitably develops a closed-ended, teacher–student communication pattern. It also prevents the child who has a particular piece of information to offer from entering the discussion if the teacher does not ask "the right question."

Effective Prompts (Questions)

When we ask questions after a group lesson we can enhance the students' achievement, but good social studies teachers know that they *must* use a variety of the right kinds of questions if they are expected to serve as effective prompts for productive discussions. It is my contention that beginning teachers should not be burdened with any one of several question-asking schemes that group questions into a number of categories. It is much more important to have an intuitive feel for types of questions that stimulate either *closed* or *open* thinking. It seems reasonable to conceive of classroom questions as being on a continuum. At one extreme of the continuum, questions require students to simply recall or retrieve facts. We call these questions *closed questions* because once the child answers the question, further dialogue is closed off; there is nothing more to say. Although the use of closed questions has been abused, such questions have several benefits: (1) they help us determine whether students are learning the material; (2) they help us detect what misconceptions students have about the material; and (3) they help students monitor their own learning to determine whether they understand the information. Conversely, *open questions* involve more complex, higher mental activity; these questions deal with personal feelings and opinions and have several possible answers. Students are usually more involved in class discussions when open questions are used because open questions are more thought provoking.

Check your ability to distinguish open from closed questions by categorizing each of the following examples:

1. What was the name of the ship that brought the Pilgrims to Plymouth Colony in 1621?
2. The Pilgrims at Plymouth Colony had very strong religious beliefs. How do you think these beliefs affected their ability to survive the hardships of the first winter?
3. When spring came, many Indians visited the Pilgrims. What questions do you suppose the Indians wanted to ask the Pilgrims? What questions do you suppose the Pilgrims wanted to ask the Indians?
4. What Indian tribe visited the Pilgrims at Plymouth Colony that first spring?
5. Squanto taught the Pilgrims how to plant corn. What did he show them to use as fertilizer?

Did you select questions 2 and 3 as *open questions?* They are open because students are encouraged to think of their own answers or explanations. Discussion usually is extended because students elaborate on each other's responses. Questions 1, 4, and 5 are *closed* because they deal either with the recall or organization of specific information. There usually is only one correct answer and, once the answer is given, discussion usually comes to a halt.

Although teachers have been urged over the years to ask a greater number of open questions in their social studies classrooms, we have to ask if there is research evidence to indicate that this is actually a better way of questioning for dynamic social studies programs. In a summary of research on questioning, Woolfolk (1995) uncovered some surprising results. She found that closed questions were positively correlated with learning in the primary grades but that open questions led to achievement gains, too, if they are used appropriately. Based on these research findings and subsequent recommendations offered by Woolfolk, the best pattern of questioning in dynamic social studies classrooms for younger students and lower-ability students of all ages appears to be simple questions that allow for a higher percentage of correct answers, ample encouragement, help when the student does not have the correct answer, and praise for positive effort. For older students and high-ability students, the successful pattern includes harder closed and open questions and more critical feedback.

Question-Asking Skills

Knowing that questions can enhance the learning process is only one step in leading group discussions in the social studies. Teachers must also be skilled at sequencing or "patterning" the questions so their students can be systematically guided toward intended learning outcomes. How to pattern questions is a personal decision, for two teachers may view the same material in completely different ways. Let us look at two questioning sequences planned by different teachers who wanted their classes to understand how the migration of white settlers changed the lives of the Plains Indians. One teacher initiated a questioning sequence by asking, "The Plains Indians had deep respect for nature and the land. Do you think the white settlers shared this point of view?" In other words, an *open question* was asked to set the discussion in motion. The students offered several responses, and the teacher continuously asked them to support their suggestions with accurate data: "What evidence do you have to support your contention?"

The second teacher, by contrast, set the discussion in motion with this question: "In what ways did the white settlers upset the Plains Indians' way of life?" The teacher added each fact to a growing list of data of how the settlers caused conflict with the Plains Indians. Immediately after the students pulled out their knowledge, the teacher suggested, "Let's look at all the information you gathered. Why do you suppose the white settlers had such a different attitude toward nature and the land than the Plains Indians? Do you think Americans today have the same attitude as the white settlers?"

The first teacher used an *open question* to start the discussion and asked the students to support their personal ideas with relevant information. The second teacher did just the opposite; she asked the students to recall specific information and then draw their personal conclusions from the data. Which approach is best for elementary school social studies instruction? Both are acceptable; instead, the major considerations of question schemes should be thought of in terms of Dewey's (1933) popular "art of questioning" guide, as proposed over 60 years ago:

- Questions should not elicit fact upon fact, but should be asked in such a way as to delve deeply into the subject; that is, to develop an overall concept of the selection.
- Questions should emphasize personal interpretations rather than literal and direct responses.
- Questions should not be asked randomly so that each is an end in itself, but should be planned so that one leads into the next throughout a continuous discussion.
- Teachers should periodically review important points so that old, previously discussed material can be placed into perspective with that which is presently being studied.
- The end of the question-asking sequence should leave the children with a sense of accomplishment and build a desire for that which is yet to come.

Learning to lead and sustain thought provoking class discussion is a key part of direct instruction.

Many questioning patterns are possible following a social studies learning experience, but the main consideration should be to think carefully what you want to achieve with any plan. Random questioning with no end in mind gives the illusion of teaching, but wise teachers know that questions with the most positive impact on the learning situation are planned so that one leads to the other throughout a logical sequence, deliberately provoking deeper thought or creating new understandings.

Framing Questions

In addition to following Dewey's suggestions, it is important to give students adequate time to respond to the questions we ask. This is referred to as the technique of *framing questions*. The basic approach for framing questions is: (1) ask the question, (2) pause for 3 to 5 seconds (wait time I), (3) call on someone to respond, and (4) pause for 3 to 5 seconds again while the students think about the response (wait time II). The process is grounded on the principle that students attend better to questions if a short pause follows. There are several benefits when we pause after asking questions. First, students think more deeply about the question, so a greater number of students volunteer answers. Second, their responses tend to be longer, more complex, and more accurate. Third, a pause provides the teacher with a little time to "read" nonverbal clues from the class. With experience, teachers can readily observe such signals as pleasure, fright, or boredom. Fourth, teachers who pause after asking questions become more patient waiting for student responses. Therefore, when you allow at least 3 seconds to elapse after asking a question, you will discover that children make longer responses, offer more complex answers, interact more with one another, and gain more confidence in their ability to make worthwhile contributions.

Concept Mapping

Although there may be more effective ways to help children understand social studies content, the fact remains that teacher's guides to textbooks—the most frequently used instructional tool in social studies classrooms—emphasize question asking. Many beginning teachers seem to have been conditioned to think that asking questions is the preferred way to learn in the social studies classroom. However, in recent years, concept mapping has been gaining increased popularity as a strategy to help students arrive at important understandings. Concept maps, used with good questioning strategies or by themselves, are simply visual representations that help students organize information from any learning experience. The visual representation takes the form of a diagram or map (charts, graphs, and drawings, for example) based on the students' personal interpretation of the content. Some of the most popular maps are described in the sections that follow.

Webbing

A key diagram designed to help students organize complex ideas is the *concept web.* Webs help children discern a central idea as well as its distinguishing attributes. Webs have three basic elements: *core concept, web strands,* and *strand supports.*

The *core concept* serves as the focus of the web. All the details, facts, and information generated in a discussion should relate to the core concept. An example of a core concept is "Daily Life in New England Colonies." Place the core concept in the center of a growing matrix (see Figure 5–8) and begin a group discussion with a thought such as, "Think about the kinds of activities that made up daily life in the homes of New England colonists."

As the students suggest categories such as jobs, school, food, furniture, entertainment, religion, and sports, place their suggestions at various points around the core to represent different categories of information related to the concept. These points are referred to as *web strands.*

The specific facts or details the students use to support each web strand are called the *strand supports.* The strand supports extend from each web strand, summarizing important information and relationships. Here the students should select such particulars as *hornbook, doll, New England Primer, stool ball,* or *oxen.*

FIGURE 5–8
Web: Daily Life in Colonial America

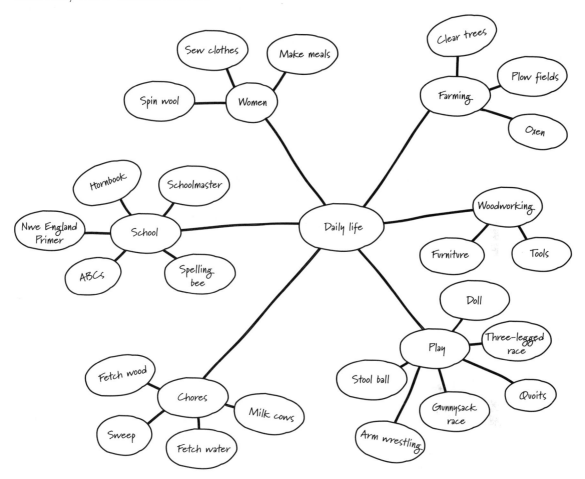

Venn Diagrams

Venn diagrams are illustrations that students can use to compare and contrast two distinct cognitive elements. Introduce Venn diagrams by drawing two overlapping circles on the chalkboard. Suppose you have just read a selection contrasting life in the city and in the country. One circle represents the city and the other represents the country. The attributes unique to each environment are listed inside the separate circles, and the attributes common to both are listed in the area where

FIGURE 5–9
Event Chain

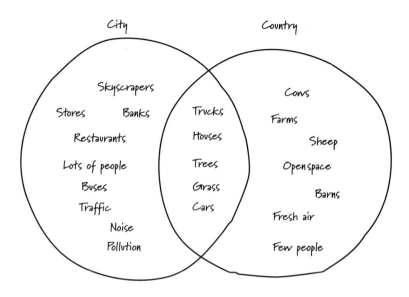

the two circles intersect (see Figure 5–9). When students know how to use the Venn diagram, you can use it to guide comparisons of a variety of topics; for example, two (or more) people, forms of government, communities, nations, religions, and so on.

Cycles

This type of map helps students visualize a series of connected events that occur in sequence and produce a repeated result. The months and seasons of a year, for example, exemplify a cyclical process, as do the water cycle and the metamorphosis of a caterpillar. One cyclical phenomenon treated in most social studies programs is recycling. Help your students understand this process by first writing the word *recycling* on the chalkboard. Discuss why so many communities in the United States recycle their waste and why our country has developed a wasteful attitude throughout its history. Then have the students complete a cycle graphic organizer as they read about the process of recycling. See Figure 5–10 for a sample.

Concept maps are fundamental to concept development in social studies. They provide the stimulus for learning in ways that questioning and discussion strategies alone cannot do. Clearly, you should offer opportunities to develop and use concept maps whenever you ask children to seek relationships among ideas.

This teacher's description can be useful as you plan to introduce the use of group maps (or individual maps) in a constructive way:

FIGURE 5–10
Cycle Organizer

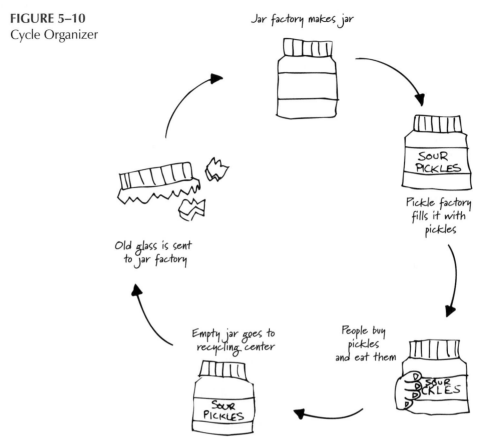

Jar factory makes jar

SOUR PICKLES

Pickle factory fills it with pickles

Old glass is sent to jar factory

People buy pickles and eat them

Empty jar goes to recycling center

SOUR PICKLES

SOUR PICKLES

Classroom Vignette

After my students have finished reading *The Tongue-Cut Sparrow,* a Japanese folktale, I would hold a general discussion of the story. I would then initiate instructions such as the following:

"I would like everyone to map their ideas of the story on a sheet of drawing paper. A map is a drawing that shows your personal understanding of the story. There are many ways you might do this. Some people like to draw pictures; others use words, circles, squares, and other shapes. Here are some samples. (At this point I would display maps of simple, familiar stories the children should know, such as *Little Red Riding Hood* or *Three Billy Goats Gruff.* I will show my samples for only a short time, for I want students to refresh their understanding of what maps look like but I don't want them to develop the idea that their maps should look like mine.) A story map can look any way you want it to look. Now, create your own map from the story ideas in your mind.

But, please don't look back at the story right now." After the students complete their maps, I will invite them to hold them up so they can see what others have drawn. I will allow several moments for the children to look at each other's maps. Next, I will invite individual students to share their maps with the class. As the map is shared, I will use prompts that encourage the students to explore story characters, events, and story plot. For example, "Why did you decide to put an arrow pointing from the old man to the small chest?"

FUNCTIONAL APPLICATION

Functional application is the final phase of a direct instruction lesson. During this phase, activities offer students the opportunity to practice the new skill or concept on their own through special projects or independent activities. Because functional application experiences often tend to focus on creativity and choice, they may include such things as painting, group murals, writing, construction projects, drama, puppetry, and music. In all of these activities, students use what they have learned to construct deeper meaning. The key to success in this phase of direct instruction is the same as was stressed for success in all of social studies teaching—variety. Vary your activities to keep interest and motivation high. Because social studies topics differ in their complexity, some strategies will be more useful than others for enriching, reinforcing, or extending the targeted concepts or skills. Some examples of functional application strategies follow.

- Dramatize the firing of the "shot heard around the world."
- Cook and eat Harry S. Truman's favorite food, brownies.
- Give first-person accounts of famous historical events as if the students were the actual characters.
- Select geographic locations students would like to "cheer" about. Then design paper pennants with colors and illustrations that have some meaning to the location.
- Choose a favorite scene from an historical era to recreate in a three-dimensional model, or diorama.
- Write an historical period news magazine. Students write their articles in the style of popular news magazines using headlines, drawings, eyewitness statements, and so on.

By reading professional journals and organizing an idea file, you will begin to accumulate ideas for functional application experiences. Remember, though, that these experiences must be an *integral part* of the total direct instruction lesson, not "extra busy work" or cutesy "icing on the cake" activity. This final activity extends the learner response phase by taking the discussion or concept mapping activity and using the information generated to discover potential relevance or usefulness. The following example shows how both activities fit together much like two puzzle pieces.

Classroom Vignette

Gary Mack's third graders had just completed reading the first chapter of the book *Sarah, Plain and Tall* by Patricia MacLachlan, a story filled with adventures about a pioneering family pushing west into the Great Plains. Sadly, the family's wife and mother passed away shortly after the family moved westward. Chapter 1 ends with Papa reading a letter in which Sarah, a woman from Maine, expresses interest in becoming a mail order wife and mother for the family. Mr. Mack knows that the first chapter of a book helps explain the characters, problem, and setting, so he focused attention on the element that was developed most deeply—characterization. He helped his students organize the information they learned about the main characters into character webs (see Figure 5–11). He began an "Anna" character web by placing her name in the middle of a large sheet of chart paper and asked the students to offer descriptive words for Anna. The students were required to cite evidence from the story to support each descriptive word. The descriptive words and supporting story evidence were placed around "Anna" and connected to her name with straight lines. The class was then divided into groups to finish Anna's web and to make character webs for Papa, Caleb, and Sarah. The webs indicated to Mr. Mack that his students had a firm grasp of Chapter 1, but he wanted to personalize the information to a greater degree. So, he told the students that Anna, Caleb, and Papa each writes back to Sarah. Keeping the character descriptions in mind, he asked each group, "If you were (Anna, Caleb, or Papa), what would you write in your letter? What questions would you ask? What things would you tell Sarah?" Mr. Mack asked the groups to record their ideas on the character web charts. Once this outline of ideas was complete, Mr. Mack directed each student in the character groups to write a letter to Sarah using the friendly letter format they had been learning about in language arts class. Children who wished to do so read their completed letters to the class. Others were displayed on a bulletin board.

It is important to note that the concept web provided the context for, or "set up" the letter-writing activity. It offered the substance which the students applied to the job of writing an original letter back to Sarah. It is logical that since the chapter ended with Sarah's letter to the family that the family members write back. However, what

FIGURE 5–11
Character Web

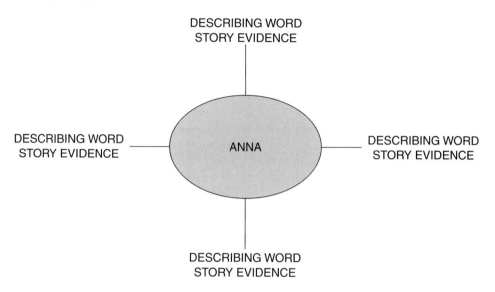

do they write? Students gain that insight by studying the characters deeply through their character webs. The connection between the *learner response* and *functional application* phases is apparent. However, think how many times a classroom teacher might be tempted to follow up the character webs with a request for the children to make a mural of the prairie. Although the mural is a nice idea, it has no relationship to the learner response activity and stands out alone rather than as part of a cohesive whole.

AFTERWORD

Children learn through a combination of physical and mental activity; they "mess about" and naturally want to get into or try out everything. They may come to elementary school knowing a little bit about a lot of things, but one characteristic they all share is a thirst for experiences that will help them find out more. When these enthusiastic, energetic youngsters come to school they expect to learn about all that interests them in much the same way, through activity and involvement. They are not greatly interested in memorizing information or in confining activities such as completing ditto sheets or workbook pages. They want to try things out.

Helping elementary school students classify and categorize information into meaningful concepts is one of the foremost challenges to social studies teachers. A major part of this challenge is to help learners attach their backgrounds to the learning experience and to organize new information into appropriate schemata. As Piaget and Vygotsky emphasize, meaningful learning takes place only when learners are able to bridge the gap between the unknown and the known. To that end, this chapter has described a system of instruction through which the teachers are able to effectively direct their students through the process of concept reception.

Developing a teaching plan is a complex professional responsibility involving a great deal of knowledge, hard work, and skill. As a new teacher, you may wonder whether the results are worth the effort. In effect, you may say, "Why bother? After all, the textbook and teacher's manual were written by experts in the field who really know the social studies." To an extent, you are correct. Manuals can be very helpful, especially for student or beginning teachers. As guides, though, they must be viewed as suggestions, not as prescriptions. You will probably want to start your career by using the teacher's guide closely, but as you gain experience, you will adapt it to the changing needs of the different groups of children you will teach each year. The constructivist approach described in this chapter allows you the flexibility to constantly change your teaching approach within a framework of sound planning.

References

Ausubel, D. P. (1961). In defense of verbal learning. *Educational Theory, 2,* 16.

Bodrova, E., & Leong, D. J. (1996). *Tools of the mind: A Vygotskian approach to early childhood education.* Upper Saddle River, NJ: Merrill/Prentice Hall.

Boyer, E. L. (1982). Seeing the connectedness of things. *Educational Leadership, 39,* 384.

Bruner, J., Bruner, J. S., Olver, R. R., and Greenfield, P. M. (1966). *Studies in cognitive growth.* New York: John Wiley.

Dewey, J. (1933). *How we think.* Boston: D. C. Heath.

Douglass, F. (1845). *Narrative of the life of Frederick Douglass, an American slave, written by himself.* Boston: Anti-Slavery Office, 9–15.

Freedman, R. (1987). *Lincoln: A photobiography.* New York: Clarion.

Hunter, M. (1982). *Mastery teaching.* El Segundo, CA: TIP Publications.

Metcalf, F. (1992). When rice was king. *Social Education, 56.* In unnumbered special section following p. 408.

Ormrod, J. E. (2000). *Educational psychology: Developing learners.* Upper Saddle River, NJ: Merrill.

Piaget, J. (1964). Three lectures. In Ripple, R. E., & Rockcastle, U. N., *Piaget rediscovered.* Ithaca, NY: Cornell University Press.

Remy, R. C. (1990). The need for science/technology/society in the social studies. *Social Education, 54,* 204.

Ryan, K., Burkholder, S., & Phillips, D. H. (1983). *The workbook.* Upper Saddle River, NJ: Merrill/Prentice Hall.

Woolfolk, A. E. (1995). *Educational psychology.* Boston: Allyn & Bacon.

6 Problem-Centered Learning

S tudents in Sheila Kirtland's fifth-grade social studies class have been investigating various customs of people around the world. One day, as they were watching a video on diverse cultural traditions, the students became particularly interested in the historically popular Islamic greeting of keeping the palm of one's hand open and touching the breast, forehead, and lips, signifying endearment in heart, thoughts, and words. Traditional greetings of various cultures around the world became the rage in Ms. Kirtland's class as the students engaged in a spontaneous and productive discussion of the many interesting ways people greet one another: "The people of Japan bow politely when greeting someone." "The French and other continentals 'kiss' one another on both cheeks." "Boy Scouts in America shake left-handed—the hand nearest the heart—with the three middle fingers extended to the other person's wrist." The greatest reaction surfaced after some students demonstrated the handshaking styles of contemporary athletes. The main area of interest seemed to be, "Who started all of this?"

Ms. Kirtland encouraged her students to pursue their interest by helping them locate and explore various resources that might assist them with an answer. Some searched the World Wide Web, using Yahooligans (*http://www.yahooligans.com/*) as a starting point. Others looked through books and magazines. A few students thought they might uncover information by interviewing parents, teachers, and other adults. Their investigations uncovered some fascinating information, such as that as far back as the 16th century the Ashanti of Africa had a handshake and an expression of greeting: "Five must lie within five." The Dutch of old ended business negotiations with a ritualistic hand slapping, high and low, which served as a basis for the expression "striking a bargain." As far as identifying the source of these modern "shaking" variations of contemporary athletes, the students discovered that, although they are rooted in African-American culture, the evidence was inconclusive as to who actually started the practices: Basketball player Magic Johnson claimed to have originated the "high five" at Michigan State in 1980; long jumper Ralph Boston argued that the "slap five" began among African-American track athletes on the international track circuit prior to 1968; and some insist that revolutionary handshakes started prior to the 1940s when African-American musicians greeted each other with a special shake accompanied by the jive phrase, "Gimme a little skin, man."

Ms. Kirtland allowed her students to pursue this interesting fact-finding investigation because she is convinced that the most productive learning experiences in social studies happen when children work at what *they* want to know. Her teaching philosophy is based on a belief that elementary school students are naturally inclined to dive headlong into whatever excites their curiosity and that teachers should be a little more flexible, a little more spontaneous, and a little more willing to depart from direct instruction in order to help students explore their personal interests. In this case, Ms. Kirtland knew that the students' attraction to the topic was very strong and that it was important to take the risk of temporarily departing from the regular curriculum to help them search for information to resolve the issue at hand. "If teachers are reluctant to take risks, how will they ever inspire these ideals in their students?" she asks. "It is the job of social studies teachers to value, not suppress, the natural curiosity their students bring to the elementary school classroom. The students are often their own best teachers."

Ms. Kirtland demonstrated how social studies teachers inspire and nurture learning through their students' natural interest in the world, often getting a good idea of what to do by listening to and watching them while looking for clues about what students find interesting and puzzling. Perhaps this natural spirit to learn was demonstrated best by a fifth grader who was asked what it would be like to know everything. "Awful," the child replied. "Why?" the teacher asked. "Because then there would be nothing to *wonder about!*" A dynamic social studies classroom encourages its students to look at the social world with awe and wonder.

WHAT IS PROBLEM-CENTERED INSTRUCTION?

Teachers in dynamic social studies programs do not commit themselves to any single method or source of information as the exclusive "right way to teach." They realize that there are many ways to unlock the mysteries of life and that the heart of solid social studies instruction is balance and proportion. Sometimes, as direct instruction teachers, they will "bring the action to the children," demonstrating and explaining in order to help their students construct concepts or refine skills. At other times, as problem-centered teachers, they will allow the children to "initiate the action"—supporting students emotionally and intellectually as they independently strive to explain a puzzling question or problem. Good teachers realize that the key to effective social studies instruction is *flexibility*—looking at options, weighing the advantages and disadvantages of each, determining the relative value of the choice, and selecting the most appropriate approach for the situation at hand.

The role of the teacher changes dramatically as one moves from direct instruction to problem-centered instruction. Instead of taking on the role of a "technician" who employs carefully prescribed patterns of instruction, the problem-centered teacher functions as a "facilitator" who provides something meaty for the children to sink their teeth into and then sits back to wait for and support their reactions. Facilitators do not check the teacher's manual to see what to do next; they go to where the action is. Only one thing matters: a teacher who is tuned into and encourages the exploration of the interests and curiosities of the students. Problem-centered instruction allows students to learn content and process at the same time. Therefore, dynamic social studies becomes not only a body of information to be learned or a set of skills to be mastered, but also a way of thinking and acting. Students are accepted as what I like to call *young social scientists*—not learning about the world only through direct-instruction teachers who make facts, concepts, and skills interesting and understandable, but also through problem-centered teachers who encourage students to use strategies employed by real social scientists—addressing questions or problems, gathering data, and searching for answers.

Since the topic of this chapter is problem-centered learning, it would be unwise to go any further unless we have a common definition of the term *problem-centered*. The term has been used by various writers to cover a number of things we do in social studies, from putting puzzles together to coming up with solutions to world crises. However, no description of problem-centered instruction has had more impact on elementary school social studies instruction than the one advocated by John Dewey (1916) during the early part of the 20th century. At that time, Dewey proposed that not only social studies, but also the entire elementary school curriculum should be based on problems, situations that he described as *anything that creates doubt and uncertainty in learners*. Because that definition has been validated over the years, it will serve to guide our discussion. Problem-centered instruction, therefore, will be defined as anything that creates doubt and uncertainty in learners.

Bridging Direct and Problem-Centered Instruction

If your students have had little or no previous experience with problem-centered instruction, you must make sure that its more student-centered processes do not run against the grain of what they have grown accustomed to. There must be a transition period during which students participate in activities designed to move from teacher direction to more independence. Student-centered learning does not emerge spontaneously in elementary school classrooms; it is a product of helping students work together in new ways. It may take several weeks or months to reach the point where you and your students work together to investigate problems productively. I have heard teachers give up after only one unsuccessful experience, saying, "I knew it! Children just can't think for themselves," and go through the rest of the year doing the children's thinking for them. But I doubt that these teachers would ever say, "These children cannot add and subtract by themselves," and thereafter remove any opportunity for them to learn those skills.

Before you make this mistake and end up crying, "Bring back the textbook! Give me back my worksheets! This business of promoting autonomous learning just doesn't work," you must know that, even though new stimuli may be needed in your social studies program, children must move slowly. You must give them many opportunities to grow and learn; the skills needed for problem solving will not be achieved after a single experience. The presence of any new element in a classroom is potentially distracting. Introducing new expectancies all at once produces a situation in which the children's cognitive systems collapse under an overload of input. Time, patience, and your belief in the importance of autonomous learning are the key ingredients of a successful transition from teacher-directed instruction to student-centered learning.

The foundation for problem-centered instruction in dynamic elementary school classrooms lies in problem-solving activities connected with direct manipulation and observation of objects. Children require rich opportunities to interact with real things before they can begin to conceptualize more abstract ideas. As they interact with objects in their environment, children become well able to establish relationships (comparing, classifying, ordering) between and among the objects they act upon. Such relationships are essential for the emergence of the logical thought processes required for problem-centered instruction.

Teachers help students learn independently by offering early problem-solving experiences that consist of the following elements: (1) *designing captivating classroom displays*, (2) *discussing the displays*, and (3) *encouraging children's questions*.

Classroom Displays (Mini-Museums)

Classroom displays, or what I like to call "mini-museums" (others have variously called them *interest areas, curiosity centers,* and *theme tables*), are basically exhibit areas in your classroom designed to stimulate children to explore, comment, and ask questions. Today's exhibit might be a compass, origami, Chilean rain stick, foreign coins, butter churn, powder horn, shark's teeth, campaign button, tape recording of city sounds, or a sombrero. Whatever items you select for the mini-museum, they should be treated like exhibits in the best public, child-oriented museums—not with a "hands-off" policy, but with a policy that invites touching, exploration, and investigation.

Most elementary school students have visited museums during a school trip or with their families (if they haven't, you should take them), so they will have many ideas on how to create a museum in their classroom. Even upper-grade and middle school students enjoy designing exhibits and sharing their interesting collections of objects. You must remember, though, that these collections should not be haphazardly displayed, nor should they be briefly explained and then put on a shelf and forgotten. In essence, the practices involved in setting up a regular museum display should be employed when setting up mini-museums. Monhardt and Monhardt (1997) suggest that the best place to start is to have the children share their past museum experiences. The authors suggest a discussion guided by questions such as these:

- "What are some different museums you've visited?"
- "What museums were most interesting?"
- "What made them so?"
- "What is involved in setting up a museum display?"
- "How can we find out more about museums?"

Introduce the idea of a social studies mini-museum to your class. One teacher modeled the process by placing a seemingly strange cultural artifact on a table—a stiff brush used by dog groomers. On the wall above the table was a sign that read, *Classroom Mini-Museum*. Almost instantly the children began looking at the brush, touching it, and talking about what it might be. The teacher watched and listened, occasionally asking open-ended questions and making comments to stimulate the children to think more deeply about the object. One child tried to use the brush on her own hair. She was surprised to see just how stiff the bristles were. Naturally, the other children had to try, too.

Rosa eventually identified the object. Her mother was a veterinarian and Rosa often helped around the office. She was obviously thrilled to share her knowledge with the class; their interest became much deeper as they listened to the expert description by their friend. To capitalize on this growing fascination with the brush and what veterinarians do, the teacher invited Rosa's mother to visit the class to talk about her job.

You can see from this example that it is a good idea to plan what you display at the mini-museum rather than selecting items in hit-or-miss fashion. You want something to happen at the center—interests to grow and concepts to deepen as the activity extends in the direction of the children's interests.

Discussing the Displays

Rather than making the mini-museum as an informal observation/conversation place, as Rosa's teacher did, some teachers prefer to introduce the mini-museum during planned meeting times that can involve a small group or the entire class. If you prefer this approach, inform the students about how a social scientist works, as well as about how the mini-museum will operate. The following dialogue illustrates one teacher's initial meeting. That teacher, Roland Comegys, opens the meeting by calling the students' attention to a large study print of an archaeologist examining some artifacts.

Classroom Vignette

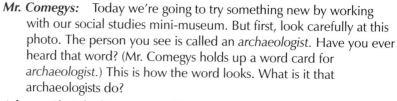

Mr. Comegys: Today we're going to try something new by working with our social studies mini-museum. But first, look carefully at this photo. The person you see is called an *archaeologist*. Have you ever heard that word? (Mr. Comegys holds up a word card for *archaeologist*.) This is how the word looks. What is it that archaeologists do?

Adam: She's looking at something.

Mr. Comegys: Yes, she's looking at, or examining, something. What kinds of things do you suppose archaeologists examine?

Denise: Fossils?

Tamara: Yeah. To see what Earth was like a long time ago.

Mr. Comegys: Yes, they examine old objects and use that information to describe what people were like very long ago. That's a good start. Does anyone else have an idea?

Patrick: They dig for old tools and things. Even bones.

Mr. Comegys: Good, they might study fossils, bones, tools, paintings, clothes, furniture, and other objects. These things are called *artifacts*. Studying artifacts can help us understand the lives of people. Have you ever seen an archaeologist examine artifacts?

Nelson: I saw a picture of an archaeologist in a museum. It was in an exhibit about China.

Lucinda: On a TV show once, I saw some archaeologists looking in old pyramids for mummies.

Mr. Comegys: Yes, archaeologists study many things to learn about the lives of people. Artifacts can tell about the games they play, the tools they use, and even the way they eat their meals. From now on, this table will be called the mini-museum. It is a place where you will come to discover things about the lives of people by doing some of the important jobs archaeologists do—examining them and keeping a record of observations. (Mr. Comegys brings out a cornhusk doll.) I've brought something for you to examine today. Look at it carefully and try to make a discovery. Amour, what are some of the things you notice? What is the item made of? What parts does it have?

Amour: It looks like a toy—a doll, maybe?

Mr. Comegys: What makes you think it's a doll, Amour?

Amour: It looks like it has arms and legs. The top is like a head.

Mr. Comegys: It is shaped like a doll. What do you notice, Raphael?

Raphael: It looks like it might be made out of corn. It feels like the dried corn plants we put out at Halloween.

Mr. Comegys: That part of a corn plant is called a *husk*. Here, look at this. It is a cornhusk before it was made into a doll. (Shows cornhusk.)

Moira: How do people ever make dolls out of cornhusks?

Mr. Comegys: Here, I'll show you. It's quite simple. (Takes a few minutes and demonstrates how to construct a cornhusk doll as the children follow his lead to make their own. Mr. Comegys is demonstrating that people can be a very important source of information. He emphasizes the importance of checking other sources, too, calling the children's attention to a set of five books, each of which contains information about cornhusk dolls.) What does the doll tell you about the people who used it? (The craft activity continues for several more minutes.) You've been examining our artifact very carefully today. You made some interesting discoveries. Archaeologists don't stop with discoveries, though. They must keep careful records of everything they do so their findings can be shared with other people. We will use a special way of recording our discoveries—*observation sheets*. (Mr. Comegys passes out the observation sheets.) There's a place for your name and the date, and then the paper says, "What I Know About . . ." followed by a long blank. What words should I put in the blank?

Martha: Cornhusk dolls!

Mr. Comegys: That's right. There is also a big box on the paper for you to draw an illustration of what you examined. Please make a careful drawing there. Finally, at the bottom you will find some space to write. What are some things archaeologists might write about this cornhusk doll?

Denise: The doll is made from dried cornhusks.

Amanda: It was made in a place where people grew lots of corn.

Martin: It could have been a gift for a boy or a girl.

Louise: There weren't many stores around selling toys then. People had to make their toys from things around them. (The suggestions flow forth for several minutes. The children then complete their observation sheets. Figure 6–1 shows a completed observation sheet.)

Mr. Comegys explained to the class that he planned to display something in the mini-museum every few days and that it would be directly related to the thematic unit under study. The children were told that they would be using their skills of observing and recording to catalog information.

FIGURE 6–1
"Archaeology Attic" Observation Sheet

"ARCHAEOLOGY ATTIC"
OBSERVATION SHEET

NAME _Jesse_   DATE _4|6|94_

WHAT I KNOW ABOUT _Corn husk dolls_

The doll was made from corn husks.
It is a toy for a young child. It
probably came from a place where corn
is plentyful, like Nebraska.

As important as they are, observational experiences by themselves do not guarantee the acquisition of problem-solving skills. As you have read, Mr. Comegys used carefully worded questions or comments to help children gain meaning from the mysteries that confronted them. Because they often have relatively limited backgrounds of experience, children could misinterpret new experiences if an adult's guidance is not offered. Skillful guidance must be provided to help children make accurate observations. You can guide their investigation with questions like the following:

- "What do you see here?"
- "How do you suppose it is used?"

- "I wonder what would happen if . . . ?"
- "If we try it again, do you think the same thing will happen?"
- "Is this like anything you've ever (used, seen, tried out) before?"
- "How can we find out more about . . . ?"
- "Can we find out if we watch it carefully?"
- "What makes you think so?"
- "Who do you think might use this?"
- "Where do they live? What makes you think so?"
- "What can you tell about the people who use this?"
- "What do you think of the people who use this?"

Some of these questions and comments help children look for specific things; others are more open, encouraging higher thought processes such as predicting and discovering relationships. Through such experiences, children develop the rudimentary scientific observation skills required for more sophisticated data collection activities.

Encouraging Children's Questions

Observing and talking about interesting things under the teacher's direction are important components of early problem-centered learning and make an excellent transition from direct instruction strategies. A third important component of this process is encouraging children to ask their own questions about the experience. Children often have a lot of questions of their own to ask. For example, Thomas Edison's last day in school came when he asked, "How can water run uphill?" after he noticed that a river in Ohio did just that. Young Tom was then expelled for expressing himself in ways that were unacceptable at the time. Children come to us with a strong need to ask questions. Respect that fact and guide them in exercising this childhood gift. Here are a few questions and comments I remember while working with children: "I heard something pretty disgusting. I read in my book about a guy who *chewded* wood. And about a guy they found alive in a whale's stomach!" "Are babies born with brains?" "Where does the water go when I flush the toilet?" "Look! That man has a hole in his hair (he was balding)." When children blurt out such comments, ask yourself, "Am I listening as carefully and sensitively as I am able?" When you do listen, you communicate to the child that he or she is a worthwhile individual whose honest thoughts are valued. This is something the children need to know, for question asking is an indispensable part of the problem-solving process.

The goal of these preliminary experiences is to help children acquire the skills needed for self-directed learning. Self-directed learning emerges when students are aware that a problem or question exists and attack it in a systematic, thorough manner by understanding what they need to know and devising strategies to find out what they don't. Self-directed learners have the ability to find out something by themselves by using these three questions whenever they encounter a perplexing problem or situation: (1) What do I already know about this? (2) What else do I need to find out? (3) Where can I find this information?

INQUIRY AND CREATIVE PROBLEM SOLVING

Problem-centered learning is a general classification label that includes two teaching strategies appropriate for dynamic social studies classrooms—*inquiry* and *creative problem solving*. Although each involves different processes, they hold one very important similarity—helping students solve perplexing problems. In keeping with the spirit of unraveling baffling problems or questions, the reading material that follows will help you answer these important questions: What are the characteristics of inquiry and creative problem solving? How are these two strategies similar? Different? How do they contribute to learning? What roles do teachers and students undertake?

Inquiry

The inquiry strategy is designed to bring students directly into the scientific process. The expression *inquiry* has achieved widespread popularity in recent years among social studies educators. However, as use of the term has spread, so has confusion about what it really means. Some people think of inquiry as an approach to instruction that is indeed new. Others consider it to be at least as old as ancient Greece and that it has previously traveled under such names as the *Socratic method, problem solving, critical thinking, scholarly investigation,* and *scientific thinking.* Despite this lack of agreement, the characteristic that sets inquiry apart from other teaching strategies is that it is the only strategy designed to teach students how to investigate problems independently through a systematic process of gathering and analyzing data. Therefore, for the purposes of this text, inquiry will be considered a *systematic process for solving problems.* Descriptions of the inquiry process vary from source to source, but the most common pattern in elementary school social studies classrooms seems to be the time-honored pattern recommended by John Dewey (1916):

1. The students identify a problem or question that can be investigated.
2. The students generate hypotheses, or tentative answers that can be verified.
3. The students collect data.
4. The students analyze the data and form generalizations that can be applied to this problem and to similar ones encountered in their lives.
5. The students share their results with an audience.

Identifying Problems or Questions

In order for inquiry to work, students must have an interesting, motivating problem to solve. The problem must hold a degree of mystery and appeal—children will not want to investigate anything they do not care about. I vividly recall one insightful youngster reacting to a teacher's suggested problem: "Research is boring when you

Interesting collections of objects bring an important dimension to social studies instruction.

don't care about what you're learning about!" What are good social studies problems? The best are those that capture the interest of the students. Herein lies the real challenge of problem-centered instruction: "How do I find out what interests my students?" First and foremost, children will be interested if they are able to attach themselves to the problem; they must be convinced that the problems they encounter are worth thinking about. The problems must be clear, understandable, and meaningful, and involve a high degree of mystery. Additionally, the problems must lie within the students' sphere of competence. Problem situations that generate the most interest are those that offer just the proper level of mystery to challenge previously established ideas. If the problem is foreign to previously established ideas, students will not be motivated to solve it; they have no personal stake in it. In contrast, if the problem is easily unraveled by the children's current level of knowledge, it will be considered too routine, and the students will quickly lose interest.

There are two basic types of inquiry problems in social studies: (1) problems generated by the students themselves, and (2) problems sparked by a teacher's presentation of a question or discrepant event.

Problems Generated by the Students

To illustrate this type of problem, let's examine one social studies teacher's use of inquiry.

Classroom Vignette

Jennifer Puksta's fourth-grade students returned from lunch upset about watching their peers unload huge amounts of food from their trays into the trash containers. The developing discussion inspired a searching "why" from one of the students. Antonio suggested that most children didn't like meat loaf, the meal served that day. There were competing views; however, most thought that the overall quality of food needed to be improved. Ms. Puksta sensed that student interest in this situation was deep and that their concerns centered about two questions: "What is the least favorite food served for lunch?" and "What is the favorite food served for lunch?" Ms. Puksta could have held a brief class discussion about the matter and then dropped it to go on to spelling class, but self-directed learning was an important part of her teaching repertoire. Therefore, she seized the opportunity to use their deep interest as a springboard for putting their emerging problem-solving abilities to work. Rather than have the children engage in a "heated exchange of ignorance" about why so many lunches ended up in the trash, Ms. Puksta decided to help her students substantiate their claims about what the children "really liked" by turning their problem into a fact-finding investigation.

Problems Sparked by a Teacher

Often, problems arise spontaneously during the course of the day. At other times, problems can be arranged by the teacher, especially when a particular topic needs to be covered.

Classroom Vignette

Michael Wang's sixth-grade class had been working at the sand table (borrowed from the kindergarten classroom) digging up and brushing off replicas of artifacts their teacher had hidden beneath the soil he had substituted for the sand. Acting as "apprentice archaeologists," the students divided the surface of the site into squares with grids made from yarn. Then they carefully removed the soil from their own section, layer by layer, to dig up the remains from an unknown past civilization. As they dug, students carefully recorded the exact location of everything they found. Jenny uncovered what looked like an old coin. She examined it closely and searched for any information that might tell where the coin was from. She observed what she thought was a likeness of a Roman emperor wearing a laurel crown. The word AUGUSTUS was printed above the likeness.

The problem, often stated as a question, is the initial spark for further discussion. For the problem to be effective, it must be closely attached to what the children already know or care about. If children are confronted with a problem with which they have little or no previous experience, they often reject it because of its unfamiliarity. However, the problem should not simply be an opportunity for children to "rehash" information they already know. Instead, the problem situation should be balanced between what students already know and what is new to them. That balance seems to stimulate elementary school children's greatest needs to explore and search for answers. This first step of the inquiry process is crucial; if children cannot accurately define a problem based on previous experiences or if they are not interested enough to pursue it, the subsequent steps of the process are futile.

Stating the Hypothesis

Once the problem is clarified, the students are in business. They are ready for action; it is now time to work systematically toward solving it. I hesitate to use the word *hypothesis* to describe what is done next because future teachers often become intimidated by such jargon, feeling there has to be an intricate scientific mission to accomplish. But, for elementary school students, the hypothesis is simply an "educated guess" or a "feeling," "hunch," or "suspicion" based on incomplete evidence. You can help your students come up with a hypothesis by asking questions such as these: "What do you already know about this situation? What have you already learned that we might be able to use now? How could this information help us come up with an answer to our problem? What ideas can you suggest based on what we've just discussed?" The purpose of these questions is to help children associate information they already know. Formulating hypotheses involves a certain amount of risk to the child, so you must be especially careful to value individual contributions. Remember that these educated guesses are nothing more than guesses. Definitive solutions and opportunities to reject hypotheses arise when children become involved in gathering data.

Classroom Vignette

Ms. Puksta's fourth-grade students offered responses such as, "Spaghetti is one of the favorites. The other kids hardly throw any away." "Tacos are a favorite because everybody likes tacos." "No one likes meatloaf. It ends up in the trash container every time it's served." and "Chicken fingers are awful. They don't even look like chicken!"

In Mr. Wang's class, Jenny used her observations from step one to formulate a reasonable hypothesis. "That's it! That's how I'll prove to Mr. Wang that this mystery coin is from ancient Rome!" The excitement in Jenny's voice was as clear as the pleasure reflected in her teacher's face. She saw what she thought looked like a Roman emperor wearing a head decoration that looked like leaves. Based on this quick observation, Jenny guessed that the coin could have come from ancient Rome.

Now that the hypothesis has been stated, it will guide the students through the next phase of the inquiry process, data gathering.

Gathering the Data

The data collection process can take any of several forms, depending on the nature of the problem. In most elementary school classrooms, data are collected in one of three ways: *survey, descriptive research,* or *historical research.*

Surveys

Surveys involve a systematic collection of data that helps determine the views of a particular group of people. The first step in conducting a survey is to design precise questions that will yield the specific data required to confirm or not confirm the hypotheses.

Classroom Vignette

For example, in the case of Ms. Puksta's class, the students want to find out exactly what their peers prefer for lunch as the cafeteria staff prepares it. So their survey question could be, "Of the following, what is your favorite lunch served in the lunchroom?" or "Of the following, what is your least favorite lunch served in the lunchroom?" However, a question such as "What do you think of the food served at school?" is too general and would not yield the necessary data.

Once the clear-cut polling question has been determined, the actual survey instrument must be constructed. Should the students poll their peers orally and keep a running record of the results, or should they distribute questionnaires similar to the one illustrated in Figure 6–2? Regardless of the choice, the students must next decide on the sample population for their investigation.

Do they need to question every child who eats lunch in the cafeteria? Should only a small portion of the student body be surveyed? Since it is unnecessary to

FIGURE 6–2
Sample Survey Questionnaire

Lunchroom Survey

Put an "x" in front of your favorite lunch served in the lunchroom.

_____ pizzas	_____ hot dogs
_____ tacos	_____ chicken fingers
_____ spaghetti	_____ meat loaf
_____ hamburgers	_____ sloppy joes

---

gather data from every child eating lunch in the cafeteria, the students would be wise to choose some type of *sampling* strategy. Perhaps they will interview every third child who gets through the lunch line. They may want an equal boy/girl distribution; if so, they could poll every third boy, every third girl, and so on. If neither of those techniques appeals to them, they might want to randomly select 50 students by drawing names out of a hat. Whatever the choice, the students must do all they can to ensure that their survey data is gathered from a representative sample of the larger population.

Direct observation is a frequently used data gathering technique in the elementary school classroom.

Descriptive Research

Descriptive research is probably the most frequently used of all data-gathering techniques in the elementary school social studies classroom. It involves two major strategies: *direct observation* and *indirect observation*. *Direct observation* includes all the real-life experiences that involve children in touching, handling, trying out, and viewing objects or events. A trip to the commuter train station to observe the "crunch" of rush hour, a trip to a dairy farm to see how a cow is milked, and a visit by the town's mayor to question her about the need for a traffic light near the school are all examples of direct observational experiences offering data for problem solving.

Indirect observation involves the use of information sources other than direct experiences. These sources often include library resources such as encyclopedias, informational books, magazines, newspapers, pamphlets, almanacs, catalogs, dictionaries, travel brochures, atlases, guides and timetables, posters, films, videos, photographs, and even the phone book.

Classroom Vignette

In Mr. Wang's class, Jenny put to use indirect observation as she searched through a series of reference books arranged beforehand by her teacher and learned about the history of Roman coins. She also consulted an electronic encyclopedia available on the classroom computer. She found that Roman coins were decorated with the likeness of the emperor in power at the time. She also discovered that laurel crowns were used to adorn emperors and also that the title AUGUSTUS was given to the popular Roman leader Octavian when he succeeded Julius Caesar following his death at the hands of Brutus. There was no date on the coin, but Jenny compared her information about Roman coins and emperors with what was on the coin she uncovered and made an educated guess that this coin was made in Rome between 31 B.C. and A.D. 14.

Historical Research

When children delve into conditions of "long ago," they must gather and evaluate relevant traces of the past. These may include physical remains (artifacts, relics, and other "accidental survivors") and oral or written records (stories and documents). Michael Wang's students were involved in both historical and descriptive research because they were unearthing information from the past using real objects (replicas) as well as print sources. Historical research was treated with detail in Chapter 2.

FIGURE 6–3
Tally Sheet

Our Favorite Lunch							
Pizza	Tacos	Spaghetti	Hamburgers	Hot Dogs	Chicken Fingers	Meat Loaf	Sloppy Joes
‖‖‖ ‖‖‖ ‖‖‖ 11	‖‖‖ ‖‖‖ ‖‖‖ ‖‖‖ 11	‖‖‖ ‖‖‖ ‖‖‖ 1	‖‖‖ 11	111	‖‖‖ ‖‖‖ ‖‖‖ ‖‖‖ 1		11

Displaying and Analyzing the Data

Once students have gathered the data, they must organize and interpret it. This is a major task, and teachers must be prepared to help the students through the process, for a variety of displays require the ability to construct accurate tables, charts, graphs, and other visual displays. When Ms. Puksta's surveys were completed, for example, the children tallied the responses and were helped to organize the data into a summary graph. Figure 6–3 shows the tally sheet and summary graph used by the class.

After they had recorded the information, the children examined and analyzed the data to assess their hypotheses on the basis of the data. Naturally, you don't impose a complicated chi-square analysis, but you do want them to interpret the data simply and accurately.

Classroom Vignette

The findings of the lunchroom survey drew the following interpretations, or *generalizations,* from the students:

> Most of the children like tacos for lunch.
> Chicken fingers are almost as popular as tacos.
> Meatloaf, sloppy joes, and hot dogs are the least favorite.

Discussing data in relationship to the hypotheses is an important part of the inquiry process, so Ms. Puksta asked the students to examine their findings in relationship to what they had originally thought. "Look," said Alex, "we were right about the meatloaf. Nobody likes it!" "Yeah," added Laura, "and most of the kids liked tacos. Just like we thought!" "You guys are right," interrupted Caesar, "but before you get too carried away, take a look at the chicken fingers. We said they were awful, but a lot of kids like them."

Discussing the data is a valuable experience for the students, and they should be encouraged to examine and argue the results.

Classroom Vignette

Jenny was somewhat sure her research evidence supported her initial hypothesis about the Roman coin but, working with an open mind, she stood ready to alter her initial feelings should further evidence demand it. She compared the artifact she uncovered with one found by a classmate, Russell. He had dug up a miniature banner with the letters SPQR. Russell's research indicated that these were the initials for Latin words translated, "The senate and the people of Rome." Together, Jenny and Russell concluded that they had enough corroborating evidence to substantiate a contention that these items were from ancient Rome. The question was, were these artifacts from Rome or were they left behind in foreign lands as the result of trade, travel, war, or migration? They would await the findings of their classmates before making a definitive conclusion.

Sharing Results

The final step of an inquiry episode is sharing the results with an *authentic audience* using appropriate vehicles of communication. In the adult world, much of the reward gained from research comes from having an impact on desired audiences; professors of education take great pride in sharing with their peers new ways to improve teaching, and medical researchers are eager to share new treatments for difficult-to-control diseases with others in health care professions. The resulting recognition and excitement are often missing for students whose primary audience is usually limited to the teacher and fellow classmates.

When sharing the results of the lunchroom survey with an authentic audience, who will benefit from the results (cafeteria manager, principal, school newspaper)? Ms. Puksta's students may choose to show the cafeteria manager their findings after summarizing the lunchroom survey on large charts with precise data. The students would then need to orally communicate this data with accurate, effective statements. What might constitute an authentic audience for Mr. Wang's students? As you ponder that question, keep in mind that authentic research goes beyond the traditional formal written report by engaging children in varieties of communication possibilities—oral presentations, graphics, photographs, audio- or videotapes, debates, dramatic skits, and other means.

Teachers' Responsibilities

While students are engrossed in problem-solving ventures, they acquire content and develop important processes at the same time, such as how to solve problems and think logically and critically. Interest in a problem becomes "fuel" for constructive learning; after students experience authentic research, they will want to address, independently or as a group, other absorbing questions, problems, or situations that come up both in and out of the classroom. Students will probe deeply into what interests them, searching for data to use while assessing their hypotheses. Because students need much encouragement and guidance to keep the inquiry process moving, the following suggestions should be helpful in the planning process:

• *Build a varied collection of research materials.* Do not limit your students' problem-solving resources to the textbook. Certainly, textbooks can be effectively used as one source of information, but be sure to use other books, magazines, travel brochures, audiotapes and videotapes, real objects, resource people, field trips, the computer, and other tools.

• *Investigate together.* When children have a real stake in a problem, motivation soars; they realize they own the problem and are able to use data in their own personal way to arrive at a solution. Teachers have a crucial role in enhancing this spirit. They act as models, thinking aloud with the students and practicing every behavior they want their students to use. They ask questions: "What's going on here? What do we need to know more about?" They coax and prompt students to ask similar questions as they assume responsibility for their own learning. They are open to new experiences. Children relish teachers who reach out for the different and unusual—teachers who look at life with a sense of wonder. The world is full of miraculous phenomena, a never-ending mystery. Teachers stop, look, and listen; they feel, taste, and smell. They ask, "What is it? Where does it come from? What is it for?" They perform their own careful inquiries and, through their enthusiasm for new discoveries, offer the greatest form of encouragement to their children. We must do our best in dynamic social studies programs to nurture curiosity for life; one of the best ways to do this is to be a teacher who responds to the world with a probing, wondering mind and regularly proposes, "Let's find out!" Discovery is usually accompanied by a strong feeling of elation. This is what drives people to explore and investigate: *It feels good!*

• *Provide adequate time.* If students are going to conduct authentic research, they will become deeply and personally involved in seeking solutions. They must have opportunities to work for extended periods of time. You should allot a minimum of 1 hour for each problem-solving session during which students conduct research to explore problems that interest them. However, inquiry episodes usually require more than one class period.

• *Facilitate social interaction among your students.* Students learn to solve problems by interacting with one another as well as with the teacher. Thus, you will often divide your class into cooperative learning groups (see Chapter 7) where students share the responsibility for carrying out the inquiry process. This social context offers students opportunities to experience the feelings and ideas of others and motivates them to seek agreement if conflict in perspective is encountered. Part of the inquiry effort, then, is to establish an environment where students develop both a growing consciousness of others' ideas and a need to cooperate and share with them.

Lifelong problem solvers and researchers get their start in elementary social studies classrooms as teachers arouse curiosity for stimulating problems and propel their students into research to obtain answers or solutions. This inquiry process, oriented and constantly monitored by the teacher, leads to the intellectual independence we so often find in lifelong learners.

Inquiry is but one variation of problem-centered learning. Another problem-centered model challenges students to explore problems of personal interest and to generate novel, creative solutions, rather than solutions based on hard data. We will refer to this model as *creative problem solving.*

Creative Problem Solving (CPS)

> *We once saw a tyrannosaurus.*
> *And we feared he'd end our lives faurus.*
> *But, "Look!" said my friend,*
> *"He's rubber, can bend!"*
> *Then we realized we were in Toysaurus.*

My son Jeff wrote this limerick in elementary school as part of an integrated thematic unit on dinosaurs. His teacher sent the poem home along with a note telling us that she felt the limerick was remarkably creative. Of course, Jeff's "impartial" parents thought likewise and displayed it proudly in the middle of the refrigerator door. What do you think about Jeff's limerick? Do you think it is creative? If so, what makes it so? If not, you just failed this course (just kidding)!

Creative Thinking

What do we mean by *creativity?* Like other inner-directed, deeply personal concepts such as love, patriotism, or intelligence, the term creativity means different things to different people and is virtually impossible to define precisely. However, most definitions of creativity include two major components:

- *Novel or original behavior.* Behavior that has not been learned from anyone else; it is fresh, novel, and unique.
- *An appropriate and productive result.* A suitable or worthwhile product, or an effective solution to a problem.

How does Jeff's poem reflect these two components? Is it fresh and original or did he get the idea from somewhere else? Is it an appropriate response to the teacher's assignment to write a limerick about the dinosaurs they had been studying?

What prompts some children to create in such engaging ways, while others struggle to move beyond the ordinary? Although a certain degree of intelligence was required, something more was needed. In Jeff's case, that "something more" was the ability to "play" with ideas—to see things in a new way. What that "something more" is has been debated over the years, but consensus seems to exist that an individual's level of creativity is influenced by the combination of four cognitive traits: *fluency, flexibility, originality,* and *elaboration:*

- *Fluency* is the ability to produce a large number of ideas. The child who responds to the question, "What things are crops?" with "Wheat, corn, beans, peas, and tomatoes" is more fluent than the child who responds "Wheat and corn."

- *Flexibility* is the ability to produce a number of different categories of responses. The child who responds to the question, "What things are crops?" with "Wheat, tomatoes, apples, peanuts, and tobacco" is a more flexible thinker than the child who responds "Wheat, rye, oats, and barley (all grains)."

- *Originality* is the ability to produce unusual or clever responses. The child who responds to the question, "What can you do with an empty cereal box?" with "Make a snowshoe out of it" is more original than the child who says "Store things in it." Originality is usually determined statistically; the response can be considered original if it is offered by less than 10 percent of those responding.

- *Elaboration* is the ability to expand on a simple idea to make it richer. The child who responds to a teacher's directive to draw a picture of the geographical area where the Sioux lived with a simple landscape drawing shows less elaboration than the child who includes buffalo, tipis, and Sioux farmers working in the fields.

If we hope to help students enhance their creativity in dynamic social studies class-rooms, we need to examine the kinds of classroom activities and practices that are supportive of creativity. Certainly, a major responsibility of social studies instruction is teaching the content and, up to now in this text, we have examined various ways to challenge students to explore and discover new information. However, creativity takes knowledge a step further; it "puts knowledge to work." What that means is that creative acts allow students to express what they know in unique ways. People who are creative are creative within a discipline they know, so a skills and knowledge base are essential components of coming up with new ideas. If content knowledge and skills are essential to creativity, then social studies teachers must teach in ways that allow students the opportunity to work in the same manner as those who are creative in social studies. They must approach social studies with a thinking focus— seeing things from a unique point of view. The remainder of this chapter examines how the content of social studies relates to creativity and the processes of thinking creatively.

Creative Problem Solving in Dynamic Social Studies Classrooms

Perhaps the oldest and most widely practiced technique to encourage creative thinking in social studies classrooms is the creative problem solving (CPS) method developed by Alex Osborn (1963) and Sidney J. Parnes (1981) during the 1960s and 1970s. To understand CPS, it would be instructive to contrast it with inquiry. It is important to remember that the goal of inquiry is to systematically search for facts and information to answer questions or solve problems. Some students do this well; they sort out the clues, pull them together, look for patterns, and deliberately arrive at valid conclusions. Others, however, attack problems in quite different ways. This variation of a well-known story helps illustrate the difference between the two:

> An engineering major and an elementary education major were hiking in the woods when they came across a grizzly bear. Both were terrified and quickly began to search for an escape route, each in his/her own way.
>
> The engineering major took out his pocket calculator and quickly computed the mathematical differential between his speed and the bear's. His face turned ashen as he stared at the results.
>
> The elementary education major simply took off her hiking boots, opened her backpack, slipped on a pair of jogging shoes, and took off.
>
> "Boy, you are STUPID," the engineering major yelled to the elementary education major as she sprinted down the trail. "My calculations show you can't outrun a grizzly bear!"
>
> "I don't have to," the elementary education major hollered back. "I only have to outrun YOU!"

This story illustrates how two people can respond very differently to the same problem. The two response categories in this example are generally referred to as *systematic* (the engineering major who relied on facts and logical thought) and *intuitive* (the elementary education major who relied on gut feelings and intuition). Original, creative solutions to problems come from the ability to shift directions in thought—to move beyond the obvious to the subtle. Some would say that the elementary education major's solution was more creative than the engineering major's. Do you agree? What made it so? Regardless of whether you are trying to figure out how to study for an exam when your friends want you to go to a movie or where to get money for next year's tuition hike, you are likely to rely on one of these cognitive styles.

Most schoolwork calls for systematic rather than intuitive thinking, so we find that schools tend to overemphasize logic-related skills at the expense of intuitive-related skills, giving our children an apparently "lopsided" education. Teaching for creativity does not minimize the importance of a solid background of information, however, for creativity in social studies is more likely to occur when students have mastered the content. Neither interesting problems nor their solutions often pop up without having substantial background knowledge. Teresa M. Amabile (1989) emphasizes this partnership in her acclaimed model of creativity. Her model has three components; in order for creativity to occur, all three must be in place:

1. *Domain knowledge and skills.* This component includes the technical skills and content needed to solve a given problem. A student who is to design a creative problem or make a creative contribution in geography, for example, will need to know something about geography. Creative ideas do not spring forth from a vacuum.

2. *Creative thinking and working skills.* Knowledge is but one component of creative thinking, but is not sufficient in itself. Creative thinking and working skills add that "something extra" that forms the creative personality. The four cognitive traits of creativity described earlier help explain the kinds of thinking abilities that help students take new perspectives on problems and come up with unusual ideas. These are required for creative performance.

3. *Intrinsic motivation.* Intrinsic motivation comes from within the person, not from an outside source. Intrinsically motivated students enjoy the pleasure and satisfaction of the work itself, not the external recognition or reward.

The Creative Problem Solving Model

The original CPS model is usually described as being made up of six steps, each driving the creative process. That model works well with older students and adults, but I prefer a simplified three-step procedure for elementary school students. A unique feature is that each of these steps, in turn, involves both a divergent phase during which students generate lots of ideas and a convergent phase during which they select only the most promising idea(s) for further exploration.

The divergent phase of each step utilizes *brainstorming* strategies. Brainstorming is an appropriate strategy when you want to generate a large number of ideas, particularly new and original ones. In traditional brainstorming, students work in groups of three to five members with a recorder to keep track of ideas and an encourager to monitor the rules. Some basic ground rules will help students be more effective as they brainstorm during each CPS stage. You (and your students) should become familiar with these ground rules so they will be second nature to all during CPS.

1. *Seek many ideas.* The goal of brainstorming is to find as many ideas as possible. Don't worry about quality; that will be considered during the convergent phase of each step. Large numbers of ideas, rather than small numbers, tend to generate good ideas.

2. *Accept all ideas.* Receive every idea that is offered, no matter how strange or silly it may seem at first glance. You may later discover a strange or silly idea has fascinating possibilities if modified for use. No idea, then, is to be critiqued or evaluated until all ideas have been shared. All forms of verbal and nonverbal criticism (eye rolling, face making, derisive smiles) are ruled out.

3. *Seek to combine and improve ideas.* Don't be afraid to be a "hitchhiker." Latching onto ideas shared by others can formulate exciting, new ideas. The combination of your ideas with others helps form new possibilities that are more intriguing than either by itself.

4. *Make yourself stretch.* Many students will exert only minimal effort and then proclaim, "I just can't think of anything else." However, effective problem solvers make a strong effort to extend themselves beyond this point of "idea exhaustion."

New and original ideas, however, seem to pop out after people think they have done all they can. Encourage sluggish students to keep at it.

The convergent phase of each step begins as groups restate the problem and narrow down lists of suggestions by addressing such concerns as, "Will this idea actually solve the problem? Will it create new ones? Will it work? Is it practical? Will we be able to use it in the near future? What are the strengths and weaknesses? Can any of the ideas be combined into one useful solution?" After narrowing their lists, each group works toward an agreed-upon decision. The ultimate choice might contain one idea or a combination of ideas.

The three steps of CPS are outlined below; an example of how they might be implemented in a dynamic social studies classroom will follow.

The Mess

This step involves saying all that can be said about a problem situation—factual statements as well as feelings. For example, Jorge Gonzalez created a simulated community environmental condition as shown in Figure 6–4. He divided his class

FIGURE 6–4
Environmental Pollution Problem

Timberland is a small village in rural Forest County with a population of just over 5,000. It is a friendly place to live with characteristics common to most villages—homes, stores, places to worship, restaurants, a movie theater, service stations, schools, government, and a few miscellaneous businesses. People are genuinely happy with their lives in Timberland, but a problem of great concern has surfaced recently and has grown in scope during the past year.

The problem is how to capture what the village residents describe as a "new breed of vermin"—illegal garbage dumpers. Village officials have identified 17 illegal dumpsites in the 85,000-acre forest that surrounds Timberland. They have noticed rubbish and junk accumulating along the roadsides, stream banks, and hollows.

The problem has grown in part because communities surrounding Timberland have grown in size the past few years and there is more and more rubbish to dispose of. The Departments of Sanitation of these communities have been overworked and the local landfill is nearly filled to capacity, so each household is limited to the removal of two trash containers per week.

Hoping to help solve the problem, the local Unspoiled Forest Club organized a campaign that prompted the village government to post signs threatening fines or possible arrests for those caught dumping junk illegally. The junk piles kept spreading, however, and the forest faces the danger of becoming an unhealthy eyesore.

The condition of the forest is a topic of daily conversation among the residents of Timberland. Some have suggested that more drastic actions be taken toward the solution of their problem.

into five-member groups and asked each group to line up in front of charts that were taped at even intervals around the room. The first student in line was to write a word, short phrase, or statement about the "mess" the community is in. These can be factual statements or feelings about the issue. Basically, the students were advised not to hold back, for everything about the problem should be said whether they think it fits or not. In turn, the rest of the students rotated to the chart, writing as quickly as they could the things they knew, felt, or thought about the "mess." Some groups wrote in excess of 20 entries in the 5 minutes allotted for this task.

Using each long list, Mr. Gonzalez directed the students to talk together and select one item that "bugs" them most or seems to capture the situation best. Each group then rewrote the selection as a problem statement. The Osborn/Parnes model suggests that the problem statement begin with an "IWWMW. . .?" phrase, or "In what ways might we. . .?" For example, if a concern about the pristine beauty of the forestland was the most bothersome aspect for one group, its statement might read, "In what ways might we control the junk that is being dumped in the forest?" If health concerns were the most bothersome aspect for another group, it might write, "In what ways might we avoid infestation by dangerous vermin and avoid contagious diseases?"

Idea-finding

This step involves brainstorming dozens of possible remedies for the problem. What are all the possible ways to address the problem statement? Using a clean sheet of chart paper and employing the same "relay" strategy as they used in the previous section, Mr. Gonzalez encouraged the students to write whatever came to mind, even if it seemed to have nothing to do with a solution. No idea would be critiqued until after all ideas had been written down. The students then looked over their lists and determined which one or two might best be made into an interesting solution. The students considered all possible ways that the tentative solution might be elaborated on or improved. Mr. Gonzalez prodded the students' thinking with thought-provoking questions such as those that follow:

- New Ideas
  - Can it be used in new ways as it is?
  - Can it be put to other uses if it is changed in some way?
- Adaptation
  - What else is like this?
  - What other idea does this make us think of?
  - What new twist could we add to the idea?
  - Could we change the color, shape, sound, or odor?
- Enlargement
  - What can we add?
  - Should we make it longer, wider, heavier, faster, more numerous, or thicker?

- Condensation

    What can we take away?

    Should we make it smaller, shorter, narrower, lighter, slower, or thinner?

- Substitution

    What else can we use to do the same thing?

    What other materials or ingredients might we use?

One group suggested making big signs such as "Spruce Up Around Here!" (*Spruce* was their play on words!) Another thought it would be constructive to establish a community "clean up" festival. Still a third announced that it would install surveillance video cameras in secret locations throughout the affected areas with the goal of identifying and prosecuting violators, and forcing them to pay fines or spend time in jail.

Action-Planning

This step involves expanding the interesting idea or combining interesting ideas into a statement that outlines an action plan that details the steps necessary to implement the solution: What course of action will we take? Does this plan depend on someone else's approval or support? What steps are needed? Who will do what? When must the steps be completed? The video camera group, for example, suggested that its program would need to be approved by the village council and could be paid for with a grant from the state Department of Environmental Protection.

Three major benefits result when students are involved in CPS in dynamic social studies classrooms: (1) higher feelings of self-confidence, self-esteem, and compassion result; (2) wider exploration of traditional content subjects and skills are undertaken; and (3) higher levels of creative invention in content and skills are used by the students. Therefore, our classrooms must encourage not only the systematic problem-centered efforts associated with inquiry, but also the inventive, intuitive thinking associated with creative problem solving.

Creative problems differ from inquiry problems in that they call for divergent thinking as opposed to convergent thinking as students search for solutions to problems. Both types of problems, however, are crucial to today's dynamic social studies classrooms in that they encourage children to dig into things, turn over ideas in their minds, try out alternative solutions, search for new relationships, and struggle for new knowledge.

AFTERWORD

It often happens that when the word *research* is mentioned to college students, groans and steely glares accompany the predictable questions, "How long does the paper have to be? How many references do we need? What's the topic?" Submitting students exclusively to this form of research often builds negative attitudes toward the processes of problem solving and inquiry. Written reports on teacher-determined topics using secondary sources not only generate negative attitudes from students but also present an unrealistic view of research in the real world. While this kind of research has a place, to present it as the only form of research does students a disservice. We cannot limit elementary school research to such practices; remember that most elementary school youngsters are natural problem finders. They revel in the mysteries of their world. On their own, they deftly uncover problems of interest—the important first step of research: "Why do farmers cut off the corn and leave the lower part of the plants behind?" "Could the early people of England actually invent the idea for Stonehenge?" What they need at school is to learn the methods by which these interests can be investigated. While all the particular problems children bring to school may not be particularly significant, the processes they go through and the feelings they gain about themselves as capable researchers *are*. Therefore, authentic research associated with creative problem solving and inquiry processes teaches students that they have the skill and ability to pursue knowledge in a meaningful way and that their efforts have real value now and in the future.

References

Amabile, T. M. (1989). *Growing up creative.* New York: Crown.

Dewey, J. (1916). *Democracy and education.* New York: Free Press.

Monhardt, R. M., & Monhardt, L. (1997). Kids as curators. *Science & Children, 35,* 29–32; 80.

Osborn, A. F. (1963). *Applied imagination: Principles and procedures of creative problem-solving.* New York: Charles Scribner's Sons.

Parnes, S. J. (1981). *The magic of your mind.* Buffalo, NY: Creative Education Foundation and Bearly Limited.

7

Cooperative Learning

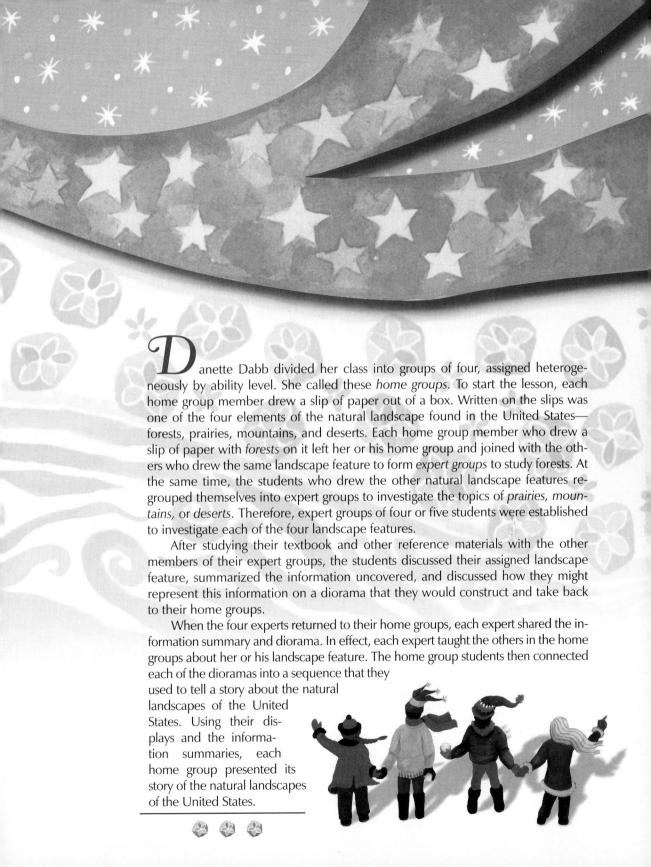

Danette Dabb divided her class into groups of four, assigned heterogeneously by ability level. She called these *home groups*. To start the lesson, each home group member drew a slip of paper out of a box. Written on the slips was one of the four elements of the natural landscape found in the United States—forests, prairies, mountains, and deserts. Each home group member who drew a slip of paper with *forests* on it left her or his home group and joined with the others who drew the same landscape feature to form *expert groups* to study forests. At the same time, the students who drew the other natural landscape features regrouped themselves into expert groups to investigate the topics of *prairies, mountains,* or *deserts.* Therefore, expert groups of four or five students were established to investigate each of the four landscape features.

After studying their textbook and other reference materials with the other members of their expert groups, the students discussed their assigned landscape feature, summarized the information uncovered, and discussed how they might represent this information on a diorama that they would construct and take back to their home groups.

When the four experts returned to their home groups, each expert shared the information summary and diorama. In effect, each expert taught the others in the home groups about her or his landscape feature. The home group students then connected each of the dioramas into a sequence that they used to tell a story about the natural landscapes of the United States. Using their displays and the information summaries, each home group presented its story of the natural landscapes of the United States.

Up to this point in our examination of flexible teaching in dynamic social studies, we have considered two major models of instruction—*direct instruction* (Chapter 5) and *problem-centered instruction* (Chapter 6). We have also learned that distinct, specialized professional strategies are associated with each. In the previous example, however, Ms. Dabb used a strategy to carry out instruction within the framework of either model—*cooperative learning*. Cooperative learning is an approach to dynamic social studies instruction where students work together in small groups to help one another maximize their own and one another's learning. Students work together to achieve common success; in other words, they apply the old adage, "All for one and one for all." It is instructive to think of this old adage as a characterization of the idea of "interdependence." Think of interdependence as the type of teamwork we find on a basketball court when one player passes the ball to another who, in turn, slams the ball through the hoop. The actions of both players were independent of each other (they both did completely different things) but the points could not have been scored without the actions of both. This is what is characterized as interdependence in cooperative learning—each student needs to perform competently for success to result. Cooperative learning happens in situations where two or more individuals work together on some task, coordinating their efforts to achieve success.

Cooperative learning, then, can truly be considered a prime vehicle of "social study." By *social* we mean that group participants are active and contributing members of a social organization that gains from their participation. By *study* we mean that the groups are involved in the systematic pursuit of knowledge and in efforts to apply their knowledge in meaningful situations. *Social study,* then, refers to a process through which students acquire new knowledge and use it within viable, collaborative groups, or social organizations.

BUILDING SOCIAL SKILLS AND TEAMWORK

Like all other classroom expectancies and routines, the ability to work with others involves a number of sophisticated skills that must be mastered if cooperative learning is going to be successful. Group work skills do not develop simply because students spend time together. Rather, group work must be made to seem attractive by making peer interaction a desirable aspect of the classroom and helping students learn how to function as group members, just as they must be taught to read or write. It is beyond the scope of this book to examine this topic thoroughly, but it is important to realize that groups begin to function effectively only after members understand their roles and feel accepted by others. Instruction on group skills such as these appears to help establish the sensitivities and behaviors required to establish cohesive, supportive cooperative groups:

- following directions
- keeping focused on the task
- completing responsibilities on time
- asking for help when you need it
- listening attentively to others
- contributing ideas when you have them
- considering the ideas and feelings of others
- offering encouragement to others
- making sure everyone has a chance to participate

Selma Wasserman (1989) recalls her frustrations with moving sixth-grade children into responsible cooperative groups when many did not have the opportunity to develop these skills: "The first weeks of so-called 'cooperative' group work was anything but. All manner of uncooperative behavior emerged. . . . They couldn't focus on the tasks; they didn't care about each other; they didn't understand 'what they were supposed to do.' In the absence of clear and specific teacher direction (i.e., 'Do this now and do it THIS way!'), they fell apart" (p. 204).

Wasserman described her biggest disappointment in watching her students become unruly: "My biggest disappointment was *not* that the children were unable to function in . . . sophisticated, mature and self-disciplined ways. . . . The killing blow was that the children wanted, asked, *begged* for a return to 'the way we did it in Grade 5' " (p. 203).

It was at that point that Wasserman realized that she needed to help the children learn the skills required to function as thoughtful, responsible, cooperative learners. She needed to provide experiences for students to gain practice in those skills. This was not an easy task either, for as she admits, "It may be a lot easier to teach children to read and spell than it is to teach them to behave cooperatively . . . with each other" (p. 204).

To engage students in cooperative learning, you must help them gain knowledge of themselves as group members and of what it takes to function effectively with others. You do this by making interpersonal functioning an important learning goal. This is good news—cooperative skills can be taught and learned just like any other skill. In addition, like any other skill, you must exercise great patience and offer meaningful opportunities to help children learn. It may be February before you are able to recognize any major shift in students' group work, but you must follow the children's lead, never moving so fast that students are overwhelmed by your efforts.

When you target a special social skill to work on, it may be most productive to begin with game-type tasks, especially those where team members must cooperate to reach a common goal. A game I have found to be particularly useful is *Puzzle Squares,* which is described below. Fortunately, many books and professional magazines contain suggestions for similar cooperative games.

Puzzle Squares

Materials

1. Cut out five heavy tagboard squares, each about 5 inches square.
2. Cut each square into three segments using the following patterns (see Figure 7–1).
3. Scramble the 15 pieces and put them into a large manila envelope.
4. Repeat the procedure for each set of five children in your classroom.

Goal

Each child in a group must use the puzzle pieces to complete a 5-inch square consisting of only three pieces.

Procedure

1. Place children into groups of five and have them select a group leader.
2. Give each group leader one of the envelopes containing 15 puzzle pieces.
3. On signal, the group leader opens the envelope and randomly passes three puzzle pieces to each group member.
4. Direct the students to examine their puzzle pieces and try to make a square from them. Signal the students to begin. Most groups will complete this task within 20 minutes; therefore, allow the proper amount of working time. If the students discover their segments will not form a perfect square, they may exchange pieces with other members of their group, but only with these rules:
   a. No talking. *The game must be played in complete silence.*
   b. No eye signals, hand signals, or gestures. *Communication of any kind is discouraged.*
   c. No taking another puzzle piece from another player, unless he or she first offers it to you.

When the time is up and several puzzles have been completed, discuss questions such as, "How did you feel when you first started working with the

FIGURE 7–1
Puzzle Squares

group? Did you find it difficult to cooperate with others as you kept working? What were some of the problems that your group experienced? How could they be resolved? What feelings did you have toward the other members of your group? What made you feel that way?"

As you share additional activities designed to help students acquire important social skills through pleasurable cooperative games, you will need to keep a number of important suggestions in mind:

1. *Define the skill clearly and specifically.* Be sure the students understand their responsibilities as distinctly as possible. For example, "When I say, 'talk in quiet voices,' I mean that you will need to use your 'foot voices.' " You can then show your students a 12-inch ruler to demonstrate how far a "foot voice" should carry.

2. *Ask students to characterize the skill.* A version of the "T-chart" can effectively demonstrate what is meant by skill characterization (see Figure 7–2). To construct a T-chart, draw a horizontal bar and write the skill above it in question form. Draw a vertical line down from the middle of the bar. On one side, list student responses to the question, "What would this skill look like?"; on the other side, list their responses to the question, "What would this skill sound like?" You then model the skill until all students have a clear idea of what its correct performance looks and sounds like.

FIGURE 7–2
T-chart Characterization of a Social Skill

Skill: Getting into your group quickly and quietly	
Looks Like	**Sounds Like**
• wait for a signal to go	• silence
• gather all materials	• quiet
• stand up and push chair in smoothly	• peacefulness
• walk slowly and softly to your group area	• stillness
	• noiseless movements
• wait for all members to arrive before you start your work	• hushed voices

3. *Practice and reinforce the skill.* You cannot teach students to work cooperatively with a single lesson or experience. Growth occurs over time with meaningful practice and effective reinforcement. One teacher's effort to help her students acquire the skill of "encouraging others to contribute" follows:

Classroom Vignette

Barbara Wertz formed small groups, which were given a short task to complete cooperatively. Ms. Wertz handed out five teddy bear counters to each group member, with a different color for each. The students were directed to place a counter in a box at the center of their table every time they spoke while completing the task. When a student had "spent" all his or her counters, he or she could speak no longer. When all the counters were in the box, each student could get his or her five colored teddy bear counters back and start again. There were several frustrated students when their five counters were the only chips in the box and a need to talk arose! This scheme usually works very well and needs to be used only once or twice to get the message across (although first graders can become strongly attached to teddy bear counters). Can you imagine this technique being used for a faculty meeting? Five teddy bear counters are handed to each teacher when they come in. When they speak. . . .

If we are to prepare students for cooperative learning in the social studies classroom, we must provide them with frequent and meaningful opportunities to function as group members. Rather than lecturing students about appropriate group behaviors, they must practice the skills that will enable them to fulfill their roles in cooperative learning groups.

ORGANIZING COOPERATIVE LEARNING

Once you are convinced that your students have acquired the skills and sensitivities prerequisite to fulfilling their roles as members of cooperative learning groups, you will be able to move on to identifying procedures that are likely to make cooperative learning more effective: *determine group membership, select a cooperative learning technique,* and *choose a reward system.*

There are key advantages to careful planning in the social studies classroom.

Determine Group Membership

There is no magic recipe to help all teachers arrive at the ideal number of students for a cooperative learning group; you will need to experiment to see what works best for you. From my own experience, I have found that the larger the group, the more skilled the members must be for the group to function well. Therefore, the best advice is to think small. Initial cooperative learning groups (for primary-grade children or older children with little prior group-work experience) should be formed of pairs, or *dyads*. Primary-grade students work best in pairs because young children more easily reach agreement with only one other person. By grade 3 or 4, as the children become more skilled at working together, you might assign them heterogeneously to three-member teams. Students through middle school can function as quartets, but they need much success working in dyads and trios before they can work in groups of four. Freiberg and Driscoll (1992) have examined the common group sizes found in elementary school social studies classrooms and detail some of the benefits of each:

> *Two-person group.* This size promotes a relationship and generally ensures participation. This is a good way to begin with inexperienced "groupies" (students who have not been grouped before). In a pair, students gain experience and skill before working in a complex group arrangement.

Three-person group. This arrangement allows for a changing two-person majority. Participation is very likely because no one wants to be the odd person out. Roles in this size group can be those of speaker, listener, and observer, and learners can experience all three roles in a brief period of time. This size group is appropriate for creating descriptions, organizing data, drawing conclusions, and summarizing ideas.

Four-person group. In a four-person group, there will likely be different perspectives. This size is small enough that each member will have a chance to express himself and can be comfortable doing so. Often this size groups [*sic*] emerges as two pairs when opinions are expressed. A group of four people requires basic communication and cooperation skills, but offers ideal practice for learning group process. (p. 277)

Determining group size will mean carefully examining your students and the maturity they have developed through informal and formal group experiences. A common mistake many teachers make is to have students work in larger groups before they have the necessary skills.

In addition to group size, the actual "coming together" of cooperative learning groups begins with the careful assignment of students to teams. For most tasks, you should use heterogeneous grouping (place "high," "middle," and "low" achievers within the same learning group). A useful technique is to arrange your students ac-

Sometimes cooperative groups can be as small as two members; oftentimes dyads are the most effective beginning cooperative groups.

cording to ability level and assign numbers according to how many teams you want. Figure 7–3 illustrates one teacher's heterogeneous breakdown of students into four-member teams. Darcee is the top-ranked student in this classroom while Donald is the lowest, so they are both members of Team 1. Warren and Robin are next at both extremes, so they are together in Team 2. In like manner, the rest of the class is divided into teams.

Notice that even though the teacher carefully divided the teams on the basis of ability, minor adjustments needed to be made as each group's composition was examined more closely: "Are all the talkative (or quiet) children in one group? Have I put together children who act up? Did I balance gender, race, or ethnic factors? In order to complete the group's goal, must I add someone with a special skill (e.g., to draw an illustration)?"

In addition to the responsibilities described to this point, another major responsibility involves assigning roles to the group members. Each member must carry out her or his role successfully for the group to function effectively. Different tasks call for different roles; you will need to reassign roles as the demands of unique

FIGURE 7–3
Establishing Heterogeneous Groups

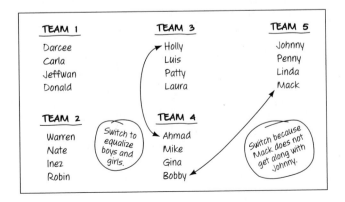

STUDENT	GROUP		STUDENT	GROUP
1. Darcee	1		11. Jeffwan	1
2. Warren	2		12. Inez	2
3. Holly	3		13. Patty	3
4. Ahmad	4		14. Gina	4
5. Johnny	5		15. Linda	5
6. Penny	5		16. Mack	5
7. Mike	4		17. Bobby	4
8. Luis	3		18. Laura	3
9. Nate	2		19. Robin	2
10. Carla	1		20. Donald	1

TEAM 1
Darcee
Carla
Jeffwan
Donald

TEAM 2
Warren
Nate
Inez
Robin

Switch to equalize boys and girls.

TEAM 3
Holly
Luis
Patty
Laura

TEAM 4
Ahmad
Mike
Gina
Bobby

TEAM 5
Johnny
Penny
Linda
Mack

Switch because Mack does not get along with Johnny.

group pursuits change from situation to situation. Some frequently used cooperative group roles include the following:

> *Group Captain* Reads the task aloud to the group. Checks to make sure everyone is listening. Makes the task as clear as possible. Coordinates the group's efforts; provides leadership.
>
> *Materials Manager* Gathers, distributes, and collects all research books and other supplies.
>
> *Recorder* Fills out forms and writes down and edits the group's report. Shares the group's result with the class.
>
> *Illustrator* Draws any pictures, graphs, charts, or figures that help communicate the group's findings.
>
> *Monitor* Keeps the group focused on the task. Makes sure each member of the group can explain the answer or information and tell why it was selected.
>
> *Coach* Sees that everyone has an equal chance to participate; offers praise and encouragement to members as they work.

Cooperative learning groups should be carefully constructed and should reflect a range of differences that make cooperative learning effective; a major purpose of cooperative teams is *social study,* which means helping students learn to work with everyone. For this reason, having students select their own groups is often unsuccessful. Self-selected groups often become homogeneous: high achievers working with other high achievers, boys working with boys, girls choosing girls, and so on. More elaborative thinking, more frequent giving and receiving of explanations, and greater perspective taking in discussing material seem to occur in heterogeneous groups, all of which increase the depth of understanding, the quality of reasoning, and the accuracy of long-term retention.

How long should the groups stay together? Cooperative learning groups vary in duration, depending on the task to be accomplished. Some groups are together only a short time—until they complete a special project, study new material, or solve a problem. Others change throughout the day (dyads, for example), especially those formed quickly when teachers ask questions that have numerous possible answers (for example, listing the possible reasons why Judaism has been able to survive and prosper over the years). Others, called *base groups,* stay together for an entire year or semester, providing a means by which students can provide a sense of support and belonging to each other.

Select a Cooperative Learning Technique

Teachers interested in using cooperative learning in their dynamic social studies classrooms can choose from among a wide variety of techniques. The following are examples of specific, tried-and-true strategies that you might find useful.

Turn-to-Your-Partner Discussions

This is one of the simplest cooperative learning techniques, most appropriately used during direct instruction. The teacher stops at natural "break" points during a lesson and directs the students to *"Turn to your partner and . . ."* (think about what makes the San Joaquin Valley a good place for farming; predict where the Mississippi River begins; tell why the Cheyenne held Medicine Dances; for example), and then give the children a minute or two to talk together to come up with their ideas. The teacher may ask a few groups to share their ideas with the class, but all groups are not required to share each time.

Numbered Heads Together

Numbered Heads Together involves a simple, four-step procedure:

1. The teacher divides the class into groups of four and assigns a number to each student: 1, 2, 3, or 4.
2. The teacher asks a question related to a topic under study. For example, while studying Medieval England, one teacher asked, "What were some of the responsibilities of the women inside a castle?"
3. The teacher tells the students to "put their heads together" and work until she gives them a signal to stop.
4. The teacher calls a number (1, 2, 3, or 4). The students with that number raise their hands, become the groups' representatives, and can be called on by the teacher to respond.

Kagan (1989) reports that positive interaction and individual accountability are both included in this *Numbered Heads Together* technique since the high achievers share answers because they know their number might not be called, and they want their team to do well. The lower achievers also listen carefully because they know their number might be called. Therefore, students work together and help one another find answers, and all students learn what is being studied.

Jigsaw

Jigsaw is a cooperative learning technique developed by Elliot Aronson and his colleagues (1978). The process starts when the teacher assigns students heterogeneously to groups of four, called their *home group.* Each home group works on content material that the teacher has broken down into sections. Each home group member is responsible for one of the sections. Members of the different home groups who are responsible for the same sections leave their home groups and meet in *expert groups* to study and discuss their specialized sections. These students then return to their original home groups and take turns sharing with their teammates about their respective sections.

To best understand the *Jigsaw* technique, it is best to describe it being used in Jack Emerson's social studies classroom. Mr. Emerson and his fifth graders use *Jigsaw* frequently in social studies. To introduce the activity, Mr. Emerson divided his students into home groups of four members each. The home groups followed this procedure for a lesson on Helen Keller:

Classroom Vignette

1. The class started out with five home groups, each consisting of four members.
2. Mr. Emerson gave each member of a home group a sheet of writing paper with a different-colored adhesive dot (red, yellow, blue, green) at the top. The dots signified membership in different expert groups.
3. Students left their home groups and went to a section of the room designated by sheets of colored construction paper that matched their dots, and joined the other members of their expert groups. Mr. Emerson had four expert groups, each focused on a different aspect of the life of Helen Keller: *Helen's childhood, Helen's early accomplishments, Helen's adult life,* and *Helen's influence on history.*
4. While in their expert groups the students studied the contents of several resources and selected important information, becoming "experts" on that part of Helen Keller's life. Finally, they summarized the information on their writing paper so it could be used to teach the members of their home groups about what they learned. The experts then returned to their home groups, and there was an expert on every section in each home group.
5. Every expert shared information on his or her part of Helen Keller's life with other members of the home group. Because students were able to learn about other parts of Helen Keller's life only by listening carefully to their teammates, they were motivated to attend to one another's work.
6. Each home group planned a special presentation to share its collective learning about Helen Keller.

Pick Your Spot

Pick Your Spot is a useful cooperative learning technique that helps students identify and declare their opinions on an issue and then explore them with like-minded classmates. Developed by Ellis and Whalen (1990), the technique has the following four steps:

1. Ask the students a question from which they are to select an answer from among alternatives: for example, "What is the one national symbol that best reminds us that we are a nation of free people?"

2. Identify "spots" in the room that offer alternatives. Word or picture cards attached to the wall can identify alternatives such as "The American Flag," "The Bald Eagle," "The Liberty Bell," "The Statue of Liberty," or "The Star Spangled Banner." Ask students to quickly pick a spot and quietly congregate there.

3. Tell students at each spot to discuss their choice to see how many good reasons they can generate for their position. Have a large sheet of chart paper available so the students can record their responses.

4. After allowing sufficient time to build a strong case, call on one student from each group to share the group's results with the rest of the class. After each group has reported, ask if any students would like to convert and pick a new spot. If they do, ask which argument persuaded them. (Including questions or challenges from other groups can expand this step.)

What determines a group's effectiveness is not who the members are, but how well they work together.

Student Teams–Achievement Divisions (STAD)

Robert Slavin (1995) developed a technique for cooperative learning that works very well with direct instruction. One of the most popular of all cooperative learning techniques used in schools today, the process starts when the teacher presents or explains the content under study. Next, the class is divided into teams of mixed ability, gender, and ethnicity. Four-member teams are ideal, but five can also work effectively. The teacher calculates a base score, or *individual learning expectation* (*ILE*), for each team member. This score is computed by averaging the student's grades or test scores from previous work. The ILE represents the student's average level of performance.

After presenting the direct instruction lesson, the teacher distributes study sheets that help students focus on important information. The study sheets can include questions, outlines, charts, hierarchies, and the like. Students work together in their teams to prepare for quizzes that are taken individually, just as in traditional classrooms. Based on quiz scores, each team member can earn from zero to three points for the group. Table 7–1 shows how comparing the student's current test score to his or her ILE score awards points. As you can see, every student has an equal chance to earn the maximum number of points for the team, and every student (not just the most able) has reason to work hard. Slavin feels that students should be judged on their own abilities, and not compared to others.

Each week the team earning the greatest number of points is declared the winner. All team accomplishments can be recognized in a newsletter; with a certificate, a group photo on a "wall of fame," a button to wear around school; or with special privileges. The teams should be changed every few weeks so all class members have a chance to work with each other. Likewise, ILE scores should be recomputed every 2 weeks by averaging grades on recent tests with the old base score.

These are among the most popular of all techniques of cooperative learning. As you search the professional literature, you will discover variations of these and many new ones. Like Gerald Livingston below, you might also someday create a cooperative learning technique you can call all your own. Mr. Livingston's cooperative learning technique for writing in social studies is outlined for you.

TABLE 7–1
Awarding Individual Points

QUIZ RESULT	POINTS EARNED
Perfect score	3 points
10 or more points above ILE score	3 points
5-9 points above ILE score	2 points
1-4 points above ILE score	1 point
Below ILE score	0 points

Classroom Vignette

1. *Bring in a number of objects related to a topic or theme of instruction.* For example, after reading about the traditions of Hanukkah during a thematic unit on Winter Holidays, Mr. Livingston brought in six objects: a menorah, a dreidel, latkes, an oil lamp, a replica Torah scroll, and a model Jewish temple. One object was displayed on each of six tables spread throughout the room. Taped to the wall near each table were two large sheets of chart paper.

2. *Divide the class into cooperative learning groups of four students each.* Mr. Livingston explained that the objects were to be used for a writing activity that would take place in stages, each lasting for about 5 minutes. One group would start out at each of the tables, brainstorming words associated with the object on the table and writing them down on one of the sheets of chart paper.

3. *Rotate each group to the next table.* Each group will now have an object different than the one it had brainstormed. The menorah group, for example, moves to the oil lamp table, the oil lamp table moves to the Torah table, and so on. The groups must write a short story about their new object on the second sheet of chart paper, using the brainstormed words of the previous group.

4. *Rotate again to the next table.* This time, each group revises the story written by the previous group—adding to, deleting, or revising ideas.

5. *Rotate once again.* Now each group edits the work of the previous groups, checking for such mechanics as spelling, punctuation, and grammar.

6. *On this fourth rotation,* students examine the edited piece and compose a final copy on a clean sheet of chart paper (leaving room for an illustration).

7. *Rotate still one more time, assigning each group the responsibility to create an illustration for the final written piece.*

8. *On the last rotation, each group should be back at its original position,* examining the story written for its object and sharing its reactions. The separate pages could be read aloud, and then bound together as a chapter in a class-generated book on Winter Holidays.

Choose a Reward System

Slavin (1984) advises teachers to provide clear incentive for children to work together, such as *rewards* (stickers, bonus points, snacks) or *recognition* ("super team," "great team," or "good team"). Under these conditions they are driven toward productivity by reinforcing one another's behavior. He notes that students' accomplishments should be recognized whenever cooperative learning groups succeed at a task—academic or social. All group members must receive a reward for the accomplishment; everyone is rewarded or no one is rewarded. For example, every group member gets 10 minutes extra recess time when group members correctly match all United States capital cities with their states. The rewards should be both enticing and inexpensive. Ways of rewarding groups include the following:

- Bonus points added to all members' scores when a team achieves an academic task.
- Nonacademic rewards such as free homework passes, stickers, erasers or pencils, or extra recess time.
- Social rewards such as smiles or verbal praise. Noticing special accomplishments becomes sufficient reward in most cases ("You improved your previous best test score by five points!").

Is cooperative learning worth the time and hard work to establish in your dynamic social studies classroom? Or, is cooperative learning a fad that will come and go like so many others in recent years? In short, research seems to indicate that the benefits of cooperative learning activities, done well, help students improve their learning and experience joy and satisfaction in assisting one another. Watching students take to cooperative learning is something like watching survivors on a lifeboat. They quickly realize that they'll either sink or swim together. They learn to be patient, less critical, and more compassionate. If they see a teammate in need, they go to his or her aid. Cooperative learning seems to be here to stay because evidence indicates it does what people say it will do. Teachers can use it with confidence in a variety of ways because the benefits of the approach are considerable.

Every successful teacher has been fortunate to discover his or her unique teaching personality.

There are no shortcuts to learning how to use cooperative learning in your social studies classroom; gaining expertise requires years of effort and long-term commitment to self-improvement. Seek help from colleagues, attend professional workshops, exchange ideas, and read widely.

AFTERWORD

Of all the instructional options available, teachers seem to prefer whole-group direct instruction to any other in social studies classrooms. Among the reasons for this situation is that (1) teachers are more familiar and more comfortable with direct instruction because this is how they themselves were taught in elementary school and high school; (2) most classrooms are crowded, so teachers find whole-class instruction more efficient than other forms of instruction; and (3) with so many subjects to teach each day, elementary school teachers have little time to plan specialized social studies instruction for small groups and individuals.

For today's classrooms and those of the future, however, teachers must subscribe to the notion that children learn best when exposed to a *variety* of instructional strategies. Teachers must offer their students a wealth of organized activity: Children may compare and contrast ideas with Venn diagrams, investigate interesting problems, or collaborate in groups with their peers. Sometimes you may choose to use direct instruction, but not always. Sometimes you will challenge students with interesting problems, but not every day. You realize that because elementary school students learn in many different ways, they must be offered a wide range of learning opportunities. In dynamic social studies, this includes whole-class instruction, cooperative group work, and opportunities for problem solving and inquiry. In effect, you must be an experimenter, trying a wide range of techniques to reach each child, while sensitive to the diversity present in every classroom.

References

Aronson, E., Blaney, N. T., Stephan, C., Sikes, J., & Snapp, M. (1978). *The jigsaw classroom.* Beverly Hills, CA: Sage.

Ellis, S. S., & Whalen, S. F. (1990). *Cooperative learning: Getting started.* New York: Scholastic.

Freiberg, H. J., & Driscoll, A. (1992). *Universal teaching strategies.* Boston: Allyn & Bacon.

Kagan, S. (1989). The structural approach to cooperative learning. *Educational Leadership, 47,* 12–15.

Slavin, R. E. (1984). Students motivating students to excel: Incentives, cooperative tasks and student achievement. *The Elementary School Journal, 85,* 53–62.

Slavin, R. E. (1995). *Cooperative learning* (2nd ed.). Boston: Allyn & Bacon.

Wasserman, S. (1989). Children working in groups? It doesn't work! *Childhood Education, 5,* 204.

8

Planning for Instruction

$\mathcal{A}$pril Warren, a student teacher impatient to spread her teaching wings, had spent her first 2 weeks in Mr. Burkhart's third-grade classroom observing the daily routine, correcting papers, taking lunch count, assisting individuals in need of special help, and getting to know the children. Finally, the day she had been longing for arrived. April's cooperating teacher assigned her to teach her first lesson— a social studies lesson! The class had been involved in a cross-curricular study of Russia, and the students were having a wonderful time discovering some of the unique aspects of Russian culture. To enrich their understandings, April's job was to prepare a lesson about Russian money. In a twinkling of an eye, her eager anticipation turned to sudden alarm. "Russian money?" April's mind shrieked silently. "Why, I don't know anything about Russian money. How does Mr. Burkhart expect me to teach something I don't know anything about? Even worse, what kind of activity can I possibly use to teach information I know nothing about?" Trying to maintain a coolness that wouldn't betray her inner panic, April swallowed hard and choked out the words, "I'd love to teach tomorrow's lesson about Russian money!"

April was jolted into a sudden realization that dynamic social studies teachers need a lot of information—much more than they can store in their minds. She learned that to be effective, she must be ready to say many times a day, "I don't know. I'll look it up." So April went back to her college library, sat down at a computer, and searched through a CD-ROM encyclopedia for the information she needed for her lesson.

When April felt comfortable with the content that would serve as the focus of her lesson, she took on the next challenge of locating a specific instructional strategy. Knowing she had only 40 minutes for the entire lesson, April needed to employ a time-efficient model. Therefore, she decided to plan a direct instruction lesson beginning with a short explanation of the Russian monetary system, followed by a game designed to reinforce the concept.

Mr. Burkhart required his student teachers to write very detailed daily plans. "With a plan in place," he advised, "you'll feel better prepared to face the students." Therefore, April went to work composing a general sketch of the content as well as the specific procedures she planned to carry out during the lesson. After an hour of jotting down and organizing ideas, a lesson finally took shape (see Figure 8–1). April needed only one more hour to make the needed game materials and gather other necessary supplies! "Who said teaching was easy?" joked April. It became increasingly apparent that the self-satisfaction and confidence received from taking on and carrying out a demanding professional challenge far outweighed the hard work April found herself contentedly involved with. "I *can* do

FIGURE 8–1
April's Lesson Plan

Theme: Russia

Grade: 3

Teacher: April Warren

Goal

The children will understand the Russian monetary system.

Objectives

1. The student will identify Russian coins and state their value.

2. The student will solve problems that require the regrouping of kopeks into rubles.

Materials

1. Duplicate for each of five groups: 1 game card; ten 1-kopek coins, ten 10-kopek coins, and one ruble.

2. One die for each group.

3. A set of four small toys such as a plastic car, box of crayons, ball, and spinning top.

Procedure

1. Divide the class into five groups of four students each.

2. Give each group an envelope containing Russian coin cards and ask them to empty the contents on their table.

3. Ask: "Does anyone know the country these coins are from?" "What are they called?" Have them examine the coin cards, calling special attention to the Cyrillic letters.

4. Explain that the coins depicted on the cards are from Russia and that a ruble is much like the dollar in the United States, a 10-kopek coin is like a dime, and a 1-kopek coin is like a penny. (There are also 2-, 3-, 5-, 15-, 20-, and 50-kopek coins and 3-, 5-, 10-, 25-, and 100-ruble bills.)

5. Invite the students to examine the money for a short period of time. Then introduce them to the Kopek/Ruble Game. In each group:

   a. Players take turns rolling the dice.

   b. Count the sum of the dice and ask for the amount in kopeks.

   c. Take the designated amount of kopeks and place them on the kopek section of the game board.

   d. When a group has 10 kopeks in the first column, it must exchange them for a 10-kopek coin and place it in the second column. When a group has ten 10-kopek coins, it exchanges them for a ruble.

   e. See which team can be first to trade ten 10-kopek coins for a ruble.

6. To bring closure, tell the students that you will show them a small toy and the price of the toy in Russian money. Each child in the group numbers off from one to four. Assign each child a toy and ask him or her to select the duplicated coins necessary to buy the assigned toy. Ask, "What are the important Russian coins and bills?"

Assessment

1. Observe the students during the game to see if they are selecting the appropriate coins and whether the coins have been correctly regrouped.

2. Check to see if each student has selected the correct Russian coins to pay for his or her toy.

3. Students write in their journals about Russian money.

Kopeks/Rubles		
100	**10**	**1**
	Must trade ten 10-kopek coins for 1 ruble	must trade ten 1-kopek coins for one 10-kopek coin

it," thought April, "It's not easy, but like *The Little Engine That Could,* I think I can do it."

"It worked!" April whooped as she returned from school the next day. "The lesson actually went well. It wasn't perfect, but it went well. I lost my train of thought once or twice, but the lesson plan kept me on track. I told the children about the Russian monetary system and they listened. We played the Kopek/Ruble game and they loved it. When they got excited and I told them to calm down, they actually did. I feel like a real teacher, like I'm ready to step in and take over my own classroom. After all the course work, and all the worries, and all the dreaming, I think I'm ready to take the next step."

Can you imagine a lawyer going to court without a legal summary (brief) of the case? How about a football coach going into the Super Bowl without a game plan? Would you allow a builder to erect your house without the help of a complete set of blueprints? I think you would agree that the results in each case would be catastrophic. Yet whenever the topic of planning for instruction comes up in methods courses, students are likely to whine, complain, gripe, and mutter their most plaintive cry: "Do we *have* to write plans?"

One of the reasons why so many pre-service teachers underestimate the value of careful planning is that a great portion of their lives has been spent as students in classrooms. From this student-oriented perspective, teaching can look so smooth and effortless that the hours of background preparation and planning is indiscernible. However, effective teaching is not a hit-or-miss process; teachers do plan ahead—formally or informally—to create an environment that sets learning in motion. Careful planning brings about such a command of content and mastery of teaching strategies that a well-prepared teacher's delivery looks instinctive. This message comes across loud and clear to skeptics who stand in front of the classroom for the first time. They are often shocked to realize that when the spotlight is on them, the classroom (which they saw previously as an orderly domain) suddenly starts spinning with chaos and pandemonium lurking at every turn.

Luckily, April Warren realized that the framework on which successful social studies teaching rests is deliberate planning. After she experienced just how much the planning process contributed to the success of her social studies lesson, active planning became a high-priority task in April's professional life. "Plans help keep you on course by serving as a roadmap or guide, plotting out the logical flow of instruction," counsels April. "A friend of mine went into a lesson stone cold and the whole thing ended up a joke. The kids were bored and confused, and he kept wandering off track. I know he learned his lesson. I'm not so naïve to think that a plan in itself guarantees success, but it does give you a good guide for instruction and serves as a record of what should be kept for the future and what should be improved on next time."

To be fair, planning can be a tiring and time-consuming process; however, all good social studies teachers plan their daily lessons. Although experienced teachers will not spend the time and energy writing out the step-by-step, formal plans that you will at this early stage of your career, they nevertheless understand the importance of planning. They think deeply about what needs to be done, even though their efforts may seem unsystematic (for example, jotting down their ideas on a note pad or in the small squares of large weekly planning books). They can record their lesson plans so briefly because their experience has helped them sense what methods and materials work best in certain situations. You await this background of firsthand experiences; therefore, you must be much more deliberate in outlining step-by-step plans.

Effective social studies teachers do not teach from day-to-day, but their planning starts by getting the big picture through laying out the entire year. They open up the textbook and examine the important topics to be covered during the year. They examine the district curriculum guide for an idea of targeted concepts and skills that need to be covered. They become familiar with professional, state, and local standards that specify what students should know, understand, and be able to do. These sources are like a steering wheel on a car; they steer teachers in the right direction and guide the planning process. The planning process itself is carried out on two different levels. On one level, we have long-range plans called unit plans. The year's social studies curriculum is broken down into a series of 2- to 3-week topical segments, or units of study. The plans detailing how those segments are to be carried out are called unit plans. Lesson plans detail how the daily instructional experiences will be carried out. Unit plans describe long-term preparations; lesson plans are short-term instructional designs.

PLANNING UNITS
OF INSTRUCTION

The unit plan is a traditionally popular technique for planning long-term social studies instruction. *Unit plans* are extended frameworks for instruction created by teachers around a central idea; they contain an orderly sequence of lessons that provide a sense of cohesiveness, or unity, to social studies instruction. The actual format of unit plans varies from teacher to teacher; for some it could be merely a chapter from the textbook supplemented with a few additional activities jotted down in the margins of the teacher's guide, but for others it may be a detailed written outline for instruction incorporating content and processes from across subject lines (from math, science, literature, creative writing, music, art, and others). For many teachers, integrating other subjects into social studies units makes sense; it is an idea that centers education in a true "child's world." Literature-based thematic units have become especially popular in the past few years, in which instruction is centered on a large collection of books all focused on the central unit topic.

A five-step format provides the structure upon which most unit plans are built:

1. Select a topic for study.
2. Formulate goals and objectives.
3. Organize the content.
4. Select a rich variety of learning experiences.
5. Assess the degree to which the goals and objectives have been met.

Selecting the Topic

Although it may appear deceptively simple on the surface, topic selection is one of the most delicate of all responsibilities associated with planning a social studies unit. Among the many factors that make it so are a school district's written curriculum and the multifarious standards movement. The amount of conformity required for standards-driven curricula varies from district to district, so it is possible to encounter any number of freedoms or constraints. Consider the following examples:

• School district A has developed a district-wide, textbook-based social studies curriculum in its effort to meet state standards. It expects every teacher to follow the curriculum so closely that all teachers must be on the same page of the book at the same date; at times a subject area supervisor may check plan books to see that this is being done. There is very little room for individual planning; the district feels that the textbook alone most effectively ensures achievement of state standards.

• School district B supplies a comprehensive curriculum guide and a set of textbooks for each teacher. Teachers are permitted to extend and enrich the specific topics as long as the basic subject matter and sequence of topics do not change. Conformity to specific content among all teachers is considered the most effective way to achieve state standards.

• School district C furnishes a curriculum guide and textbook program, but considers these more as guides for attaining standards than as "the final word." Standards guide the curriculum, but the formality associated with a text-based program is missing. Topics and instructional strategies are modified as growing interests and needs dictate. This is a middle-of-the-road approach, moving away from textbook-dominated instruction, although not yet completely.

• School district D has developed a philosophy that the social studies curriculum should be planned by the teacher in response to state standards, but content is guided by student interests and backgrounds. Based on the standards, teachers formulate their own goals and objectives, learning activities, and assessment strategies. However, a great deal of emphasis is placed on integrating the curriculum and on the use of a variety of instructional resources, including individualized instruction, learning centers, self-paced materials, learning packets, and specialized research projects.

In other words, you will find yourself fortunate enough in many school districts to have the freedom to select a unit topic, but how do you go about selecting it?

The primary consideration is that it be of high interest to both you and your students. What fascinates you? What trips have you taken? What are you enthusiastic about? What do you enjoy reading about or watching on television? Do you have a rich knowledge of a particular culture? Likewise, what are your students excited about? What are some of their favorite stories, movies, television shows, and things to talk about? What questions do they ask? What are their cultural backgrounds? In addition to these major considerations, make sure that the topic is rich in literature, both fiction and nonfiction, and that it has natural links to other areas of the curriculum.

A lot of a teacher's energy and time go into selecting a topic, much more than first meets the eye. Nonetheless, the value of a personalized social studies program is obvious: It is chosen with *your* children in mind. And who knows *your* children better—you or the developers of a textbook series or curriculum guide?

Take Michael Bowler, for example. He developed a fifth-grade thematic unit, *Customs and Traditions of Japan,* for five good reasons: (1) "Japan" is a mandatory topic of study in his school district's fifth-grade curriculum; (2) the textbook treatment of Japan is quite limited and needed to be enriched; (3) he took a trip to Japan last summer and was fascinated by the land and its people; (4) he had collected assorted souvenirs, photos, videotapes, and slides during his stay; and (5) his students were enthusiastic learners and very interested in studying a variety of cultures from around the world.

One helpful way to think of actual unit construction is to imagine yourself as a "tour guide" and your students as "passengers." You are about to escort this group as they embark on a most thrilling journey—a delightful excursion to exciting new worlds. Choosing a unit topic for a thematic unit answers the question, "Where are we going?" It provides you and your "passengers" with a general destination—"We're going to Japan!"

Formulating Goals and Objectives

A tour guide and his or her passengers cannot expect to embark on a pleasurable journey without having a deeper understanding of the trip's intent: "So, we're going on a trip to Japan? Great! But, what are we going to do when we get there?" Likewise, without careful long-term planning, you run the danger of flitting about from one activity to another without getting much out of any, or of spending so much time on one or two things that you run out of interest, time, and energy for the others. Listing a series of learning goals and objectives helps you pinpoint what you want to achieve from the unit, and helps prevent the instructional sequence from wandering aimlessly or from becoming stalled in one place too long.

Goals

Goals, first of all, are broad, general statements of intended educational outcomes; in fact, they are such all-embracing statements that most will not be achieved during the unit or even during the school year. Rather, students work toward accomplishing goals throughout all 12 years of schooling, and maybe throughout their entire lives. An appropriate goal statement, for example, is, *"The students shall acquire understandings about the human experience from the past."* An inappropriate goal statement is, *"The students shall understand the positions of the Republicans and the Federalists on the French Revolution."* Can you see that the second statement defines *specific* understandings while the first is broader and more sweeping? Words that are commonly used in goal statements include terms such as *understand, know, appreciate, synthesize, value, create, evaluate,* or *apply.* Because goals are so comprehensive, it is best to limit the number for any unit of study to three to five.

For the sake of illustration, we will look at the goals Michael Bowler wrote for the expanding sample social studies unit, *Customs and Traditions of Japan:*

Classroom Vignette

1. The students shall understand that the culture in which a person lives influences thoughts, values, and actions.
2. The students shall understand that human beings throughout the world are more alike than different.
3. The students shall value the diversity of cultures around the world.
4. The students shall understand how geography affects the lives of the people in different locations.

Objectives

The goals you have just read provide a general idea of what we might do with the unit topic, but they are much too general for us to attach much unit-specific meaning. To extend our trip analogy, the tour director might inform us that we will be visiting certain important people and places in Japan, but does not tell us specifically whom we will meet or what we can expect to see or do at the planned locations. Telling us that we will be visiting an elementary school in the heart of Hiroshima is helpful, but we are still confused about what we can expect to do or see at this special place. Likewise, our unit planning responsibility now shifts to making known the specific learning outcomes expected with clearly stated, detailed statements.

Writing objectives helps us inform. Objectives are statements that target the specific outcomes students will accomplish as a result of experiencing the unit of instruction. There are many recommended formats for writing unit objectives, but Norman Gronlund (1991) developed a style that seems particularly useful for our purposes. Gronlund's approach, or adaptations of it, is perhaps the most popular among curriculum writers and teachers today. In this text, I have chosen to use Gronlund's approach, which starts with writing a general goal and then clearly breaking it down into its more specific objectives. To illustrate, one of the goals for Michael Bowler's expanding sample unit, *Customs and Traditions of Japan,* is more precisely detailed below:

Goal

The students shall understand how geography affects the lives of the people in different regions around the world.

Specific Objectives

1. The students shall locate Japan on a map and on a globe.
2. The students shall identify Japan's major landforms, such as islands, mountains, and major bodies of water, from a map.
3. The students shall describe the seasons, climate, and weather of Japan.
4. The students shall identify the major cities and regions of Japan from a map.
5. The students shall compare the size of Japan and their state.
6. The students shall explain how living on a Pacific island affects the way the people of Japan design homes, use the land, build cities, deal with environmental phenomena such as earthquakes and tsunami, and use natural resources.

It is beyond the scope of this book to break down the remaining general goals in Mr. Bowler's unit plan, but the same process would be used to clearly detail each. Although this process is time-consuming, it is important to begin your unit plan with carefully selected and clearly articulated educational goals and objectives. These statements pinpoint what students should be able to know or do as a result of taking part in the unit and also serve as handy guides for addressing the planning responsibilities that follow—organizing the content, selecting instructional activities and materials, and assessing the effectiveness of instruction.

An Objectives–Professional Standards Connection

It is important you realize that goals and objectives are a vital part of the "real world" of teaching today. Their use is certainly an inescapable part of the educational scene. Educational goals were highlighted on the national scene when President Clinton signed *Goals 2000: Educate America Act* (see *National Education Goals Report,* 1995). The document listed such broad goals as, "At the fourth, eighth, and twelfth grades, students will demonstrate competency in . . . history, geography, civics, and government; they will be prepared for responsible citizenship and productive employment."

The public stood up and demanded a more rigorous curriculum in response to these goals, so professional groups met to develop curriculum standards that would identify precisely what students will be taught, how they will be taught, and in what ways their learning will be assessed. Several disciplines, boosted by the endorsement of the federal government, developed statements of desired outcomes; the social studies standards emerged in 1994 (see Chapter 1). If you examine the social studies standards carefully, you will note that the way they are stated corresponds quite closely with Gronlund's format. Each of the 10 general standards functions much as a general objective. For each standard, a set of precise student performance expectations is specified. These performance expectations correspond very closely to Gronlund's objectives. To illustrate, take a close look at the following excerpt taken from the *Curriculum Standards for Social Studies* (National Council for the Social Studies, 1994):

Standard III

Social studies programs should include experiences that provide for the study of *people, places,* and *environments* so that the learner can:

Early Grades
a. construct and use mental maps of locales, regions, and the world that demonstrate understanding of relative location, direction, size, and shape;
b. interpret, use, and distinguish various representations of the earth, such as maps, globes, and photographs;
c. use appropriate resources, and geographic tools such as atlases, data bases, grid systems, charts, graphs, and maps to generate, manipulate, and interpret information. . . . (p. 35)

Therefore, in one way or another, you will confront general as well as specific statements describing what the public or the profession view as the end targets of education. If you do not encounter them as goals and objectives while writing unit plans, you will certainly become acquainted with them in the context of *curriculum standards* that are used to guide curriculum planning.

Organizing the Content

After stating the goals and objectives (or citing appropriate standards), the process of building a unit of instruction continues with a thorough search of the content required to help accomplish the goals and objectives. For that reason, the requirements of this stage cannot be fulfilled adequately unless you are tuned in to subject matter. You must be broadly educated, well informed, and have a powerful drive to learn. When you teach, you teach with all you know and can find out about a topic. To be sure that your students will construct meaningful and accurate concepts, you must have much information at your fingertips. Each concept related to a topic under study has its own body of facts; each has a precise set of data used by social science scholars. You must respect this knowledge and use it to support and deepen children's learning.

Factstorming

A useful process to use while selecting and organizing the unit content is called *factstorming*. Virtually any theme can be factstormed by a group or an individual. All that is required is to create a web of potential content for the study (see Figure 8–2).

FIGURE 8–2
Factstorming Web for Social Studies Unit

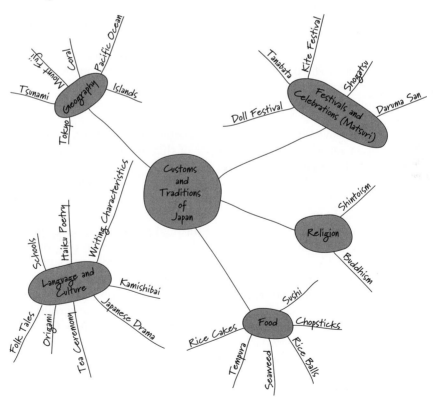

First, write the topic selected for your unit in the middle of a large sheet of paper and draw a circle around it. Then, to extract what you already know, begin by asking questions about the topic (for example, "What would it be like to go to school in Japan?"). Write your responses on the paper, and draw lines to connect them to the topic. Repeat the process with each potential subtopic. The web is only the beginning. Most teachers do not know enough at this beginning stage of the content selection process to single out all the information needed to carry out the unit, but are only brainstorming "idea starters," or representative examples to help judge whether the chosen topic has possibilities. If the topic appears to be either too overwhelming or too narrow in focus, it should be adjusted or discarded altogether. If the topic seems appropriate, you will need to move on and expand and enrich the web.

Expanding the Content

Should you decide to continue with the topic, the next step is to expand the content. As a teacher of elementary school social studies, you must have much information at your fingertips, and must also have access to resources to supplement that knowledge (the computer and a wide variety of print sources are the places to start). Check with your school or public librarian for references on a topic you wish to pursue. Constantly search for information to bolster what you already know; you should find yourself saying, "I'll look it up," many times a day. Books and the computer will provide most of the needed information, but you will also want additional resources. Consult specialists either in person or by telephone, visit museums and other sites, view films or filmstrips, listen to audiotapes, and seek other opportunities to broaden your background. You will need to spend a great deal of time uncovering and organizing information: What is the key information necessary for a satisfactory understanding of this topic? What are the concepts, skills, and big ideas that might emerge?

The background material you uncover should be substantial. You will not use all of it, but when you teach elementary school children it's best to learn all you can find out about a subject (often learning more than you can ever expect to cover in the unit). As they progress through the unit, children will ask many questions requiring information on specific points. It certainly is okay for a teacher to reply, "I don't know. Let's find out together," to children's questions each day, but the teacher who can answer a great many questions is the one who gains his or her students' confidence and respect. Study the key facts related to the topic, and think about the concepts that should emerge. This important background material should be outlined at the beginning of your unit. The content outline is usually preceded by a short summary paragraph, like the one below that will be used in Michael Bowler's expanding sample unit, *Customs and Traditions of Japan:*

Classroom Vignette

Summary Paragraph

Japan is an island nation located off the coast of China about 5,000 miles from the United States. It consists of four major islands and many small islands. Of the major islands, the largest is Honshu, the furthest north is Hokkaido, and the remaining two are Shikoku and Kyushu. Tokyo is Japan's largest city. There are many mountains sprinkled with thick green forests in Japan; Mount Fuji is the highest. Some of the mountains, including Mount Fuji, are inactive volcanoes.

Japan is a very old country, with customs and traditions that go back thousands of years. Japanese customs, traditions, and family life will be the focus of this unit, with added emphasis on its climate and geographic features.

Following the drafting of a summary paragraph, you will normally prepare an outline of the content or display the completed web. For the content outline, it helps to record the primary concepts that will be highlighted throughout the unit and then list the key information below each. The example below displays the content outline for two concepts central to Michael Bowler's expanding unit, *Customs and Traditions of Japan: geophysical features* and *traditional clothing.* If space permitted, this outline could continue for each targeted concept: possibilities include *festivals, typhoons and earthquakes, school, meals, music, art, religion, pets or animals, games, industry, fishing, language,* and so on.

Classroom Vignette

Content Outline

Concept: Geophysical Features
- Japan consists of nearly 3,000 islands.
- Japan has four major islands: Honshu (the largest), Hokkaido, Shikoku, and Kyushu.
- If all the islands were put together, they would constitute an area about the size of California.

- Japan is located 5,000 miles from the United States, off the coast of China.
- Tokyo is the capital of Japan and is its largest city.
- Japan's climate is much like that of the Middle Atlantic States.
- The tallest mountain is Mount Fuji, which is nearly 12,500 feet high. It is an inactive volcano.
- Many other high mountains make up the island; several are inactive volcanoes.
- Thick green forests cover the mountains.
- In the summer, Japan may experience *typhoons,* very strong storms with high winds and heavy rains.
- Japan has a great deal of rain during the *tyusu* (rainy season) that comes in early summer.
- Rain is good for the rice crops, a favorite food in Japan.

Concept: Traditional Clothing
- Contemporary Japanese wear clothes that are just like the clothing worn in the United States.
- Japanese people traditionally wore layered robes called *kimonos.* Today some do so only on special occasions.
- A long piece of cloth is tied around the waist of the kimono. It is called an *obi.*
- Japanese children may wear brightly colored kimonos. Women's silk kimonos and obis may be blue, red, or green, often made with colorful designs. Men's kimonos are not as fancy. The cloth is usually cotton and the color is brown, black, or gray.
- Japanese wear shoes, but some like wearing *getas,* slippers with high wooden soles and a strap going between the big toe and the next toe. *Zori* are similar, but often have straw soles.
- Geta and zori are worn over *tabi,* or split-toe cotton slippers that allow the straps to slip between the first and second toe.
- Many people of Japan have fancy silk umbrellas called *kasa* because it rains so much in Japan.
- Some children in Japan wear uniforms to school. Each school has a different uniform.

Selecting the Learning Experiences

Once teachers target and outline the content, they face the next big challenge: "How can my students most effectively learn this content?" Learning experiences are the instructional activities we employ to achieve the content goals and objectives, the actual experiences that involve children in puzzling, wondering, exploring, experimenting, finding out, and thinking. Not all learning experiences need to be new or unique; highest priority should be placed on balance and variety. Do not choose activities because they are "cute" or gimmicky; select them because they stand the best chance of accomplishing your goals and objectives.

This is often the section of a unit plan that creates the most confusion in the minds of pre-service teachers. Some instructional models (called *direct instruction*—see Chapter 5) require the teacher to guide all students through the same experiences at the same time. Other models (called *problem solving/inquiry*—see Chapter 6) invite students to actively pursue individual interests. Which model works best? That is a question that has baffled educators since the onset of formal education, and one that has not yet been convincingly answered. After examining the pros and cons of the models, however, Jarolimek and Foster (1997) offered some perceptive advice on the issue: "The act of teaching is so complex that it is nearly impossible to demonstrate that a specific way of teaching is superior to other ways for all purposes, with all teachers, with all children, for all times, and in all circumstances. . . . We are forced to conclude that there are many good ways to teach" (p. 146).

The nature of dynamic social studies calls for flexible, adaptable teachers—informed professionals who eagerly anticipate the challenge of selecting instructional methods and materials that are most appropriate for any specified learning outcome. However, responsive teachers realize that instructional decisions are highly complex and make decisions only after thoughtfully considering a number of factors. Orlich and his associates (1990) identify these factors as: (1) the teacher, (2) the students, (3) the goals and objectives, and (4) the physical environment.

Every *teacher* has a unique set of personal and professional strengths that serve as a firm basis of instruction. It takes some time to discover how these strengths can be transformed into an individual's "teaching personality," however, so new teachers often benefit by beginning their professional careers using direct instruction strategies. In this mode, classroom management challenges are minimized because children are not placed into potentially disruptive situations. This does not mean, however, that you should shy away from student-centered strategies; once you have developed a sense of confidence in the classroom, it is important to try new things in an effort to continually expand your teaching repertoire.

The mode of instruction must also match the *students'* experiences and backgrounds. Just as every teacher's special combination of likes, dislikes, abilities, strengths, and needs contributes to a teaching personality, so do student preferences and unique skills influence "student personality." Some students interact freely with peers while others are shy and withdrawn. Some students prefer to be "turned loose" to attack a problem of personal interest while others would rather sit and listen to an interesting lecture by the teacher. Some students require direct help with specific subskills before attempting a certain task while others readily achieve success if left to their own devices. The effective social studies teacher uses key information about students to make appropriate decisions related to classroom instruction.

Additionally, all procedures must have a purpose; these purposes, the desired outcomes of the learning experience, are specified in the *goals and objectives* portion of the lesson plan. Obviously, procedures must tie in to the goals and objectives of the lesson, so I like to encourage students to make this connection by using the logic required to complete *if-then statements.* For example:

- *If* we want students to understand how to use latitude and longitude to locate places, *then* the teacher would direct the students through the process with the aid of a large classroom globe.
- *If* we want students to investigate their community's past as a historian would, *then* they could search town records to create a "living history" of their community.

The final consideration influencing the choice of learning experience is the *physical environment.* Because instruction depends on where the learning will take place, you must ask the question, "Is the environment the best one to ensure learning with the procedure I've chosen?" A classroom in which student desks are arranged in rows will not enhance cooperative learning; a school located near a large museum should schedule at least one field trip there. Both in-class and out-of-class environments must accommodate the selected instructional strategies and activities.

You may be surprised to know that not all contemporary teaching methods are original; most are adaptations or reworked versions of older, more traditional methods. The crux of successful teaching lies not in whether a method is new or old, but in flexibility; the younger the children, the greater the variety should be:

- Sometimes you will lecture to your children, but rarely; the younger the child, the less you should talk.
- Sometimes you will use good questioning and discussion techniques to hold discussion sessions with the entire class.
- Sometimes you will lead whole-group learning experiences from the textbook or another common source of information; such experiences offer balance and proportion to the program.
- Sometimes you will bring to class objects such as jewelry, clothing, dolls, toys, books, catalogs, containers, tools, and other realia. These items help to form classroom connections to other people, times, and places.
- Sometimes you will choose books, computer applications, videos, slides, tapes, filmstrips, records, pictures, bulletin boards, and other learning aids to provide a variety of essential learning materials.
- Sometimes you will encourage children to solve problems and search for answers to their own questions; an independent quest for information is a lifelong asset.
- Sometimes children will work alone; meeting personal interests and needs must assume high priority in all classrooms. At other times you will

encourage children to work together cooperatively; children learn a great deal from one another.

Phases of Planning the Learning Experiences

There are many ways to structure the learning experiences for the unit; the organized outline that follows is only a suggested guide. It should be adapted or restyled to suit your personal preferences and the needs of your students.

Phase One: Introductory Activities

Establishing classroom conditions that motivate children to learn is a major mission during this initial portion of the unit. What kind of "bait" can you cast out to "hook" students on the topic so they will participate as eager and involved learners? Usually encompassing the first day or two of instruction, the "hook" for a social studies unit can be cast out in a number of ways.

There are many good introductory experiences (field trips, community visitors, videos, realia, and the like); however, in my opinion, one of the most important ingredients is a good book. I call books that are used during the introductory phases of social studies units *literature launchers*. Their purpose is to sweep up the students and carry them off into captivating, enchanting worlds of curiosity. As powerful as they are for this purpose, however, good books should not be limited only as the introductory experiences. They are powerful instructional tools when used at strategic points throughout the unit, also. A good unit is rich in literature, both fiction and nonfiction.

Phase Two: Developmental Experiences

Following the introductory activities, you enter into what is often described as the "brass tacks" of the unit. Activities may be done independently, in small groups, or by the whole class, but what really matters is that you stay within the periphery of the children's interests. You must choose worthwhile experiences that are neither mere entertainment nor busywork, but are rich and powerful in potential learning (see Chapter 10). You must be constantly ready and eager to reach out and seize the most gripping experiences available. As you set this phase in motion, the students go to work; you may retreat a bit, since your initial teaching responsibilities as a stimulator and arranger are now over. You will have more teaching to do, but in other ways; you must now analyze the unit objectives and ask, "What learning experiences will most effectively help me achieve this objective?"

The developmental experiences, then, offer students opportunities to wrestle with real problems. The whole point is to help them comprehend the interrelationships among phenomena that can make this nation and the world better places to live. We do this not only from the social studies alone, but by utilizing experiences that cut across all subject lines.

Phase Three: Culminating Activity

While the preceding phases of the unit were content or process specific, the culminating activity allows students to review, summarize, or bring closure to the topic. This concluding portion of the unit usually takes the form of a whole-class project that gives students an opportunity to apply or extend what they have learned. The culmination might be a time during which group projects are shared; a festival where dance performances, creative skits, and cultural meals are enjoyed; a "readers forum" where written reports are read; or a construction project, such as a model community, where children represent the major concepts learned.

Sometimes the unit that follows is such a natural transition that neither a culminating activity for the first unit nor an introductory experience for the second is necessary. If one unit deals with "The First Americans" (Native Americans), for example, and is followed by "Settling the Land" (early settlers), continuity from one unit to the next need not be broken.

You must carefully orchestrate the three major phases of activities, because each activity flows from preceding experiences and furnishes the foundation for those that follow. Think of the entire collection of activities as being much like a beautiful symphony that is made up of separate movements which, when woven together, shape a grand masterpiece.

Thematic Blueprints and Daily Lesson Plans

Teachers usually describe the specific learning activities they have selected for the unit using either (or a combination) of two formats: *unit blueprints* and *daily lesson plans*. A *unit blueprint* gives brief descriptions, in paragraph form, of all the daily activities a teacher intends to provide throughout the unit. *Daily lesson plans*, by contrast, are much more detailed; they expand on the blueprint's brief description by clearly and comprehensively detailing how things will be done. The material below describes Michael Bowler's blueprint for his expanding unit on Japan. Any of the descriptive paragraphs can be easily expanded into a lesson plan (a process that will be detailed in the next section).

Classroom Vignette

Phase One: Introductory Activities

1. Prepare a "cultural box" with items that reflect many aspects of both Japanese tradition and contemporary life. Obtain items from parents, friends who travel, restaurants, gift shops, or local import stores. Wrap the box in brown wrapping paper and address it to the class with a return address from Japan. Add a stamp (if possible, an authentic Japanese stamp) and a postmark. Ask a colleague or

staff member to deliver the box to your classroom at a designated time. Looking surprised, gather the students in your group conversation area and say, "What do you suppose is in this special delivery package? Where did it come from?" Examine the box together for clues. Open the box and use these questions as you share each item: "What is it? How do you think it is used? What do we have in our culture that compares to this?" Lastly, pull out a manila envelope containing a round-trip airline "ticket" to Japan for each child. Challenge students to find Japan on a large world map. Ask them to name the ways people can travel to Japan, tracing the various routes.

Have the students prepare for their trip by completing "passports" (samples are available at local post offices).

2. Prepare for an airplane flight to Japan by distributing 18 × 24" sheets of white construction paper. Have students cut out large red circles from construction paper and paste them to the center of the white paper, modeling the Japanese flag. Keeping one long side open, staple or glue an 18 × 24" sheet of red construction paper to the white. Attach handles on the top to look like the handles of a suitcase. Have the students make travel journals that they will fill out to summarize their travels through Japan. The suitcases make handy containers for their journals, as well as an excellent portfolio case where students can store work samples throughout the unit.

3. Arrange the students' chairs in sections of three to resemble the interior of an airplane. Process the students through "customs," checking their passports. After everyone is safely on board, the flight attendant (you, the teacher) offers a greeting and explains critical pre-flight and in-flight information. Play a sound-effects tape or CD of a jetliner taking off. As the simulated flight levels off, explain how long the flight will take and that an in-flight video (*Rim of Fire*) will be shown. Darken the room and start the VCR. After watching the tape, invite the "families" to fill in the first page of their journals describing their pre-flight and in-flight experiences.

Phase Two: Developmental Experiences

1. A Japanese *kamishibai* storyteller (you, the teacher, in costume) is the first person the children meet. Kamishibai, a unique form of Japanese street storytelling, is an authentic folktale strategy. The kamishibai storyteller, who was also a candy seller, rode from neighborhood to neighborhood on a bicycle equipped with a small stage for showing story cards. He entered a neighborhood, striking together two wooden clappers, and everyone knew it was story time. The children ran from their homes so they could buy candy—those who did were able to sit closest to the stage. The storyteller illustrated his stories with 12 to 16 story cards, on the back of which were written the corresponding parts of the story. He was sure not to read all the cards during one visit, for he wanted to capture enough intrigue by reading only three or four in order to entice the children to

come back again. The teacher will use a number of illustrated kamishibai cards to tell a traditional folktale, *The Tongue-Cut Sparrow.* (Authentic kamishibai cards can be ordered from: Kamishibai for Kids, P.O. Box 20069, Park West Station, New York, NY 10025-1510; telephone and fax, (212)662-5836.)

2. Read *Tree of Cranes* by Allen Say. It is a tender story of a young Japanese boy that helps bridge Japanese and American cultures. A large part of the plot has to do with paper cranes symbolizing love, peace, and sharing. Discuss and chart the features of the book that highlight the traditional way of life in Japan. Afterward, the children will enjoy learning how to make origami cranes from folded paper (this is a centuries-old art form in Japan).

3. Simulate a day in the lives of Japanese schoolchildren. Pretend to be a teacher (*sensei*) who will be teaching the children some basic school-related information, such as Japanese writing characters and number characters. Detailed lesson plans for this learning experience follow.

4. Folktales help illustrate Japanese beliefs and values. This activity uses a Japanese folktale to depict the extended family household, common in Japan even today. Read *The Boy of the Three-Year-Nap* by Diane Snyder (Houghton Mifflin). Discuss the ways decisions were made in the extended family. Then, show the children an outline of a large Japanese house drawn on a piece of chart paper. Have the students draw, inside the house, all the family members and extended family members who would live there. Each student should write a sentence or two about who these members are.

5. Read the book *How My Parents Learned to Eat* by Ina R. Freedman. Discuss the contrasting eating habits of Japanese and American culture. Give each child a set of chopsticks (readily available from Japanese or Chinese restaurants). Show the children how to use the chopsticks: hold one like a pencil; slip the other chopstick under the first one, leaning it against your third finger; move the second chopstick up and down to pick up crumpled pieces of paper. Have the students practice for awhile. Set up a model Japanese restaurant with a low table, pillows for sitting, menus, kimonos for waitstaff, and pre-prepared, simple Japanese food such as kiwi fruit, vegetable tempura, rice balls, or rice cakes. Show students pictures of food normally enjoyed in Japan as well as pictures of authentic Japanese restaurants. Read the book *Everybody Cooks Rice* by Norah Dooley (Carolrhoda Books).

6. Students will enjoy learning about Japanese holidays. On the fifth day of May, for example, Children's Day (*Kodomo-Hi*) is celebrated. On this day a tall bamboo pole is erected in front of homes. At the top of the poles fly brightly colored cloth or paper streamers in the shape of carp (because of the carp's energy and strength). One carp flies for each boy in the family with the largest one for the eldest and the others ranging down in size to the youngest. Although the carp streamers fly only for the boys, the real purpose of the holiday is to teach all children the importance of being good citizens. Make tagboard fish patterns for the children to trace around. Show the children how to make a carp streamer: trace around the pattern onto any of a variety of colors of construction paper; decorate one side with paint, sequins, glitter, tissue paper streamers from the tail, beads, wiggle eyes,

and so on. Wait until this side dries, then flip the carp over and repeat. Hang the carps from a pole outside the classroom or mount them inside.

7. Plan a field trip or invite guest speakers into the classroom:

Visit a martial arts studio or ask someone to give a classroom demonstration.

Visit a local gardening shop to examine bonsai trees, or invite someone to come to school to talk about bonsai trees.

Invite someone of Japanese heritage or someone who has traveled to or lived in Japan to speak about the culture.

Phase Three: Culminating Activities

Michael Bowler chose to celebrate the customs and traditions of Japan by engaging the children in the following culminating activities:

Display the journals summarizing each family's tour through Japan.

Have the students act as *kamishibai* as they share pictures and related information about various customs and traditions of Japan.

Unit blueprints describe the general flow of the unit activities; the lesson plans shed light on the explicit way those activities will be carried out. The following examples illustrate two specific plans Michael Bowler designed to carry out experiences related to a visit to a Japanese classroom. Notice how these lessons fit together. Lessons should not be thought of as separate daily experiences unrelated to each other, but as related episodes that flow together smoothly to form a seamless sequence of classroom events.

Classroom Vignette

Lesson Plan A

Topic: Customs and Traditions of Japan
Grade: 5
Teacher: Michael Bowler

Goal
The students shall understand what it is like to grow up in other cultures.

Objectives
1. The students shall demonstrate how Japanese students greet their teacher.

2. The students shall reproduce the Japanese ideographs representing the numbers one to four.
3. The students shall learn to count from one to four in Japanese.

Materials

1. A chart depicting the Japanese ideographs representing the numbers one to four.
2. One thick-bristled paintbrush per child.
3. Black tempera paint.
4. One sheet of manila drawing paper per child.
5. Photos of children in Japanese elementary schools.
6. Chart displaying rules of a Japanese classroom.

Procedure

1. Show photos of children in Japanese elementary schools. Discuss the children's observations.
2. Inform the students that they will be simulating a visit to a Japanese elementary school. Begin by instructing them to stand behind their chairs; demonstrate how the children bow to the teacher (*sensei*) and offer the greeting, "O-hayo gozaimasu," which means, "Good morning." Practice this routine several times and then leave the room. Upon return, the children should stand and extend the greeting.
3. Read and discuss the rules of a Japanese classroom (displayed on a class chart):
   - When the teacher enters, rise and bow. Greet the teacher in unison.
   - Come to school prepared with pencils, erasers, pens, paper, and textbooks.
   - Raise your hand to answer questions. Stand before answering.
   - Be quiet during class.
   - Keep to the right in the hall. No running.
   - Help clean the classroom, hall, and yard. Pick up trash, sweep the floors and yard, and stack the chairs before anyone leaves.
4. Discuss how these rules are like or unlike their classroom rules.
5. Bring the students' attention to the display where Japanese numerals from one through four are illustrated. Ask, "What do you see?"
6. Inform the children that they are going to learn to count in Japanese from one to four. Pronounce each number name and encourage the children to recite them.

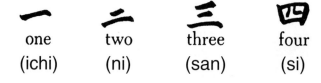

one	two	three	four
(ichi)	(ni)	(san)	(si)

7. Give each child one container of black tempera paint. Show the children how to lightly tear the edges of their drawing paper to resemble handmade rice paper. Demonstrate the *suni-e* style of painting: Hold the brush between the thumb and fingertips in a vertical position and move the arm to make the basic two strokes. To create *gung* (wide strokes), apply more pressure; for *shey* (finer strokes), slant the brush slightly and ease up on the pressure. Have students write the numerals on their drawing paper.
8. Arrange the students' ideographs in a classroom display.
9. Ask the students to think about how their school is similar to a Japanese school and how it is different. Complete a Venn diagram to help organize their responses, recording the unique features of Japanese schools in one circle, the unique features of their school in the other, and features common to both where the circles intersect.

Assessment

1. Observe students during the simulated classroom activity to judge whether they have made progress toward the desired behaviors.
2. Examine the ideographs to check the accuracy of the kanji figures.
3. Determine whether student journal entries accurately and completely incorporate the information shared in the lesson.

Lesson Plan B

Theme: Customs and Traditions of Japan
Grade: 5
Teacher: Michael Bowler

Goal
The students shall understand what it is like to grow up in other cultures.

Specific Objectives
The students shall create a traditional Daruma-san doll as a representation of a traditional Japanese custom.

Materials

1. Souvenir Daruma-san dolls from Japan.
2. One blown egg with an enlarged hole for each child.
3. One small lead fishing sinker for each child.
4. Masking tape.
5. Papier-mâché strips.
6. Paper towels.
7. Various colors of tempera paint.

Procedure

1. Review the previous lesson by having the students greet the teacher in traditional Japanese style.

2. Review the Japanese kanji figures from the previous day. Inform the students that they will experience a sample craft lesson today; they will be making a toy that has been appreciated for generations in Japan. Discuss some of their favorite toys. Are any of their toys like the ones their parents enjoyed as children?

3. Display the souvenir Daruma-san dolls. Explain that legend has it that these dolls are named for a Buddhist priest (*san* means "mister," a title of respect) who sat in a red robe with his arms and legs crossed for 9 years while contemplating serious problems. A roly-poly doll that, no matter how it is tipped, bobs back upright represents Daruma-san. Traditional Daruma-san dolls (popularly called "roly-polies" in the United States) are painted red with a face having no eyes. The dolls are given to someone who is trying something new. When that person sets a goal, one eye is painted in; when the goal is accomplished, the other eye is completed.

4. Help the students make their own Daruma-san dolls. To make the dolls, give each student a blown egg with an enlarged hole. The children put a small fishing sinker through the hole and cover the hole with masking tape. Next, they coat their eggs with four layers of overlapping papier-mâché strips. After they dry, the students can paint on a face, leaving off the eyes. The clothes could be painted in many colors, but the most traditional is red (symbolizing a red robe).

5. Encourage the students to set two goals for the remainder of the school year. After one goal is accomplished, they should add one eye. When the other goal is accomplished, the remaining eye should be colored.

6. Discuss the patience and determination it takes to reach an important goal.

Assessment

1. Check the completed dolls to see if they accurately represent the characteristics of Daruma-san dolls.
2. Read student journals to see how students summarized what they learned.

Assessing Instruction and Learning

The final section of a unit plan deals with assessment. Teachers must constantly assess their children's work in order to determine to what degree they are achieving the intended instructional outcomes. To accomplish this crucial responsibility, teachers commonly employ two overall assessment strategies: (1) *formative assessment* and (2) *summative assessment.* Formative assessment occurs during the unit and is used to identify what the students are learning; summative assessment occurs at the end of instruction to inform the teacher of the degree to which learning took place.

Because formative assessment takes place daily and is part of all lesson plans, teachers are able to quickly identify areas in need of attention and make appropriate adjustments. If students seem to be having difficulty, the content, activities, or pace of instruction can be altered; if students are breezing through the work and seem a little impatient to move on, modifications can be made. In the elementary school, formative assessment is based on such evidence as homework, seatwork, and classroom projects.

Summative assessment occurs at the end of a unit of instruction and is part of all unit plans. It is designed to summarize student achievement for grading purposes. Although most people think of testing in connection with summative assessment, social studies teachers today have many other options.

Although many alternative measures can be used to provide assessment feedback, traditional practices have focused on student test scores. One type of test, the standardized achievement test, has become a particularly influential index of instructional effectiveness. Standardized tests are those official-looking booklets purchased by school districts and administered to students to obtain a score that can be compared to the scores of students who will be taking the same test in school systems across the country. These tests are designed to measure how much the student has learned in specific content areas, including social studies. Results help teachers determine a student's overall strengths and weaknesses, and thus serve as a catalyst to determine what adjustments, if any, should be made to the instructional process.

Although they have great potential for improving the educational process, standardized tests have been widely misused. Getting kids to perform well on the test is *the* top priority in many districts.

Despite the fact that some groups are working to increase the significance of standardized testing, even to the point of developing a national examination, many professional groups are advocating a cutback in its use. In their search for alternatives to standardized tests, professional groups now endorse the use of "authentic assessment."

Authentic Formative Assessment

Proponents of authentic formative assessment suggest that children's learning must be demonstrated in situations that address three criteria: (1) students must apply knowledge they have acquired, (2) students must complete a clearly specified task within the context of either a real or simulated exercise, and (3) the task or completed product must be observed and rated in accordance with specified criteria. Instead of recalling information from rote memory, authentic assessment is based on major sources of information: real-life tasks and teachers' observation.

Authentic formative assessment places students into situations where they must do something, make something, offer solutions to problems, or compose a critical response to an issue. It assumes that there is much more to learning than being good at recalling facts; students must try to become good at personalizing the information. For this reason, the assessment exercise closely resembles the behavior targeted in the objective as the desired learning outcome. Authentic assessment can often be carried out while observing typical learning tasks—written products, demonstrations, group projects, integrated art and music activities, construction projects, dramatizations, museum displays, and so on.

Social studies educators frequently argue that the primary mission of social studies instruction (producing students who are able to think for themselves) is contradicted by standard fill-in-the-blank, multiple-choice, or true-false tests requiring specific answers. A major goal of authentic formative assessment is to better assess what social studies educators value.

Contexts for Authentic Formative Assessment

Authentic formative assessment begins by re-examining the specific lesson objectives and determining the degree to which those objectives have been met. Assessment is directly tied into the lesson objectives. For example, the way to find out if Lamont can meet a targeted outcome of reciting by heart the Pledge of Allegiance is to have him recite it. In other words, if a lesson objective is to teach children how to write a *haiku* (form of Japanese poetry) then asking them to create an example would be considered an appropriate assessment strategy. However, if you want to determine whether Belinda understands what a traditional Japanese Tea Ceremony (Chanoyu) would have been like, it may be appropriate to have her write a creative piece about the challenges of the ancient practice, such as a letter to a friend back in the United States telling about this tradition.

Much authentic formative assessment is, at least in part, similar to regular classroom activities: games, oral reports, creative dramatics, debates, art projects, or writing tasks. By gathering evaluative information from situations and contexts similar to those in which students normally learn, their assessment may be more meaningful to you, to them, and to their families.

Criteria for Authentic Formative Assessment

The second step of authentic formative assessment is determining precisely what degree of learning you consider minimally acceptable. Suppose you designed this formative assessment activity: Write a letter to a friend back in the United States describing a traditional Japanese Tea Ceremony. The goal of this assessment situation is to measure the students' understanding of the tea ceremony, so you must first establish criteria for measuring that understanding. That is, you would expect students to express in the letter ideas similar to the following:

- Guests sit upon cushions.
- To authenticate the ceremony, guests wear kimonos.

There is far more to authentic assessment than giving students grades on tests.

- Guests are served small rice cakes or sweets.
- Guests are served powdered green tea (ocha).
- Guests bow and sip the tea.

This task is important because the criteria form your basis for evaluation—your expectations may be that every student must address in their letters each of the five ideas listed. With such definitive guidance, you can be more consistent when judging the relative effectiveness of students' efforts. Criterion lists furnish you with a set of guideposts that help focus assessment on the entire response rather than on any particular components. However, because authentic formative assessment requires students to personalize the way they present the information, end results must not all look alike, as they normally do on traditional tests. This does not mean, however, that students need not include accurate facts and information as they complete an instructional task. Should a student's letter, for example, contain references to lobster tail, it would reveal doubtful links to the traditional ceremony. In contrast, references to red silk napkins, cushions, kimonos, or rice cakes would indicate a student's basic understanding of Chanoyu.

What to Do With the Results

The third step of authentic formative assessment has to do with the issue of what to do with the results—pointing to what the student must do to improve his or her subsequent learning. In contrast to testing techniques that often seem to search for deficiencies in learning, authentic assessment helps determine the students' areas of strength, as well as what must be done to improve. For example, assessment may indicate that Eugenia has difficulty keeping Japanese matsuri (holidays) in proper chronological order; therefore, to boost this skill, you could ask her to use a timeline to organize the popular holidays through the seasons.

Some contexts for authentic formative assessment, then, entail the students' actual work samples, but another major category includes *observations of students at work*. Teachers who use authentic formative assessment consistently observe students as they interact with curriculum materials, other students, and adults. They observe, but realize that their observations cannot be done in a hit-or-miss fashion. To get the most from an observational experience, teachers must know specifically what they are looking for and then organize their observations systematically. Two helpful organizational tools are checklists and rubrics.

Checklists. Because they are easy to use, many teachers find it convenient to employ checklists for observational purposes. Checklists are particularly helpful when you wish to assess the degree to which students have demonstrated specific skills, behaviors, or competencies. To determine these, you must first list the specific outcomes to be assessed, and then record the occurrence of each. A sample checklist of behaviors considered important for cooperative learning is shown in Figure 8–3.

FIGURE 8–3
Checklist of Cooperative Learning Skills

	Sometimes Present	Mastered
Assists co-workers when needed		
Follows group-established rules		
Does fair share of group work		
Respects group decisions		
Shares materials willingly		

Directions:

(√) Place a check in the "Sometimes Present" column if the characteristic is occasionally observed.

(+) Place a plus in the "Mastered" column if the characteristic occurs habitually.

( ) Make no mark if the characteristic is observed seldom or not at all.

Rubrics. Teachers often use rubrics to identify important strengths and weaknesses in student learning. Rubrics are constructed by breaking down the performance associated with each educational objective and assigning a number to indicate the quality of work. For example, to assess written reports on topics related to Japan, a teacher would list the types of behaviors associated with various qualities of work and assign it a numerical score. The following rubric may give you a sense of how this system is used to assess the written reports.

3 The written report is complete. It clearly communicates the content, provides accurate and relevant details, and allows the reader to grasp the writer's message easily.

2 The written report is partial, but is fairly well organized. Although the information selected includes mostly accurate details and ideas, some confusion is encountered from time to time.

1 The written report is sketchy, inconsistent, and incomplete. It includes many random details not organized into cohesive statements, or contains a number of irrelevant or erroneous ideas. The reader has no idea about what the writer is trying to say.

0 There is little or no written material, or it is illegible.

The skills to carry out authentic formative assessment are not easy to apply. Because related judgments are deeply personal, you must be careful to be fair. It is important to establish specific criteria and clear standards of expectation. However, like any other sophisticated professional ability, experience and effort will combine to improve your skills in this complicated process. As you find your teaching style in the social studies, you will also need to key in these new evaluative strategies used to assess student learning. Today's teachers are learning to look more and more at student performance while making decisions about educational growth and needs.

Formative assessment, then, occurs during instruction for the purpose of informing the teacher about the children's current level of performance. Formative assessment guides the teacher's planning, providing the data that leads the way to appropriate instructional adjustments. This method relies on such day-to-day information as teacher observations and demonstrated performance on specific learning tasks. Good assessment begins with formative assessment—the teacher's evaluation of what students are gaining from instruction every day. Such feedback helps teachers make informed decisions about student progress and about any adjustments that need to be made in the curriculum.

Summative Assessment

Summative assessment is carried out at the end of a unit of instruction. It is considered a summary of what the student has accomplished and, because instruction has been completed, is more useful for assessing final achievement (usually for assigning a grade) than for initiating changes in instruction.

Standardized Tests

One of the most hotly debated summative assessment strategies is called the *standardized test.* Standardized tests are commercially prepared, machine-scored, "norm-referenced" instruments. Norm-referencing means that the test is administered to a sampling of students selected according to such factors as age, gender, race, grade, or socioeconomic status. Their average scores then serve as "norm scores" and become the basis for comparing the performance of all the students who will subsequently take the test.

How do we use the results of standardized achievement tests in social studies? What do they offer that we can't get without them? Comparability in the context of the "big picture" is the major thing. It isn't very useful, for example, for one teacher to compare his or her students' results with those of a classroom next door and then to make decisions about instruction based on that comparison. To many, this process is too limited; it is more important to get a broader, more comprehensive picture: "*In general,* are my fourth graders learning basic social studies content as well as other fourth graders in our school district?" "Compared to fifth graders throughout the United States, how well are those in our school district doing in map skills?" "How does our district's social studies curriculum compare with others in the state?" Despite the fact that standardized tests can determine only a small fraction of any student's total achievement in social studies, the public assigns standardized test scores great weight: "Are our students making 'normal' progress?" "Is their overall achievement above or below average?" Shepard (1989) explains what happens when such sharp public interest becomes focused on the results of standardized tests: "When the scores have serious consequences—and they often do—teachers will teach to the test. . . . But teaching to the test cheapens instruction and undermines the authenticity of scores as measures of what children really know, because tests are imperfect proxies even for the knowledge domains nominally covered by the tests; and they also omit important learning goals beyond the boundaries of the test domain" (p. 5).

One major concern about standardized testing, then, is that teachers will overreact to the tested areas at the expense of the "real world" of social studies. Seldom in the "real world" is one limited to using memorized information from social studies; applying knowledge to the construction of concepts, the solution of problems, and the formation of attitudes or personal beliefs are required much more of active citizens. "Test-generated" instruction, however, often leads to repeated drill and practice on decontextualized skills and content because teachers are pressured to prove that what they are doing produces desired outcomes. Challenging activities such as problem solving and creative thinking are deemphasized; learning centers, literature, art, and music are eliminated to make more time for daily drill and practice. One first-year teacher described the stress produced by such pressure: "I was petrified that my class would do so poorly that I wouldn't be back next year. So I taught what the other teachers recommended to get them ready for the test. After the test I started teaching, good teaching. The class enjoyed it, and I think they learned more the last three weeks of school than they did the first six months, because I was more relaxed, the students were more relaxed, and I was able to hone in on those areas where they needed help." (Livingston, Castle, & Nations, 1989, p. 24)

With all this criticism, you may want to ask, "Can standardized testing be a good thing for dynamic social studies programs?" The answer is a qualified yes—if the tests are not asked to do too much. They can show differences among groups in a school or district, reveal how students in a school or district compare to other students across the country, indicate whether a school is increasing or decreasing in general social studies achievement, provide support for grouping and placement decisions, help identify students in need of special services, and aid teachers in making curricular decisions. However, school administrators must resist all temptation to view the test as a "district report card," thereby forcing teachers to resort to methods having only one overriding goal—raising the test scores.

Teacher-Made Tests

Of all the sit-down, paper-and-pencil types of tests, teacher-made tests have traditionally been the most common form of unit assessment in social studies. The reason for their popularity is obvious: When teachers expend the great time and energy needed to originate a comprehensive thematic plan, they want to know if their efforts have paid off.

Constructing Teacher-Made Tests. Teacher-made tests are referred to as *criterion-referenced tests* and are used to measure the mastery of specific instructional objectives. Rather than comparing a student's score to other students as we found with standardized tests, criterion-referenced tests measure student performance against a specific guidepost, or criterion. Students are not compared to anyone else; they are judged only by what they can or cannot do. Criterion-referenced tests, then, are used to describe individual performance. Besides this immediate feedback, teacher-made tests also serve other functions:

1. They support information on which grades can be based.
2. They support diagnostic decisions about student needs—their strengths and weaknesses in social studies.
3. They help determine a student's progress.
4. They allow teachers to modify course objectives in accordance with student needs.
5. They suggest ways teachers might alter instructional strategies, techniques, and resources.

Teacher-made, criterion-referenced tests are used to measure minimum competence in achieving targeted instructional objectives. If mastery has been achieved, instruction proceeds; if not, the objective will need further attention. Thus, the results of criterion-referenced tests offer teachers meaningful information that they can use in planning classroom instruction. These tests, however, must not encourage the kind of narrowly focused instruction that standardized tests often invite.

The Link to Objectives. To serve as legitimate evaluation sources for social studies instruction, items on teacher-made tests must be directly linked to unit objectives. This is the only way we can effectively assess the degree to which students have benefited from classroom instruction. Although teachers do not have the test construction expertise that developers of standardized tests have, they must still be aware of the basic steps in designing test items.

The first step in constructing a teacher-made test is reviewing the stated instructional objectives. Whether they originated as part of a textbook series, school district guide, state-mandated curriculum guide, statement of national standards, or were collected from various sources, these objectives set the course for the unit by clearly communicating what the students are expected to do after the unit has been completed. Before any outcome can be measured by a teacher-made test, then, there must be a clearly stated objective pinpointing the desired performance. Let us say that an objective for Michael Bowler's expanding unit on the customs and traditions of Japan was stated: "The students shall understand the basic staples of the Japanese diet." Which of the following test items would be most appropriate—A or B? (The students must examine the lists and circle the word that is not conceptually related to the other words. Then they must explain why.)

| A. rice | fish | ostrich | vegetables |
| B. tatami | kanji | money | gohan |

If you picked A, you are correct! It is directly related to the instructional objective. Although the word "gohan" in item B means "cooked rice," the primary purpose of the question is merely to see if children can pick out Japanese words. As obvious as this appears, the biggest problem in teacher-made tests is that items are not related to instructional objectives.

Portfolios

Portfolios have recently come to the forefront as an alternative and valued technique of summative assessment. Paulson, Paulson, and Meyer (1991) describe a portfolio as "a purposeful collection of student work that exhibits the student's efforts, progress, and achievements in one or more areas. The collection must include student participation in selecting contents, the criteria for selection, the criteria for judging merit, and evidence of student self-reflection" (p. 60). There is no single list of items recommended for all portfolios. They may include such items as student writings, art products, photographs, independent research reports, projects, favorite books, and other work samples from the social studies.

Michael Bowler liked the idea of using portfolios to assess his students' progress in social studies, but felt they required a model before they were able to put together their own. His introductory plan is described below.

Classroom Vignette

Mr. Bowler wrote the word *portfolio* on the chalkboard. He asked if anyone had an idea of what a portfolio might be. Anticipating that not many would have a clue, Mr. Bowler brought out a box containing items that, in effect, created a biographical sketch. He informed his students that portfolios tell a story and that he chose objects that helped tell a story about himself as a person.

The first item in Mr. Bowler's portfolio was a photograph of his family. "I love my family," he proclaimed for all to hear and showed that his wife and children were in the picture (along with Barkley and Betty Basset Hound, the family pets). Next came a diploma from college: "I was the first member of my family to graduate from college," he said proudly. Several ribbons followed—they were awarded to him as a youngster for winning several swimming championships. Mr. Bowler then held up a favorite book and explained, "In my spare time, I enjoy reading." The most fascinating items came next—a photo of Mr. Bowler taken as a fifth grader, and his report card from the same grade. "I wanted to show you the best report card I ever received," he explained. "I had a fabulous fifth-grade teacher." The last article he removed from the box was a children's book, *The Little Engine That Could.* "This is one of the favorite stories my parents read to me as a child. It taught me that a person could accomplish almost anything if he or she tried hard enough."

Comparing his collection of personal memorabilia to a social studies portfolio, Mr. Bowler asked the class what might be included in their portfolios for the unit, *Customs and Traditions of Japan.* What would show their effort and learning in social studies? He accepted suggestions: daily assignments, drawings, illustrated maps, their best writings and those reflecting problems, group projects, individual projects, journal entries, audiotapes of oral reports, and so on. The class then discussed the format of a good portfolio. They decided it should be housed in a suitable container—boxes, file folders, folded sheets of construction paper, binders, and so on. Their "suitcases," all agreed, made excellent portfolio containers. The class also decided the portfolio should be neat and include a table of contents. Furthermore, each item should have with it a short personal statement about why it was important to the learner. Following this discussion, Mr. Bowler gave his students 1 week to organize their portfolios.

There are countless ways to organize portfolios; the important consideration is that the students take an active role in selecting material for and maintaining their portfolios. Of course, the portfolios must address instructional objectives. When students create their portfolios, their exhibits become a means through which you may provide evaluative feedback and monitor progress. You should hold individual conferences with the students during which you guide portfolio review with such questions as the following:

- "How has your work in social studies changed since last year (or last month)?"
- "What do you now know about _____ that you didn't know before?"
- "What are the special items in your portfolio?"
- "What would you most like me to understand about your portfolio?"
- "How did you organize the items?"
- "What are the strengths as displayed in the portfolio? What needs improvement?"

The conference should focus not just on subject-matter accomplishments, but also on planning strategies, personal reflections, and evidence of progress. Adams and Hamm (1992) address this important aspect of portfolio use:

Learning requires communication—with self, peers, and knowledgeable authorities. It also requires effort and meaningful assessments of these efforts. Since students need to be involved actively in evaluating and pro-viding examples of their own learning, they must document the probing questions they are asking, identify what they are thinking, and reflect on their understandings. In this way students can create, evaluate, and act upon material that they and others value. Assuming active roles in the learning process and taking responsibility for what students are learning goes beyond simply recognizing that they have made a mistake to imagin-ing why, getting feedback from others, and finding practical ways to do something about it.

Portfolios provide a powerful way to link learning with assessment. They can provide evidence of performance that goes far beyond factual knowledge and offers a clear and understandable picture of student achieve-ment. (p. 105)

Portfolios are a rich source for individual assessment in the social studies. They yield much concrete evidence that allows teachers to evaluate the progress their students are making. While this is important, however, the greatest asset of portfolios may be in self-evaluation. Portfolio assessment offers students the opportunity to set in-dividual goals, select the items for evaluation, and reflect on their work. In this way it encourages pride in learning and helps students develop the motivation to im-prove.

An important point to remember about overall assessment is that no single in-strument or technique can adequately measure the range of performances and be-haviors in social studies. For this reason, educators today strongly favor using port-folios containing a variety of evidence. Despite these popular endorsements, some teachers have been slow to use portfolios in their classrooms. They are skeptical of whether the results justify the time required to evaluate multiple measures. Admin-istrators must provide teachers with the time and support required to effectively eval-uate student portfolios. School district and building administrators hold the key to helping teachers endorse this form of assessment. If teachers can see portfolios as manageable and rewarding, they will be inclined to add them to their already full workload and to evaluate them with enthusiasm.

The assessment section brings to an end our discussion of unit and lesson plan-ning. Planning at both levels is a crucial element of dynamic social studies instruc-tion. Planning, either formally or informally, is a continuous process for everyone, for there is a constant need to keep materials, activities, and techniques up to date. As a new teacher, you will work hard to accumulate a rich teaching repertoire; ideas can be found in magazines and journals, and ready-made file cabinets (virtual file cabinets) are out there on the Internet, free for the taking. So, dive into the planning process, examine the many planning sources that are popping up regularly, and ex-perience the joy and satisfaction of creating a plan for a topic that has held your fas-cination. Perhaps you will make an important professional contribution that will be-come a rich source for future generations of teachers and students in your school district to enjoy.

AFTERWORD

Creative teachers clearly demonstrate a key ingredient of first-rate instruction—a characteristic I call "stick-with-it-ness." This is a persistent, intense devotion to what they are doing. Teachers with stick-with-it-ness are thrilled with their professional responsibilities. Teaching is not only their job—it is their passion. It leaves them virtually starry-eyed and eager to devise experiences that activate children for learning. All children must believe that their teachers are captivated by what they are doing in the classroom. In social studies, this means that teachers view their world with fascination and inspire their children to accept theirs as a never-ending mystery. To do this, teachers must plan significant learning situations in which there is a little mystery, a bit of magic, and a dash of magnificence to confront the children. Elementary school children respond to these things; that is what makes their classrooms different from those for any other age.

Teachers achieve magic in the elementary school social studies program when they help *each child* become challenged by the activities and emotionally involved in the subject matter. They deliver the best for each youngster and make the most of their time every day. Teachers adapt instruction to meet the special needs and interests of all their students; they fully understand and are willing to work toward fulfilling the principles and assumptions that underlie learning carried out by small groups or individuals. Teachers use a large variety of learning tasks to meet their students' needs. Although thematic planning has become extremely popular in recent years, many elementary school social studies teachers continue to use a number of techniques that have been a part of the educational scene for years. We will learn more about these techniques and about the kinds of activities most appropriate for individual pursuits throughout the rest of this book.

References

Adams, D. M., & Hamm, M. E. (1992). Portfolio assessment and social studies: Collecting, selecting, and reflecting on what is significant. *Social Education, 56,* 105.

Gronlund, N. (1991). *How to write and use instructional objectives.* New York: Macmillan.

Jarolimek, J., & Foster, C. D., Sr. (1997). *Teaching and learning in the elementary school.* Upper Saddle River, NJ: Prentice Hall.

Livingston, C., Castle, S., & Nations, J. (1989). Testing and curriculum reform: One school's experience. *Educational Leadership, 46,* 24.

National Council for the Social Studies. (1994). *Curriculum standards for social studies: Expectations of excellence.* (Bulletin 89). Washington, DC: National Council for the Social Studies.

National Education Goals Report. (1995). Washington, DC: U.S. Government Printing Office.

Orlich, D. C., Harder, R. J., Callahan, R. C., Kauchak, D. P., Pendergrass, R. A., Keogh, A. J., & Gibson, H. (1990). *Teaching strategies: A guide to better instruction.* Lexington, MA: D.C. Heath.

Paulson, F. L., Paulson, P. R., & Meyer, C. A. (1991). What makes a portfolio a portfolio? *Educational Leadership, 48,* 60.

Ryan, K., Burkholder, S., & Phillips, D. H. (1983). *The workbook.* Upper Saddle River, NJ: Merrill/Prentice Hall.

Shepard, L. A. (1989). Why we need better assessment. *Educational Leadership, 46,* 5.

9 Literacy-Based Social Studies Instruction

Nancy Smith has been reading folktales from around the world to her first graders with the overall goal of helping them see that a culture's folklore echoes its traditions, beliefs, and values. Today, she selected *The Mitten: A Ukrainian Folktale* by Jan Brett (Putnam). The story is about a little boy who lost a fancy mitten while in the woods looking for some firewood for his grandmother. The animals found his lost mitten and crawled inside to keep warm. The parade to the inside of the mitten started with a field mouse, who was joined by a frog, owl, rabbit, fox, wolf, wild boar, and bear. To the children's surprise, all were able to fit inside the mitten until a tiny cricket joined them. Then the mitten exploded!

To start the reading experience, Ms. Smith reached into her story basket (a large picnic basket holding surprises connected to the books she reads). Filled with anticipation, the children begged Ms. Smith to open the basket so they could see what special surprises were waiting inside. Capitalizing on their curiosity, Ms. Smith surreptitiously drew out a beautiful shirt from a Ukrainian folk costume that she had borrowed from a friend. The children were fascinated by the intricate embroidery and ornate stitchery. Ms. Smith explained that this folk costume was from Ukraine, a large country in Eastern Europe, and together they located Ukraine on a globe. She showed the children a photograph of a Ukrainian family dressed in similar traditional attire. Other items were removed in a similiar manner.

Giving the impression of being on pins and needles in anticipation of what else was in the bag, Ms. Smith reached in one last time and carefully took out a pair of fancy mittens. She and the children talked about the reasons why people wear mittens and how the mittens feel on a cold winter day. Ms. Smith called the children's attention to the illustration on the front cover and asked the children to think about what might go through an animal's mind if it saw a mitten like the one on the cover lying on the ground in the forest. She elicited several predictions and then read the story.

After the story was finished, the discussion began. The children did not raise their hands in response to the teacher's questions, but joined in a natural, spontaneous conversation about the book. As they talked, Ms. Smith listened and helped the children make connections. However, the children (not Ms. Smith) determined the flow of the conversation.

Following the discussion, Ms. Smith involved the children in a readers' theater skit. To begin, the children recalled the animals that squeezed into the mitten. They then suggested the movements each animal would probably use while approaching the mitten (e.g., springing like a frog, hopping like a rabbit, flapping one's arms like an owl, and lumbering like a bear). Ms. Smith selected volunteers to play each of the animals as the story was reread. She placed an elegant construction paper glove covered with sequins and beads in the middle of the group meeting rug. The players dramatized the story as volunteer readers read each part. The readers used their voices to bring life to the animals they were reading about, while the players entered and left the area around the glove according to the script. The emphasis was not on the quality of the production, but on the interpretive qualities of the children's voices and bodily movements.

Ms. Smith embraces a philosophy of dynamic social studies instruction that is popularly referred to as *literacy across the curriculum.* To understand what this means, we must first understand what is meant by the term *literacy. Literacy* was used in the past to refer exclusively to reading, but now has been broadened to include many skills and processes. However, for the purposes of this book, literacy will be considered as those skills and processes we commonly associate with reading and writing. Thaiss (1986) has used that concept of literacy as a base from which to explain what is meant by the phrase, *literacy across the curriculum:*

> "[Literacy] across the curriculum" means basically two things. First, it means that gaining power in all the modes of language . . . must take place in every school course and at every school level, if this growth is to be deep and substantial. This meaning rejects the notion that the diverse uses of language are best learned in specific "skills" courses in, for example, [spelling, grammar, reading, or writing]. Second, "[literacy] across the curriculum" stresses the interrelationship of the modes: One learns to write as one learns to . . . read. Each ability, therefore, improves to the extent that all are exercised.

This second meaning rejects the teaching of, for example, writing or reading in relative isolation from the other. Ultimately, these two meanings of [literacy] across the curriculum come together in a third: the inseparableness of language, thinking, and learning. If we do not apply the full range of our language resources to our learning of any subject, then we stifle thought, conscious and unconscious, and so deprive ourselves of more than the most superficial understanding. (p. 2)

Ms. Smith adopted the literacy across the curriculum philosophy after her first few years of classroom teaching. At the end of her first year, while constantly thinking about the apparent connectedness among the various school subjects she was required to teach, Ms. Smith began to ask some interesting "Why" questions: "*Why* is it that when my students read a story about Sojourner Truth at 9:30 in the morning and I follow that up with a creative writing project, it is called *reading*?" "*Why* is it that when we read about Sojourner Truth at 2:30 in the afternoon and I follow that up with a dramatic skit, it is called *social studies*?" Ms. Smith looked long and hard to find answers, and finally found them when she read what literacy experts had to say. In essence, their message was, "Language is a powerful learning tool, and reading and writing are valuable ways to learn in all content areas. Effective teachers encourage students to use reading and writing in meaningful ways in theme studies so that they learn information better and refine their literacy competencies" (Tompkins, 1997, p. 32). Today, Ms. Smith operates with a firm belief that literacy should be the heart of a dynamic social studies program. Her convictions were supported by these three significant principles (Thaiss, 1986):

1. Students understand the content better when they use reading and writing strategies to learn.
2. Students' literacy learning is strengthened when they use related skills and strategies in authentic daily experiences (meaningful activities in subject areas such as social studies).
3. Students learn most effectively through active involvement, collaborative projects, and interaction with their classmates, the teacher, and their environment.

Although reading and writing are viewed as parallel processes of meaning construction, they will be described separately in this text for ease of discussion.

READING IN DYNAMIC SOCIAL STUDIES CLASSROOMS

Although students read many kinds of texts (any print source from which students derive meaning) in social studies classrooms, in this chapter we will explore the two major types of print sources that are used most often—*textbooks* and *trade books*.

Recently, there has been considerable controversy over whether textbooks or trade books should serve as the primary framework for teaching social studies in the elementary school. Goodman (1988), for example, argues that textbooks are written too unimaginatively for young children; their shortened sentences and controlled vocabulary result in dull, impersonal reading. On the other hand, Wixson (1991) states that the content and instructional design of textbooks help the student to learn. Texts, for example, have a logical flow of information, and the unity in the focus of texts helps the students acquire big ideas. Instructional supports in the texts, such as headings, graphics, and charts, relate important information in a way that guides learning. To further muddy the waters, Tompkins (1997) suggests that there is a place for both trade books and textbooks in social studies. She explains that textbooks are especially important for beginning teachers who "often rely on [textbooks] and move toward incorporating trade books in their [social studies] programs as they gain confidence in their teaching abilities . . ." (p. 203). Therefore, what should teachers of dynamic social studies do—use trade books only, textbooks only, or a combination of both? The point of view of dynamic social studies is that students need a wide variety of reading materials in their social studies program, including both well-written textbooks and quality trade books. Each has the potential to extend and enrich learning, if used properly.

Social Studies Textbooks

Social studies textbooks are graded sets of reading materials traditionally used by social studies teachers as a major source of planning and instruction. Typically published as a sequential series from kindergarten through grades 6 or 8, textbooks are developed by large publishing companies under the direction of a senior author who is usually a respected name in social studies education. Most social studies textbook programs contain an array of instructional materials, including student books and teacher's manuals.

Benefits of Textbooks

Because most social studies textbooks are easy to use and offer a convenient instructional package as well as a carefully researched, systematized body of content, they are the predominant source of information in most classrooms. Such general acceptance by social studies teachers is not hard to understand. When elementary school teachers are required to use developmentally appropriate strategies and activities in all school subjects, including math, reading, spelling, writing, and science, the thought of having specially "packaged" help in social studies greatly reduces the pressure and anxiety of daily planning. In addition, textbooks provide extensive treatment of subject matter that is organized sequentially from one grade level to the next. Each teacher from kindergarten through grade 8 knows what was done in earlier grades and what will be expected in later grades, thereby minimizing gaps or repetition. Finally, school districts and teachers appreciate the carefully researched, comprehensive nature of the teacher's manuals, which come complete with goals, objectives, lesson plans, suggestions for activities, and tests. Teachers at all levels of experience acknowledge the utility of textbook programs, but beginning teachers find them especially attractive. Jarolimek and Parker (1993) explain:

> Beginning teachers are usually most comfortable starting with [textbooks]. The teaching environment can be controlled sufficiently well to reduce management concerns to a minimum, the objectives can be made specific, the children's study materials can be preselected by the teacher, and the process can be entirely teacher directed. . . . This initial [textbook experience relies heavily] on the suggestions presented in the teacher's manual that accompanies the book. (pp. 30–31)

Although textbooks virtually take teachers "by the hand" and guide them through the instructional process, teachers must plan and execute specialized instructional strategies that help students think and learn with text.

Problems With Textbooks

Although textbooks have many beneficial features, they are not without their problems. Critics, for example, complain that textbooks are often used as *the* curriculum—the *only* source of learning in the classroom. In surveying the reactions of a dozen teachers to the question, "What do you associate with social studies teaching that is clearly not creative in the elementary school?" Solomon (1989) discovered that viewing textbooks as divinely inspired canons was a common response:

> In these classrooms, wrote one of the teachers surveyed, the "teaching [is] only from the textbook. . . . Students answer questions from the textbook or [do] nothing but worksheets, worksheets, worksheets." Another responded, the teacher is "following each page and paragraph within a textbook—giving the same emphasis to every paragraph and calling for learning." All twelve respondents had the same message: Teachers in these classrooms allow one textbook to determine what they teach, with students simply expected to learn the material in it. (p. 3)

On another note, Sewall (1988) inquired into the features of social studies textbooks that make them ill-disposed for elementary school classrooms. His findings suggest that:

1. *The physical size and weight of textbooks discourage enthusiasm for their contents.* Reviewers stressed that young children would not be drawn to "curl up and read" the social studies texts mainly because of their bulk.
2. *The prose style of most textbooks is bland and voiceless.* Even though quality varied from publisher to publisher, reviewers thought that the overall literary style of the texts fell well short of the mark: "Reviewers found textbooks generally to be more catalogues of factual material . . . , not sagas

peopled with heroic and remarkable individuals engaged in exciting and momentous events" (p. 35).

3. *Excessive coverage makes textbooks boring.* Many textbooks are most effective as almanacs, encyclopedias, or reference guides; names and dates seem to dart past like telephone poles seen from the window of a swiftly moving train.

4. *Textbook formats and graphics diminish the style and coherence of the running text.* To be competitive with television and other mass media, textbooks have lost the narrative content that made up 90 percent of their pages just a generation ago, and now substitute "endless photographs, diagrams, charts, boxes, subunits, [and] study exercises . . . especially in lower-grade-level textbooks" (Sewall, 1988, p. 35).

To be fair, it must be emphasized that most of these criticisms cannot be as closely attached to the textbooks themselves as to the ways teachers use them. Textbook publishers encourage teachers to supplement and enrich their materials with auxiliary literature sources, information books, videos, field trips, realia, computer programs, and other instructional resources. Textbooks are not meant to furnish the total social studies experience, but to serve as a single resource among an array of several possibilities. However, I'm sure your personal experiences as a student have helped you visualize the stereotyped textbook-bound teacher following the rigid routine of assigning a section to read and calling on students to answer questions. I like to call this the "dry assign-and-question" technique.

Trade Books

Disenchanted by the dryness that has often been associated with the "dry assign-and-answer" approach to textbook-driven instruction, more teachers than ever before are using a variety of texts (print sources) in social studies. Although these texts may include such sources as newspapers, magazines, diaries, primary resources, and a number of other print learning tools, this chapter will focus on the use of trade books.

Most of us shared our childhood with a menagerie of trade book companions. Some we recall with beloved memories and special delight. Their adventures and misadventures were sources of great enjoyment, and the thoughts of missing out on the likes of Winnie-the-Pooh, Strega Nona, Tom Sawyer, Laura Ingalls Wilder, or Johnny Appleseed is unthinkable. Zinsser (1990) notes that "[n]o kind of writing lodges itself so deeply in our memory, echoing there for the rest of our lives, as the books that we met in our childhood" (p. 3). First and foremost, then, good literature sources bring delight and enjoyment to children. But literature can educate at the same time it entertains. Huck, Hepler, and Hickman (1993) explain:

> The experiences children have with literature give them new perspective on the world. Good writing can transport the reader to other places and other times and expand his life space. The reader feels connected to the lives of

others as he enters an imagined situation with his emotions tuned to those of the story. . . .

How better can we feel and experience history than through a well-told story of the lives of its people and times? . . . A history textbook tells; a quality piece of imaginative writing has the power to make the reader feel, to transport him to the deck of a slave ship and force him into the hold until he chokes on the very horror of it. (p. 11)

Textbooks are primarily concerned with facts; trade books are rich both in content *and* emotion (e.g., compassion, humanness, misfortune, happiness, awe, and grief). This intrinsic value of trade books, by itself, should make it a valued part of the social studies program. Ravitch (1978) echoes this sentiment as she laments the gradual displacement of quality trade books with textbooks based on the expanding environment approach during the early- to mid-1900s:

Until expanding environments managed to push historical material out of the social studies curriculum children in the early grades in most public schools learned about . . . myths, biographies, poems . . . fairy tales, and legends. The story of Robinson Crusoe and the study of Indian life were

Fine literature is an essential component of dynamic social studies programs.

particular favorites. Stories about explorers, pioneer life, American heroes . . . , and famous events in American history were the staples of the first three grades. The line between historical literature and general literature was virtually nonexistent. Teacher guides emphasized the importance of telling stories. (p. 38)

Trade books (both fiction and nonfiction) contribute much to the social studies program; they present information about people, places, events, and times in inspiring, memorable, and relevant ways. But implementing a trade book-based social studies program means that teachers must be familiar with the books that can help students explore a theme or topic. The range of trade books is enormous, spanning all types of topics. But three types (genres) of books are especially useful in dynamic social studies classrooms: *historical fiction, biographies,* and *folk literature.*

Historical Fiction

It wasn't too long ago when history was a matter of kings and queens, wars and revolutions, conquests and defeats, explorations and inventions, all compressed neatly and tightly between the covers of social studies textbooks. History was not for true lovers of stories from the past; instead, history was a comprehensive collection of names and dates. Fortunately, contemporary history has changed. Now, with historical fiction at the center of instruction, the "story" of the past has become a part of history. No longer do students read the textbook and recite facts, but actively learn about the full meaning of the past through historical text that is inspiring, imaginative, and powerful.

Historical fiction is a term describing realistic stories that are set in the past; the facts are accurate but the characters are fictional (although they sometimes interact with actual historical figures). Historical fiction offers children opportunities to vicariously experience the past by entering into a convincingly true-to-life world of people who have lived before them. By being transported to the past through the vehicle of literature, students enter into the lives of the characters and, through mental imagery, become inspired to think as well as to feel about their condition. In *Early Thunder* by Jean Fritz (Coward-McCann), for example, 14-year-old Daniel West faces a struggle to sort out his loyalties during events leading to the American Revolution. Set in Salem in 1774, when people were either Tories (loyal to the king) or Whigs, Daniel is a staunch Tory who hates the acts of the rowdy Liberty Boys. But as the story unfolds, Daniel becomes equally disenchanted by British attitudes. Eventually, Daniel must come to terms with himself and sort out his allegiance.

This and other pieces of historical fiction invite the judgment of children upon the past, bringing to the forefront deplorable enactments of bigotry and prejudice. Mildred D. Taylor is an African-American author who has written with great sensitivity and understanding about racial injustice in rural Mississippi during the 1930s.

Mississippi Bridge (Bantam Skylark) is a tale based on a true story told to Taylor by her father. It centers on the Logan family and their determination to fight racial injustice. The story opens with Jeremy Simms, a 10-year-old white child, watching from the porch of the general store as the weekly bus from Jackson splashes through a heavy rainstorm to stop at the store. His neighbors, Stacey Logan and his sister Cassie, are there to see their grandmother off on a trip. Jeremy's friend, Josias Williams, is taking the bus to a new job. But Josias and the Logans are black; black people can't ride the bus if that means there won't be enough room for white people to ride. When several white passengers arrive at the last minute, Taylor describes the injustice that unfolds as the driver sends Josias and Stacey's grandmother off the bus. The author's words make it impossible not to attach deep emotions to the event:

> Josias stood. He picked up his bundle of clothes and he give up his seat. He took himself some slow steps to the front of the bus. I moved over to the door waiting to say my spell to him, but he still ain't got off. He stopped hisself right front of that driver and he gone to pleading. "Please, boss . . . I got to get to the Trace t'day. Please, boss. I done got my ticket. I done made all my plans. Folks spectin' me. I got t' go on this bus!"
>
> "Nigger, I said you gettin' off."
>
> "Boss, please . . . ."
>
> That bus driver, he ain't give Josias chance to say no more. He jerked Josias forward to the door, put his foot flat to Josias's backside, and give him a push like Josias wasn't no more 'n a piece of baggage, and Josias, he gone sprawling down them steps into the mud. The bus driver, he throw'd Josias's bundle after him, his ticket money too. (1992, p. 31)

Shortly after this episode, the bus skids off the bridge and all are drowned. The nightmare changes the lives of the townspeople forever.

The varied settings and conflicts of colonial America have inspired a large number of books about young children. *The Sign of the Beaver,* by Elizabeth George Speare (Houghton Mifflin), is a widely used book in many literature-based social studies programs around the country. It is the story of 12-year-old Matt, who is left by his father to survive on his own in their cabin in the Maine wilderness while his father returns to Massachusetts to bring back the rest of the family. Matt is a brave boy, but he is not prepared for an attack by swarming bees. He is rescued by Saknis, the chief of the Beaver clan, and his grandson, Attean. The boys get to know each other well; Matt teaches Attean to speak English, and Attean teaches Matt crucial survival skills. The story describes their friendship that grows through the months until Matt's family returns.

A teacher's obvious interest in and use of historical fiction strengthens the appeal of the past for children. The words of Lizzie Stanton, the diary of Colonel William Travis at the siege of the Alamo, and letters that passed between brothers fighting on opposite sides during the Civil War serve to wear away the bland acceptance of fact piled on top of fact, year after year. Suddenly, through story, the magnitude of humanity's accomplishments are revealed, arousing the imagination and interest of the reader.

Biographies

Biographies are much like historical fiction in that they are rooted in thorough research of people who lived in the past. Rather than focusing on fictional characters, however, the stories recreate true incidents in the lives of real people (i.e., nothing can be invented by the author). For children, biographies serve to keep alive interest in heroes. In contrast to the past when heroes could do no wrong, modern biographies show the human side of heroes, telling about their weaknesses as well as their strengths. Children relate more closely to heroes when they can be accepted as real people, complete with doubts, discouragement, mistakes, conflict, and frustration.

The best biographies for children characterize the subject of the biography as true-to-life, neither excessively fawning with praise nor disgraceful and humiliating. The person should be recognized as a real human being with both strong points and imperfections. Jean Fritz has been especially effective in creating remarkably true-life portraits of famous figures. In Fritz's (1979) absorbing biography of the controversial Confederate general, *Stonewall,* she characterizes Thomas "Stonewall" Jackson as a hard-working but underachieving young cadet at West Point:

> But however determined he was, no one watching him struggle would have guessed that Tom ("The General" as they called him) could have survived four years at West Point. Indeed, he could hardly get through a recitation. When called on to answer a question or solve a problem, Tom sweated so profusely his classmates joked that one day he would drown them all. (p. 26)

But, as Fritz tells, Jackson's determination to succeed served him well:

> Yet inch-by-inch Tom pulled himself through. Near the bottom of his class at the end of his first year (fifty-first in a class of seventy-two), he rose to seventeenth by the end of his senior year, and his classmates, who had grown fond of him, said it was a pity there wasn't a fifth year at West Point. Tom would have graduated top man. (p. 26)

By examining Stonewall Jackson's character at various points throughout the story, students will learn to appreciate one of the nation's most brilliant and heroic leaders as well as one of its oddest heroes. Likewise, the best biographies are intimate and memorable, presenting the facts as fairly as possible-accurate information within a frame of human emotion.

Folk Literature

As old as language itself, the vast reservoir of folktales originated in the great oral tradition of cultures. Wherever people gathered (around a campfire, hearths of homes, in market places, or while cooking or planting), these stories were told not only for entertainment, but also to pass on the values, traditions, and customs of the culture. Because folktales have been retold from generation to generation, they are an unclouded mirror of any group's beliefs, values, lifestyles, and histories. An authentic tale from Africa, for example, will include references to the land on which the people lived, their food, their homes, their customs, and their beliefs.

Take, for instance, this passage from *Mufaro's Beautiful Daughters* by John Steptoe (1989), a Cinderella tale from Africa:

> Nyasha kept a small plot of land, on which she grew millet, sunflowers, yams, and vegetables. She always sang as she worked, and some said it was her singing that made her crops more bountiful than anyone else's. (p. 4 of an unnumbered text)

Steptoe's careful research paints a vivid picture of what life was like in the region of Africa where the story originated. Throughout the book, his beautiful illustrations of the plants and animals of the rainforest bring authenticity to this outstanding work. Children not only enjoy an engaging tale, but also learn about the plants and animals of the rainforest and about the architectural wonders of the great wall that enclosed a magnificent city built on a plateau in southeastern Zimbabwe around 1100 A.D.

Elementary school children learn so many things about early African culture as they read this fascinating tale, including crops grown in gardens and the significance of such birds as the Carmine Bee-eater and the Crowned Crane. This is a major reason folktales belong in social studies classrooms; they help children learn about a culture as well as become aware of its inherent values, beliefs, and customs.

Huck, Hepler, and Hickman (1993) inform us that, despite these distinct cultural nuances, the following characteristics are common to all folk literature: *repetition* ("Fee, fi, fo, fum"); *conflict* (Anansi needs a wife but is afraid he won't find one because of his bad name), *characterization* (wolf, stepmother, or witch is a symbol of evil; innocent children are symbols of goodness), *theme* (reflects goals of people—long life, good spouse, beautiful home, plenty of food, freedom from fear, power, courage), and *motifs* (magical powers, such as the ability to spin straw into gold; transformations, such as Cinderella going from rags to riches; magic objects, such as a lamp containing a genie who grants a number of wishes; and trickery, such as in Gail Haley's *A Story, A Story*. In this Caldecott Medal winner, Anansi the Spider Man outwits a fierce leopard, swarming bees, and a fairy that no man sees to win the Sky God's stories for people around the world.

The study of folktales enlightens students about distinct cultures and contributes greatly to their understanding of people around the world. A cross-cultural study of folk literature also adds an extra dimension to helping children discover the universal qualities of all humankind.

Teaching With Trade Books

Trade books do not typically come with teacher's manuals, so teachers must establish for themselves the course instruction will take. Vacca and Vacca (2002), conceding that there is no single best way to prepare an effective text lesson, suggested a plan, called the *Instructional Framework* (IF) that offers social studies teachers a fairly representative approach to lesson organization, regardless of what text is selected. Examine the plan carefully; you will see that it corresponds very closely to the direct instruction model explained in Chapter 5. According to Vacca and Vacca's IF strategy, what the teacher does *before reading, during reading,* and *after reading* is crucial to productive learning.

Before Reading (Prereading)

An IF includes provisions for one or more of the following during the prereading phase: (1) motivating readers, (2) building and activating prior knowledge, (3) introducing key vocabulary and concepts, and (4) establishing purpose. For example, teachers often provide study guides for students—questions or activities that lead them to discover the main ideas and/or concepts developed in the test. The study guide shown in Figure 9–1 was developed by one teacher to help his students construct meaning from text dealing with Egyptian art. The guide is introduced during this phase, worked with throughout the "during reading" phase, and discussed, compared, and analyzed during the "after reading" phase.

During Reading (Reader-Text Interactions)

While most teachers recognize the most important parts of a text assignment, many students don't. Oftentimes, they read in a monotonous way; each word, each sen-

FIGURE 9–1
Study Guide Organizer

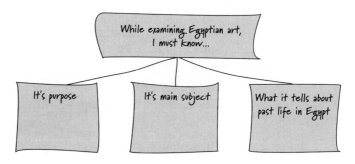

tence, each paragraph is treated with equal importance. The gap becomes especially wide while reading social studies textbooks where readers must deal with highly specialized conceptual demands. Therefore teachers have an important role to play, a role Vacca and Vacca call "instructional scaffolding." Scaffolds provide necessary supports for readers' efforts to make sense out of the text by showing them how to use strategies that they cannot do by themselves at first. K-W-L charts and other graphic representations discussed in Chapter 5 are examples of strategies appropriate for instructional scaffolding.

After Reading (Postreading)

Ideas encountered before and during reading may need to be clarified and analyzed after reading. Writing activities, illustrations, and other postreading elements are springboards to further thinking.

Two after reading strategies that appear to be most popular in dynamic social studies classrooms are (1) *discussions and response charts,* and (2) *literature response journals.*

Discussions and Response Charts

Most text-centered social studies programs value open-ended thinking in response to stories, but they operate with a conviction that the students need some support in understanding the content, too. Therefore, teachers in dynamic social studies classrooms often begin a story discussion with a series of "comprehension checks" that determine whether or not the children understand the main direction of the story, followed by a number of more open-ended questions that invite a variety of responses. Sloan (1984) has developed a helpful list of sample questions that invite "comprehension check" responses to a story:

> "Where and when does the story take place? How do you know? If the story took place somewhere else or in a different time, how would it be changed?"
>
> "What incident, problem, conflict, or situation does the author use to get the story started?"
>
> "Trace the main events of the story. Could you change their order or leave any of them out? Why or why not?"
>
> "Who is the main character of the story? What kind of a person is that character? How did you know?"
>
> "Did you have any feelings as you read the story? What did the author do to make you feel strongly?" (pp. 104–106)

Oftentimes, teachers will help facilitate student responses to these "comprehension check" questions with concrete teaching aids that help them organize and graphically represent ideas and relationships. These concrete teaching aids help students rehearse what they want to talk about. For example, Tim Jacobs read to his third graders the book *Nannabah's Friend* by Mary Perrine (Houghton Mifflin), a sensitive story of how a young Navajo girl bridges the gap between the security of home and the world outside. He thought this was an excellent book to use for character analysis, especially since the children would immediately identify with Nannabah and her plight. To help analyze Nannabah's character, Mr. Jacobs and his students made a web of her traits. See their character web in Figure 9–2.

Open-ended questions often used to initiate literature-based discussions signal a process called *grand conversations* (Tompkins, 1997). In place of the comprehension check, or as some call them—traditional "inquisitions"—during which students recited answers to factual questions teachers asked about books they had read, grand conversations shift the focus to making connections between the text and the students' own lives. Tompkins elaborates:

> Teachers often participate in grand conversations, but they act as interested participants, not leaders. The talk is primarily among the students, but teachers ask questions regarding things they are genuinely interested in

An attractive display of trade books is a great way to launch children into meaningful and productive inquiry.

learning more about and share information in response to questions that students ask. . . .

Grand conversations can be held with the whole class or in small groups. Young children usually meet together as a class, while older students often prefer to talk with classmates in small groups. (p. 259)

FIGURE 9–2
Character Web

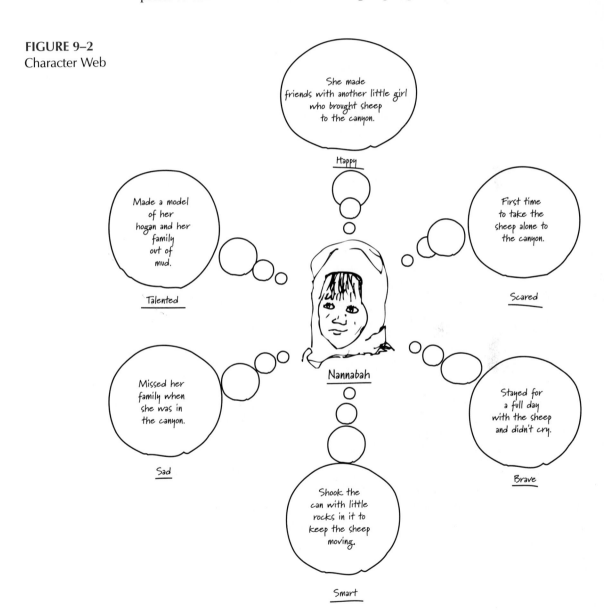

Literature Response Journals

Literature response journals are booklets or notebooks in which students write personal reactions to what they read or listen to. Tompkins and Hoskisson (1995) advise that literature response journals should not be used to simply summarize stories, but to relate students' reading to their own lives or to other literature they have read. In addition, students may also list interesting or unfamiliar words, jot down quotable quotes, and take notes about characters, plot, or other story elements; however, the primary purpose of these journals is for students to think about the book, connect literature to their lives, and develop their own interpretations.

Literature response journals are being used with children as young as kindergarten age, but most typically, kindergarten children listen to their teacher read a story and draw their response. When children start writing words to accompany their drawings, their entries usually summarize the plot. In time, the drawings become less significant and the written entries take on a more interpretive or personal nature (see Figures 9–3 and 9–4).

Journals can be powerful reading-response tools, and teachers are now using them widely in social studies programs. Cooper (1997), however, cautions teachers against their improper use:

- *Beware of the overuse of journals.* Sometimes teachers get so excited about journals that they use them in every subject. Students become bored, and the journals lose their effectiveness.
- *Don't require all students to write in their journals every day.* To get everyone to write, teachers often require a daily entry. This turns students off to journals. Therefore, allow students to use journals as a choice. Some teachers require a minimum number of entries each week, especially in middle school.
- *Make sure you know your reason for having students use journals.* Be confident about why you are using journals before you begin. Avoid using journals just for the sake of using them.
- *Remember that journals are instructional tools.* Journals should replace other, less-meaningful types of activities that students might do. They should not become just more "busy work" that is added on to what students already need to complete (p. 319).

Good books and dynamic social studies are inseparable. Through the use of various genres, students are able to enter the conditions of life experienced by people of all ages, places, and times. Using the stories that engage the minds and hearts of your students will enrich your social studies program.

FIGURE 9–3
Lower-Grade Literature Response Log

Amy L.
The BAd Gise
saiD you CAn't
pray To god. you
hAve To pray To us
not god. And The
BAd reckt Their HoUES

But that was not
nice. And A
Little Kid found
Some Oel. And
The Light StAd
for eight DAys.

FIGURE 9–4
Upper-Grade Literature Response Log

THE GIVER
by Lois Lowry

A book that made a difference for me is one called <u>The Giver</u>. It is about a twelve year old boy who lives in the future world where there is no fighting, poverty, or injustice. In this book no one can see color. All they can see is gray. When they are twelve, all children get assigned the job they have to do for their adult years (like Caretaker of the Old). The main character, Jonas, is selected by the Giver for a special job called The Receiver of Memory. This job requires one to remember the past when everyone could see color and had a choice of what they could do. But it also helped Jonas discover some very important secrets about his world.

The book made me realize that the future may not be as good as everyone thinks it will be. So I think that our time is good enough. The reason I liked this book was because it was not in our place of time. It was a future book but not with space ships and robots. It made the future look boring and very lonesome instead of fun and really high tech. Also, it showed me how one person could make a difference. Jonas left the village so that all the memories of the past would flood the minds of the people in the town in order for everyone to see color and have feelings as in the past. <u>The Giver</u> is a great book to get so check it out.

by Gayle Westover
Grade 7- Ms. Summers

WRITING IN DYNAMIC SOCIAL STUDIES CLASSROOMS

Why should elementary school children write in social studies? This question is easy to answer—children learn to write by writing. In the elementary school classroom, teachers create diverse contexts, or environments, where writing can be experienced as a functional tool of communication. These classroom contexts can be rich and varied, but must include social studies, a curriculum area "that is as appropriate for writing as the gym is for basketball" (Murray, p. 54). The social studies curriculum offers contexts in which students are able to write frequently and purposefully; an environment in which writing in different forms can be perceived as necessary and useful. In this section, we will explore how writing can be employed in dynamic social studies classrooms and examine the many ways we can enhance learning in social studies through writing.

To maximize the contributions of writing in dynamic social studies programs, we must have a clear idea about what forms the writing might take. Called *voice* or *form,* this aspect of writing describes the different types of writing used by elementary school children. Although there are several types of writing, many individuals insist that writing in social studies should take only one form: "Children should write in social studies class to record or report information about a particular topic under study. Writing is functional—it enables students to organize or express what they have learned." Certainly, the process of clearly organizing and expressing information is a valued goal of elementary school social studies instruction; indeed, practicing social scientists often write in a functional way, recording and communicating important observations and discoveries. Because social studies instruction attempts to help children search for ideas and information the way social scientists do, writing about what they learn is a very important part of the curriculum. However, communicating the content of their learning experiences is only one way children write in the elementary school classroom. Called *expository writing,* this form joins *expressive writing* and *imaginative writing* as the three writing forms most applicable for social studies.

Expository Writing

When students write to inform, explain, or persuade, they are using a form of writing called *expository writing* (also known as *transactional writing*). Expository writing is the presentation of facts, ideas, or opinions, as in the informational report, book report, or biography. Teachers have considered expository writing such an important part of the social studies program over the years that it is probably the most common type of social studies-specific writing we find today. Teachers use expository writing because they are convinced that the essence of the scientific method is precise observation, accurate data collection, orderly analysis of the data, and clear communication of applicable information. Because expository writing is inherent in each of these tasks, it becomes a functional tool that helps students operate as young social scientists.

The Drawing and Writing Partnership

Starting with the youngest children, teachers offer writing opportunities when they ask students to record ideas on paper in a variety of ways—usually with drawings accompanied by printed words. When their teacher offers a piece of drawing paper or primary-grade writing paper, most kindergartners and first graders will draw a picture of something important to them and then talk about it. Graves (1983) refers to the process of drawing pictures and talking about them prior to writing as a "rehearsal" process inasmuch as children use their drawings as a way to reflect on what they have experienced.

Classroom Vignette

For example, when Ms. Helmuth's first graders came back to school at the close of winter vacation, they were excited to share the interesting things that happened to them while they were away. Hands flew into the air during sharing time in eager anticipation of who would be the first to talk. Teri's words exploded as she told about helping her father shovel the driveway after a heavy winter snowstorm. Robert's eyes sparkled with pride as he described the new furniture that arrived at his house a week ago ("A beautiful yellow sofa, just beautiful!"). Wendy sadly recounted her family's heavyhearted task of saying good-bye to their 10-year-old beagle, who was struck down by a speeding car. Enrique happily described his seventh birthday party, and Kyle told of his family's trip to the city aquarium.

After they shared orally, Ms. Helmuth passed out drawing paper and asked each child to draw a picture of his or her special event. After the children finished their drawings, Ms. Helmuth directed them to print their "stories" in the space below each drawing. Figure 9–5 is a sample of Kyle's story about his family's trip to the aquarium. Kyle was particularly interested in the dolphins; when asked about his written piece, Kyle explained that the dolphin was saying, "Ouch!" Then he read what he had written above the drawing: "Animals should definitely not wear clothing because a dolphin might have trouble with its fin."

Ms. Helmuth explains the importance of these early writing experiences: "As the blank pages come alive with drawings and words telling of their experiences, I could see these children were showing me what they knew as well as what they needed to know."

Two types of expository writing seem to be most popularly modeled from literature in social studies classrooms—*written reports* and *biographies.* I like to add newspapers to the list because they offer a superb medium for children to express what they know and feel.

FIGURE 9–5
Young Child's Early Writing Effort

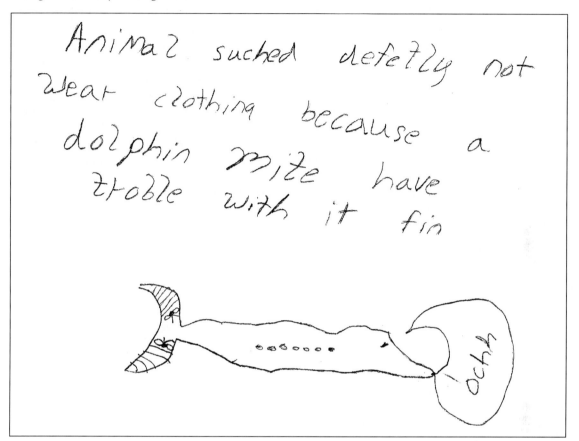

Written Reports

Written reports have had wide acceptance at all grade levels in social studies. However, you cannot expect students to compose good written reports simply by announcing, "Write a report on Sitting Bull for next Tuesday." Giving such an assignment only frustrates many students and encourages them to copy every bit of information from a written resource.

As Atwell (1990) emphasizes:

> Report writing per se, isn't the problem . . . the problem with school reports lies in our methods of assigning them. We need to put the emphasis where it belongs—on meaning—and show students how to investigate questions

and communicate their findings, how to go beyond plagiarism to genuine expertise and a "coming to know." (p. xiv)

As with younger children, writing for middle- and upper-grade children must also be supported by direct experience. Cullinan and Galda also (1994) suggest that quality sources of children's literature offer models for young authors to imitate as they evolve into more mature writers. Good models produce good results: "Literature provides a rich resource to use in any writing program. It can be used as a model for writing, as examples of interesting language used well, and as an illustration of topics to write about. There are books that illustrate unique formats: journals, letters, postcards, diaries, and autobiographies. . . . No matter what point you want to illustrate about writing, there are books to help you make it clear" (p. 399). As they read increasingly more complex books over the years, children are able to model them for their own writing projects.

Saul Greenberg realized that direct experiences and good storybook models were important, so he included both as he planned a report writing experience for his children.

Classroom Vignette

When Mr. Greenberg began his unit on "bread," he showed his children an assortment of real bread, as well as pictures and plastic models: bagels, pita, baguettes, paska, challah, tortillas, rye bread, croissants, fry bread, injera, and the like. Mr. Greenberg did this not only to provide his students with a direct experience, but also to generate enthusiasm and interest, to prime the pump for the flurry of activity that naturally follows.

Mr. Greenberg invited the children to taste the different breads and, as a class activity, marked the country of origin of each on a large map. Mr. Greenberg added another hands-on experience by giving the children small balls of prepared bread dough (available in supermarkets) and asked them to place the dough balls on sheets of aluminum foil labeled with their names. Before they put their dough into the oven, the children were encouraged to shape the balls any way they wished. The bread was baked according to directions, and the children discussed their sensory experiences as well as the physical changes they observed from start to finish. Everyone responded to this activity with enthusiasm and interest; a profusion of questions and comments followed.

To begin the actual writing phase, Mr. Greenberg formed committees around common topics of interest which, after much discussion, were narrowed down to four: bread bakers and bakeries, bread from around the world, what bread is eaten with (condiments), and homemade bread. The children wrote on a piece of paper the two topics they would most like to pursue. Mr. Greenberg then formed interest committees that would meet the next day.

The next day, Mr. Greenberg assigned to each committee the task of writing its own information booklet on its chosen topic. He specified the form the writing would take, but the children would eventually determine the content. To set the scene and familiarize the children with the form that their booklets were to take, Mr. Greenberg brought to class a model of informational writing: *Bread, Bread, Bread* by Ann Morris (Lothrop, Lee, & Shepard). The book contains exceptional photographs of various breads being made around the world. As the class surveyed the book, it paid particular attention to the photographs and text on each page. Mr. Greenberg and the students agreed that their information pages would include a drawing and a sentence with some information about the drawing—just like the literature model.

The children worked on their bread books for about 3 days during social studies period. The committees decided that their books should be about five pages long (one page for each child), but a few committees wrote more. To help the committees find information about bread, Mr. Greenberg located suitable trade books and inserted bookmarks at the proper places. At the end of the first day, Mr. Greenberg and the children sat in a circle with their papers and the books they had used. They shared what they had done and how things had gone for them. Most had gotten as far as locating something they wanted to write about and starting their drawing. The second day was spent completing the drawings and completing the associated text in their own words.

On the third day, Mr. Greenberg talked to the children about book titles and discussed how the covers of several of the books the children had been reading contained illustrations that represented the main idea of the text. The children illustrated the covers of their own books, added titles, listed their names as authors, put the pages in order, and stapled them together. The committees shared their books with one another, and the final copies were ceremoniously added to the classroom library. The "Bread From Around the World" committee compiled this book, along with appropriate illustrations, which they entitled *The Bread Book:*

Page 1: Navajos eat fry bread almost every day. It is fried, not baked.

Page 2: People from France eat baguettes. They are long loaves of bread.

Page 3: Ukrainians eat paska bread at Easter. Paska is very special.

Page 4: Mexican people like to eat tortillas. Tortillas are cornmeal bread.

Page 5: Jewish people bake challah to eat at their Friday meal. It is a white bread with braids.

The main thing about initial report writing is that children are motivated to write; you must help fuel that motivation by providing "grabbers"—objects or books that just can't be ignored. These launch the children into personal writing projects that help them both describe what they know and begin to ask questions on their own. For an example of higher grade written reports, take a return visit to Nick Bonelli's classroom on pages 101-106 in Chapter 3.

Biographies

Biographies join the written report as an especially relevant form of writing in social studies because they allow students to conduct research into a person's life and write a true account of that person. As with other forms of writing in social studies, students should begin the process by using sources from children's literature as models. Farris (1993) comments:

> Beginning any composition is a challenge for all writers; however, initiating a biographical sketch can be especially difficult for children. Because children have problems establishing a frame of reference for a setting (both time and place), their biographies often fail to describe a distinct period and locale. As a result, it is not unusual for a child's biography about a historical figure . . . to begin with "Once upon a time, there was a boy [girl] named . . . ." (p. 227)

There are many authors who could serve as exemplary models for students' biographies, but I strongly recommend the works of Jean Fritz, "a master biographer" who has written about several periods in American history, particularly the Revolutionary War. In *Why Don't You Get a Horse, Sam Adams?* (Putnam), for example, Fritz describes the colorful hero this way: "His clothes were shabby and plain, he refused to get on a horse, and he hated the King of England." Her series on the Revolutionary War contains several stories that show historical characters as real people, helping children enrich their understandings of history in an enjoyable way. Two other master biographers serve as superb models for biographical stories: Russell Freedman and Milton Meltzer. Each has published a number of fascinating biographies that breathe life into the stories of the famous and infamous throughout American history.

In writing their biographies, students must focus on (1) a realistic description of the time when and place where the person lived (biographies can be written about living people or historical figures), (2) an accurate characterization of the person, (3) a careful accounting of the significant events in the person's life, and (4) the values and interests influencing the person to act as he or she did. Cullinan and Galda (1994) suggest that social studies teachers look at biographies in two major ways. The first and most popular is the biography in terms of an historical period such as the Civil War or the Great Depression. The second contains stories of people who relate to a particular theme such as "the struggle for human rights," "great explorers," or "immigration." Whatever the choice, biographies hold an important place in the social studies writing program. Conducting research for their biographies and putting thoughts together to form a written portrait of an historically significant person help students more clearly understand the people who have made history and elucidate the strengths and weaknesses influencing the great decisions of their lives. One student's biographical sketch of Sacagawea is shown as an example of a short biography in Figure 9–6.

Newspapers

Classroom newspapers offer a stimulating expository writing medium especially appropriate for the social studies because they offer a realistic forum where students can express factual accounts of what they are learning or attempt to influence others with political cartoons or editorials. As we earlier discussed other forms of expository writing, we found out that children must be exposed to models of writing; the same is true for newspapers. A trip to a newspaper publisher and careful examination of a local newspaper can help the children understand how newspapers are planned and produced. Although newspaper publishers differ in their organization, all generally include the following positions, which many teachers have encouraged their students to simulate in the classroom:

- *Publisher.* Owner or person who represents the owner; responsible for the overall operation of the newspaper. (The teacher usually assumes this position.)

FIGURE 9–6
Biographical Sketch of Sacagawea

Mike   Mar. 0, 1991

                        Sacagawea

She was a Shoshoni Indian Woman who guided Lewis and Clark. She was helpful because her presence was a sign of peace to different tribes. they encountered. She could find Many plants and herbs to eat. She wanted to see the Pacific Ocean and the whales. Clark took her all the way to the Great Stinking Pond. The trip took two years and 6 months. Sacagawea was a brave woman,

- *Editors.* Responsible for several facets of newspaper production, such as deciding what goes into the newspaper, assigning reporters to cover events, determining where to position the stories, or writing their opinions of significant events (editorials).
- *Reporters.* Cover the stories. Columns to which reporters can be assigned include class news; school news; local, state, national, and world news; and sports; or special events. Reporters can interview others as primary sources for gathering their news or use secondary sources such as radio, television, newspapers, or magazines.
- *Feature writers.* Produce special columns such as jokes and riddles, lost and found, a version of "Dear Abby," recipes, and so on.
- *Copy editors.* Read stories for mistakes and adjust length to meet space requirements.

After assigning several students to the news gathering and processing activities, you will want to get the rest of the class involved in these "behind the scenes" departments:

- *Production department.* Put pictures and stories together and get them ready for printing. (Children who know how to use the word processor, or have a desire to learn, make special contributions here.)
- *Art department.* Create ads (with pictures) that highlight coming events, special sales, parent meetings, special assemblies, or other important school functions.
- *Mechanical department.* Run the printer or copier, collate the newspaper pages, staple them, and get them ready for distribution.
- *Circulation department.* Deliver the newspaper to the principal, other teachers, and classmates.

Newspapers may be written in either a contemporary or an historical setting. Consider Mr. Rivera's fifth-grade classroom, for instance:

Classroom Vignette

The students were involved in a unit on the Revolutionary War. They created a period newspaper for 1776 with front-page stories blaring these headlines: "British Evacuate Boston," "Grand Union Flag Unfurled Over Boston," "Redcoats Invade New York," "Congress Approves Declaration of Independence," and "Washington Stuns Hessian Fighters." Feature articles detailed such items of interest as quilting, candle-making, and the steps of drying food for winter. An advertising section offered articles for sale (spinning wheels, bed warmers, teams of oxen, pewter tableware, flintlock rifles, wigs, and the like). An employment section offered such jobs as post rider, saddler, tanner, wigmaker, tavern keeper, chandler, mason, cooper, hatter, and printer. An editorial page displayed a political cartoon showing a crowd of Continental soldiers pulling down an equestrian statue of King George III as they celebrated the signing of the Declaration of Independence in Philadelphia, and an editorial soliciting funds for the relief of widows and children of the patriots "murdered" at Lexington. There was a book review of Thomas Paine's *Common Sense,* and a sports section detailing the results of such popular events as stool ball, quoits, arm wrestling, and gunny sack races. When finished, the various sections were stapled together and distributed to the other fifth-grade classrooms.

By simulating a real newspaper, children bring more interest and excitement to their learning. Be careful, though, to make sure that all children have an opportunity to assume as many different newspaper responsibilities as possible. Let them try the various jobs so they develop a complete understanding of the many duties performed on a real newspaper.

Potpourri of Expository Writing Opportunities

Teachers of elementary school social studies should use a variety of expository writing formats in purposeful ways to avoid the dullness associated with such drab routines as copying a report from the encyclopedia. Teachers employing useful formats must share good examples from textbooks and trade books; with exposure to these models, children will learn to write quality expository pieces. The list below suggests expository writing formats that will supplement the major suggestions in this text.

- ABC books where each letter of the alphabet tells something about a topic (for example, *The ABCs of Native American Culture*).
- Calendars where each page incorporates an illustration and information (for example, a yearly calendar with each month illustrating a United States National Park).
- "Big books" where each page contains a combination of drawings and brief text about a topic under study (for example, a book about trains).
- Information books or pamphlets where groups of students write separate sections or chapters (for example, a description of life in the Southwestern United States).
- Radio, television, or movie scripts that can be performed by the students.
- Historical fiction where students write about events of the past through the stories of people who seem real to us (for example, a day in the life of a child who accompanied Lewis and Clark in their journey west).

- How-to books where students write instructions for something related to a unit of study (for example, *How to Make a Tipi*).
- Travel brochures that describe a particular location under study.
- Recipe books that describe foods eaten by particular groups of people (for example, favorite foods of our nation's presidents).

Expressive Writing

A second popular form of writing found in elementary school social studies classrooms is more relaxed and personal than expository writing. Called *expressive writing,* this form encourages children to write in much the same way as they speak. Norton (1997) defines expressive writing as "very close to speech and, consequently, very close to the writer. It is relaxed and intimate. It reveals the writer and verbalizes his or her consciousness. . . . Expressive writing is frequently characterized as thinking aloud on paper" (pp. 499–500).

Many kinds of expressive writing are employed in typical social studies programs, but the most frequently used are *journals and diaries, stories (narratives),* and *letters* written to and by real or imaginary people. Expressive writing is important in social studies classrooms because it encourages students to tell about their personal experiences, express what they already know about a topic of investigation, or communicate their personal feelings about a theme or learning experience.

Journals and Diaries

Starting with young children who draw simple pictures and scribble words to tell about interesting people, places, and things, and progressing through adult learners who write about their ideas, beliefs, and concerns, journals are a favored way to record personal thoughts. As such, they become students' dialogues with themselves that express ideas in highly personal ways. Journals can be a handmade booklet or a purchased notebook. They may be made in several styles, but the two most applicable to the social studies program are the *dialogue journal* and *double-entry journal.*

Dialogue Journals

Dialogue journals are designed specifically to promote interaction between a student and the teacher. Each student writes what she or he learned or felt about a learning experience, and the teacher composes a brief response directly on the journal page (see the sample in Figure 9–7).

Double-Entry Journal

A double-entry journal is a format by which students can summarize what they have learned and describe their feelings related to the main ideas or events. For example, students might be asked to read the book, *The Titanic Lost . . . and Found* by

FIGURE 9–7
Page From a Dialogue Journal

June 8, 2001

I read the book the Titanic Lost... and Found. The best part was when the Titanic sank. It was awesome.

"Mike!"

Your interest in historical research is certainly growing. I'm happy you enjoy it so much!

Judy Donnelly (Random House). The teacher would have blank double-entry journal pages ready for use. As the students read, they jot down brief notes about the plot in the left-hand column. Afterward, they record their personal feelings in the right-hand column (see Figure 9–8).

Diaries

Diaries are short, private records of personal observations of events and reactions to those events. Diaries unlike journals are kept private.

Imaginatively placing themselves in the shoes of real people (such as Columbus during his journey to America) can serve as a strong stimulus for diary writing. Or, your students might enjoy pretending to take part in the journey West with the Ingalls family by writing diary entries for each exciting day on the trail. The following diary entries were written by a student pretending to be a sailor on one of Magellan's ships that was unable to complete the circumnavigation of the globe in 1522:

FIGURE 9–8
Double-Entry Journal

Something Interesting	How I Felt About It
Titantic is ready to sail on its first voyage.	I would be excited.
Titantic is in icy waters off the coast of Canada.	I would feel scared in the dark.
The lookouts sees a mountain of ice.	I would panic. "Is the ship going to hit it."
The Titantic hits the iceberg.	It is going to sink!
People laugh and joke.	They still think they are on a ship that cannot sink.
Some people get into the lifeboats.	Women and children go first – they're lucky.
Titantic sinks into the black water.	I feel sad so many people died.
A ship called the Carpathia saves the people in the lifeboats.	I feel better that some people are saved.
Titantic lies many miles down under the atlantic.	I would love to see how beautiful the ship was.

Susana

Classroom Vignette

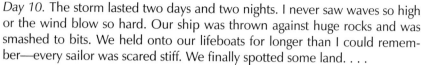

Day 10. The storm lasted two days and two nights. I never saw waves so high or the wind blow so hard. Our ship was thrown against huge rocks and was smashed to bits. We held onto our lifeboats for longer than I could remember—every sailor was scared stiff. We finally spotted some land. . . .

Day 11. Fresh water is disappearing. Our captain divided us into four groups. Each group was to go in a different direction to search for fresh water. In midafternoon the fourth group found a freshwater spring on the west side of the island. . . .

Writing is easy for children if they write about what they know (i.e., something they have read about or lived). Regardless of the source of motivation, children have much to share in the social studies classroom. Throughout the social studies program, the children use writing as a tool to understand their world and communicate their knowledge of it.

Stories

As children are exposed to a wide variety of literature throughout the curriculum, including social studies, they will become interested in using the styles of their favorite authors as a model to write their own stories. Tompkins and Hoskisson (1995) point out: "As students read and talk about literature, they learn how writers craft stories. They also draw from stories they have read as they create their own stories, intertwining several story ideas and adapting story elements to meet their own needs" (p. 347). This penchant for imitation is important because imitation plays a productive role in students' learning to write. All writers, either by accident or design, begin by imitating the style and structure of other writers. Take advantage of this imitative urge and encourage it as the students explore their "writing personalities." In other words, when students are studying the folklore of Africa they might want to compose their own "pourquoi tale" patterned after Verna Aardema's *Why Mosquitos Buzz in People's Ears* (Dial) or a "trickster tale" after Eric A. Kimmel's *Anansi and the Talking Melon* (Holiday).

When using a story related to the social studies topic, help your students organize a story map so they can visualize how the authors structured the plot. Story maps are excellent planning tools for story writing. A *story map* is a word or picture diagram that shows the main events in a story and the relationship of those events. Making a map of a story can help children identify the detail and sequence they want to use in their stories. A simple story map is shown in Figure 9–9.

FIGURE 9–9
Story Map for *Why Mosquitos
Buzz in People's Ears*

Beginning

Mosquito tells iguana a big lie.

Iguana puts sticks in his ears.

Snake crawls into rabbit hole.

Rabbit runs away.

Crow spread the alarm.

Monkey breaks limb.

Limb kills an owlet.

Mother Owl does not hoot to wake up the sun.

Middle

King Lion calls a meeting of all the animals.

End

Mosquito is found guilty.

Mother Owl is satisfied.

Mother Owl hoots to wake the sun.

Mosquito has guilty conscience.

Mosquito buzzes in people's ears.

KPAO!

Letters

Letters offer another major type of expressive writing for social studies classrooms. Children typically write two kinds of letters in social studies: friendly letters and business letters. As Tompkins and Hoskisson (1995) explain, "Friendly letters might be informal, chatty letters to pen pals or thank-you notes to a television newscaster who has visited the classroom. . . . When students write to General Mills requesting information about the nutritional content of breakfast cereals or letters to the President expressing an opinion about current events, they use the more formal business style" (p. 383). Because friendly letters and business letters are written in customary, standard styles, some teachers find it helpful to provide models (charts, samples, and so on) for the children to follow as they write. This is important, for as we explained with the other types of writing, form and content are interchangeably essential for effective communication.

Numerous experiences encourage both types of letter writing in the social studies. For example, Mr. Littlejohn, a second-grade teacher, took his children on a trip to the post office as part of their study of community workers. After they returned,

FIGURE 9–10
Paragraph From a Student's Letter

In Flagstaff it snows a lot during wintertime. Sometimes it snows as much as 2 feet. There are many winter sports in Flagstaff. People especially like to go to the Mountains to snow ski. I like to go snowboarding.
Jeffwan

he read them the book *The Jolly Postman* by Janet and Allan Ahlberg (Little, Brown). The children enjoyed hearing about the likable mail carrier who delivers letters to such fantasy characters as Cinderella and the Three Bears. Mr. Littlejohn used the story as a springboard for his children to write letters to their favorite community workers. Older children find letter writing experiences worthwhile, too.

Classroom Vignette

Wallace Kahn, a fifth-grade teacher from Flagstaff, Arizona, carried out a different type of letter-writing experience with his class of older children. It was written within the context of sending an "exchange package." Mr. Kahn asked his students to collect objects in a package that could clearly describe life in Flagstaff. The package would then be sent to fifth-grade students in a classroom in San Francisco. The students selected items such as a small chunk of volcanic cinder, a branch of ponderosa pine, a plastic fish, a cowboy hat, a pair of mittens, and a small piece of silver jewelry with turquoise stones. The students then wrote a letter that explained each of the items. Figure 9–10 shows one paragraph of the letter; it describes the significance of including the pair of mittens. The students in San Francisco reciprocated by preparing a package and explanatory letter describing their city.

Luz Ramirez, a fifth-grade teacher discussing disagreements that led to the Civil War, asked half of her class to imagine that they lived in the North but had relatives living in the South. Each "Northern" student was to write a letter to the "Southern" relative, encouraging him or her to remain loyal to the Union. The letters were randomly delivered to classmates, who were then responsible for composing a response. Ms. Ramirez made sure her students understood the issues before they presented their arguments in the letters. Aaron Wakefield asked his students to apply what they knew about Christopher Columbus by pretending to be Columbus writing a letter to Queen Isabella requesting support for his explorations. See Figure 9–11 for an ex-

FIGURE 9–11
A Student's Letter Written From the Perspective of an Historical Figure

> Dear Queen Isabella,   Oct. 1491
>   If you'll give me financial help I can prove that this world is round. I have been studying sailing since the age of 13 years old and am a very experienced sailor I could get my crew and me out of any storm we run in to.
>   I can bring you back spices, treasures and any thing else, I may find on this voyage. I have been sailing from a young age and should be successful I hope you'll consider my request.
>
>   Your Loyal subject,
>
>   C. Columbus, Navigator

ample of an imaginative letter. Both imaginative and true-to-life letters have a place in dynamic social studies instruction.

Eli Pronchik asked his students to examine the yellow pages of the telephone directory to locate the names of travel agencies. They wrote true-to-life letters to the agencies requesting information about important historical sites around Philadelphia, including Convention Hall, the Betsy Ross House, and the Liberty Bell. Writing business letters to request information and free materials is a staple of social studies instruction.

The goal of all forms of expository and expressive writing is to present ideas accurately and creatively. Salinger (1988) describes the possibilities for selecting expository and expressive writing activities during a unit on Thanksgiving:

> A unit on Thanksgiving, for example, could lead to a factual report on the Pilgrims or Indians, a letter from Plymouth colony, a description of the first Thanksgiving dinner (with drawings), a thank-you note to Squanto, a script to reenact the first Thanksgiving dinner, a report on Thanksgiving celebrations in other nations, or a recipe for cooking turkeys. A few children may find some of the assignments "silly"; others may have trouble handling a full "research report"; but because there are varied opportunities for writing, the whole class—avid storywriters to meticulous fact seekers—can accomplish something. Through assignments like these, children gain valuable practice writing in different modes, for different purposes, and to different audiences. (p. 258)

Imaginative (Creative) Writing

When social studies educators talk about writing in social studies, expository and expressive writing are the two forms that seem to dominate their suggestions. However, we must be aware that young writers enjoy writing imaginatively, too. Called *creative writing* or *imaginative writing,* this form is intended not only to communicate information, but also to express original ideas. Creative writing not only communicates information, it makes learners care about the information—it makes them feel, it makes them experience, and it "gets under their skin." Consequently, students need many opportunities in the social studies to write and to gain pleasure from sharing their creative products.

Poetry

One of the most powerful ways to encourage imaginative writing in social studies is to involve children in poetry, a form of writing that allows individuals to concoct word pictures as they express ideas in highly original ways. Through poetry, children organize their thinking about social studies content in order to convey clear images of what they have learned.

Free-Form Poetry

One of the most spontaneous forms of writing is free-form poetry. Free-form poetry need not rhyme, so some children create unintentional rhymes as they play with words, combining them in unique ways. Denman (1988) calls this process "word-smithing," or awareness that thoughts can be expressed freely on paper. The charming example shown in Figure 9–12 is a free-form poem that was written for a teacher whose early grade social studies program was focused on helping children participate as members of a compassionate classroom society. Her foundational step was to encourage children to be positive, to smile, and to feel special. The inspiring poem was one perceptive young writer's heartwarming expression of how it felt to be in that classroom—unconditionally accepted and loved for one's good (and maybe not-so-good) attributes!

The most frequently utilized way to more directly connect poetry to social studies is to focus on the content of a topical learning experience. For example, Ronald Benoit carried out a poetry experience within the context of a lesson on popcorn—a food introduced to the colonists by the Native American Indians.

Classroom Vignette

Mr. Benoit began the lesson by holding up a cup of popcorn and asking his second graders what they already knew about it. Mr. Benoit recorded their comments on a knowledge chart. They listed everything they knew about popcorn in a column labeled, "What We Know About Popcorn." Then he read aloud *The Popcorn Book* by Tomie de Paola (Holiday), a delightfully illustrated and written picture book that tells about varieties of corn and where and when popcorn was discovered. The book includes interesting information about popcorn, such as how much is eaten each year and the "popcorn blizzard" in the Midwest. After the book was read, Mr. Benoit helped the students record in the second column ("What We Learned About Popcorn") any new information they uncovered about popcorn and corrected any inaccurate ideas from the first column. Because de Paola devotes much of the book to what happens during the popping process, Mr. Benoit followed up this information-processing activity by inviting the children to pop some corn. Throughout the entire popping experience, he stressed both an awareness of the understandings the children were acquiring as well as the sensations they were experiencing through their senses. Together, they talked about the color, size, weight, and shape of the popcorn. Children suggested words to describe the aroma of the popcorn as it popped. They commented on the exploding sounds, as well as the puffy new shape the kernels were taking. More words described its feel and taste as the popcorn was eaten. As the children contributed their ideas, Mr. Benoit recorded them on a data chart.

FIGURE 9–12
Experimental Poem

A Poem,
    Me

Me me I love you me, that's just how you are.

Even though you step in tar,
Me can't drive a car, But that's just how you are.

tar

After the children exhausted all their ideas, Mr. Benoit helped them write a free-form poem, each line beginning with "Popcorn." Here is the final product:

Popcorn

Popcorn was discovered by Native American Indians long ago.
Popcorn was worn as trinkets by some Native American Indians.
Popcorn jumps up and down when you cook it.
Popcorn looks like a puffy cloud when it is done.
Popcorn tastes delicious with salt and butter.
Popcorn is yummy!

This type of poetry is called *free-form poetry* because the children select words freely to express their thoughts without concern for rhyme or structure. In social studies, it is important to create free-form poems only after the children have experienced a meaningful learning episode and brainstormed lists of words to describe what they learned.

Patterned Poetry

In recent years, teachers have found structured poems such as haiku, tanka, senryu, diamante, or cinquain to be especially useful in helping students convey ideas about social studies topics. As we discovered with free-form poetry, teachers who want children to compose structured poems must organize instructional efforts so that the children both acquire sufficient knowledge on which to build their creations and understand the literary form that the final product is expected to take.

Craig Springer understood these basic principles of writing in the social studies program, so the first phase of his fourth-grade unit involved the process of *exploring the topic*, a period of content investigation during which children were encouraged to think deeply about something new to them. Therefore, as the "Southwestern United States" interdisciplinary unit unfolded, Mr. Springer carefully prepared a variety of fact-finding experiences.

Classroom Vignette

The emphasis of the unit was on the land and the people, so one particular section dealt with early life on large Spanish cattle ranches (*ranchos*). The focus of learning was on *vaqueras* (female cattle workers) and *vaqueros* (male cattle workers), expert riders who spent most of their day on horses rounding up cattle. To help children learn about these rancho workers, the teacher selected a quality resource from children's literature: *Carlota* by Scott O'Dell (Houghton Mifflin). The students collected and organized information from the book on an information processing web titled "Life of a Vaquera." The process of collecting, thinking about, organizing, and talking about the content was an important initial step for the teacher's young writers.

The second step in creating the poem was to guide children through the process of *brainstorming what they already knew* about the topic and *organizing the data* into categories of shared relationships. In this case, Mr. Springer selected a *cinquain* as the form of written expression, so his brainstorming and data organization activities needed to correspond to the basic elements found in cinquains. Teachers have adapted cinquain patterns over the years; Mr. Springer selected one that consisted of a *first line* with one word (or short phrase) naming the theme, the *second line* having two words (or phrases) describing the theme, the *third line* consisting of three words (or phrases) describing some action related to the theme, the *fourth line* naming something associated with the theme, and a *fifth line* designating another word (or short phrase) for the theme. Because the children needed to brainstorm ideas for each line, Mr. Springer began by dividing his class into three groups to brainstorm what they knew about vaqueras. He asked one group to brainstorm all the words or phrases they could think of to *describe* vaqueras, another group to generate *action* (doing) words associated with vaqueras, and a third group to list *things* associated with these cattle workers. After about 10 minutes of active deliberation, each group selected an individual to place its word inventory on a large sheet of chart paper. Words and phrases such as *expert rider, skillful, lonely, tired, hard working, Spanish,* and *woman cowboy* were written on the "Describing Words" chart; the "Doing Words" collection contained *roping, herding, riding, camping out, tending cattle, branding,* and *rounding up;* the "Vaquera Things" chart included *reata (rope), silla (saddle), chaparreras (chaps), blanket, horse, cattle, rancho, spurs, cattle drive, stampede, hat,* and *bandanna.* By creating a comprehensive list for each grouping, Mr. Springer not only helped the students organize an extensive word bank from which to select words for their cinquains, but also reviewed and organized into meaningful categories the content under study.

After the charts were completed and displayed, Mr. Springer directed the "Describing Words" group to carefully examine the "Doing Words" list and select three words or phrases they felt best characterized what vaqueras do. The "Doing Words" group was to choose from the items on the "Vaquera Things" chart the four things they thought were most closely associated with the rancho workers. The "Vaquera Things" group examined the "Describing Words" chart to select the two words or phrases that best described vaqueras. Intensive discussion continued for some time as the children in each group argued strongly for their personal choices. Key information from previous learning experiences bubbled forth as the children sought to convince their group partners of the value of their word choices.

When each group made the appropriate number of selections from its assigned chart, Mr. Springer directed the class toward the third phase of the writing continuum—*composing a group model.* He posted a large sheet of chart paper that exhibited a writing guide, as shown in Figure 9–13.

FIGURE 9–13
Cinquain Pattern

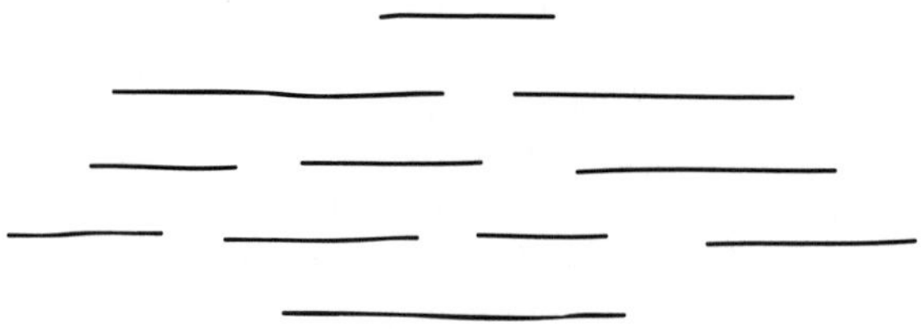

_____

_____   ___

___   ___   ___   ___

___   ___   ___   ___

_____

FIGURE 9–14
Completed Group Cinquain

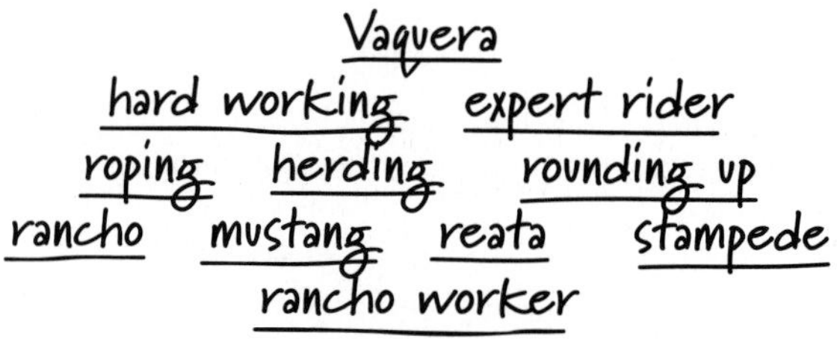

<u>Vaquera</u>
hard working   expert rider
roping   herding   rounding up
rancho   mustang   reata   stampede
rancho worker

On the top line, Mr. Spencer wrote the word identifying the topic under study—*Vaquera*. He then asked the "Vaquera Things" group to reveal the two describing words or phrases it had selected. He then wrote the phrases in the blanks on the next line—*hard worker* and *expert rider*.

In the same way, the "Describing Words" group offered its three action words or phrases for the third line (*roping, herding,* and *rounding up*) and the "Doing Words" group shared its choice of four vaquera things for the next line (*rancho, mustang, reata,* and *stampede*). "Try to think of another word for *vaquera* so we can put it in the last blank," was the teacher's final challenge. After considering all suggestions, the class decided upon *rancho worker* because the term defined these women. The group composition is shown in Figure 9–14.

During the fourth stage of the writing process—*editing and rewriting*—Mr. Springer encouraged the students to read the cinquain with him, and they revised the text appropriately. Commas were added to separate items in a list and the first word of each line was capitalized. After the corrections were made, the class chorused the selection once again.

To begin the next stage of the writing process—*individual writing*—Mr. Springer helped his students analyze the pattern of the group cinquain they had just completed: "How many blank spaces do you see on each line? What kinds of words were used on each line? How have the words helped paint a mental picture of what we have been studying about life in the Southwest?"

Mr. Springer had previously decorated a large bulletin board with illustrations depicting a vaquera twirling a rawhide reata above her head and a herd of cartoon-like cattle of the size to hold a 4 × 5″ piece of paper (the display was titled "Poetry Roundup"). He asked the students to read their cinquains and staple them to the sides of the cattle. This final phase of the writing process—*publishing*—encouraged the students to share their products with an audience. In this case, publishing took the form of oral reading and displaying the individual cinquains on the bulletin board.

Other forms of structured poetry are especially useful in the social studies. To list them is beyond the scope of this text, but they are easily located in teaching resources and literacy textbooks.

Many kinds of literacy experiences give balance to dynamic elementary school social studies programs. Through reading and writing experiences, the elementary school child learns to use language for a variety of purposes. Of all curriculum areas, students and teachers use literacy strategies more often than any other to fulfill the demands of learning in social studies.

AFTERWORD

In this text, *literacy* has been defined as an integrated approach where writing is nestled among reading experiences, resulting in a cohesive command of all language tools, more than thought possible when they were taught in isolation.

It is during infancy when we take our first wobbly steps on the journey toward literacy, an enterprise that continues throughout our lives. Vygotsky explains that it is this language process that separates humans from all other animals. Specifically, Vygotsky proposes that humans differ from animals because they are able to make and use tools in an effort to adapt to their environment and solve problems. Tools can be classified as *physical tools* (things such as saws or hammers) or *mental tools* (complex thought and language processes). In speaking of mental tools, Vygotsky emphasizes that human beings, to communicate what goes on in one's mind and how these thoughts might transform the physical environment, have created symbols. These symbolic tools might be expressed through such forms as a drawn picture, an algebraic formula, or a map, but the most influential "tool of the mind" in Vygotsky's thinking is language. And, because language appears to be so intimately associated with cognition (thinking), teachers must be aware that application of reading and writing is basic to thinking and learning in the social studies. It is through language that we learn about, and become more capable members of, our culture.

References

Atwell, N. (Ed.) (1990). *Coming to know: Writing to learn in the intermediate grades.* Portsmouth, NH: Heinemann.

Combs, M. (1996). *Developing competent readers and writers in the primary grades.* Upper Saddle River, NJ: Merrill/Prentice Hall.

Cooper, J. D. (1997). *Literacy: Helping children construct meaning.* Boston: Houghton Mifflin.

Cullinan, B. E., & Galda, L. (1994). *Literature and the child.* Fort Worth, TX: Harcourt Brace.

Denman, G. A. (1988). *When you've made it your own: Teaching poetry to young people.* Portsmouth, NH: Heinemann.

Farris, P. J. (1993). *Language arts: A process approach.* Madison, WI: Brown & Benchmark.

Freedman, R. (1987). *Lincoln: A photobiography.* New York: Clarion.

Fritz, J. (1979). *Stonewall.* New York: PaperStar.

Goodman, K. S. (1988). Look what they've done to Judy Blume!: The "basalization" of children's literature. *The New Advocate, 1,* 29–41.

Graves, D. H. (1983). *Writing: Teachers and children at work.* Exeter, NH: Heinemann.

Huck, C. S., Hepler, S., & Hickman, J. (1993). *Children's literature in the elementary school.* Fort Worth, TX: Harcourt Brace Jovanovich.

Jarolimek, J., & Parker, W. C. (1993). *Social studies in elementary education.* New York: Macmillan.

Keller, H. (1920). *The story of my life.* Garden City, NY: Doubleday.

Lapp, D., Flood, J., & Farnan, N. (1992). Basal readers and literature: A tight fit or a mismatch? In K. D. Wood & A. Moss (Eds.), *Exploring literature in the classroom: Contents and methods* (pp. 35–57). Norwood, MA: Christopher Gordon.

Murray, D. (1987). In N. Atwell, *In the middle.* Portsmouth, NH: Heinemann/Boynton Cook.

Norton, D. E. (1997). *The effective teaching of language arts.* Upper Saddle River, NJ: Merrill/Prentice Hall.

Prelutsky, J. (1984). *The new kid on the block.* New York: Greenwillow Books.

Ravitch, D. (1978). Tot Sociology. *American Educator, 6,* 38.

Salinger, T. (1988). *Language arts and literacy.* New York: Merrill/Macmillan.

Sewall, G. (1988). Literary lackluster. *American Educator, 12,* 35.

Sloan, G. D. (1984). *The child as critic: Teaching literature in elementary and middle schools.* New York: Teachers College Press.

Solomon, W. (1989). Teaching social studies creatively. *Social Studies and the Young Learner, 2,* 3–5.

Steptoe, J. (1987). *Mufaro's beautiful daughters: An African tale.* New York, NY: Lothrop, Lee, & Shepard.

Taylor, M. D. (1992). *Mississippi bridge.* New York: Bantam Skylark.

Templeton, S. (1997). *Teaching the integrated language arts.* Boston: Houghton Mifflin.

Thaiss, C. (1986). *Language across the curriculum in the elementary grades.* Urbana, IL: National Council of Teachers of English.

Tompkins, G. E. (1997). *Literacy for the 21st century: A balanced approach.* Upper Saddle River, NJ: Merrill/Prentice Hall.

Tompkins, G. E., & Hoskisson, K. (1995). *Language arts: Content and teaching strategies.* Upper Saddle River, NJ: Merrill/Prentice Hall.

Vacca, R. T., & Vacca, J. L. (2002). *Content area reading.* Boston: Allyn and Bacon.

Vygotsky, L. S. (1978). *Mind in society: The development of higher mental processes.* Cambridge, MA: Harvard University Press.

Wixson, K. K. (1991). Houghton Mifflin Social Studies (Teacher's Edition for *A message of ancient days*). Boston: Houghton Mifflin.

Young, E. (1989). *Lon Po Po.* New York: Scholastic.

Zinsser, W. (Ed.). (1990). *Worlds of childhood: The art and craft of writing for children.* Boston: Houghton Mifflin.

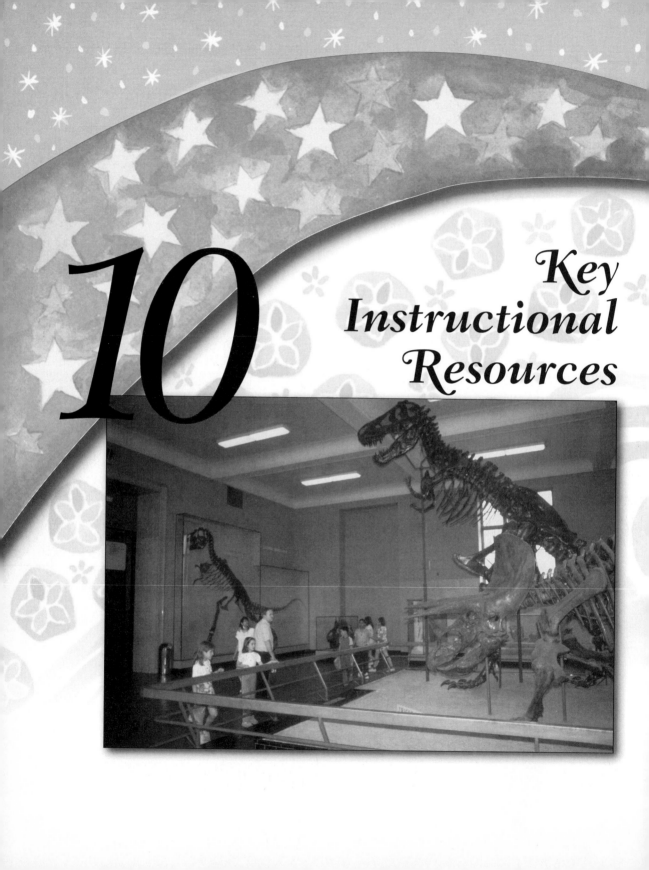

10

Key Instructional Resources

"*I* had an elementary school teacher, Mrs. Dunbar, who made social studies my favorite subject. We learned a lot because Mrs. Dunbar always had something fun for us to do. Once, in fifth grade, we were studying prehistoric life. Mrs. Dunbar asked us to clean off and bring in bones left over from dinner at home. You could probably guess that the next day we had a pile of all kinds of bones—fish, chicken, steak, you name it! Our job was to clean them thoroughly with a scrubber, boil them in vinegar water, soak them in a bleach and water solution to make them white, and put them in a large box called 'the boneyard.' Mrs. Dunbar then organized us into groups and let us choose any of the bones we wanted. We then glued or wired them together in the general shape of the dinosaurs we were studying. We gave our 'dinosaurs' their scientific names and displayed an information card next to the models. Mrs. Dunbar called us *paleontologists*. I still remember the word because it was a real thrill to be given such an impressive-sounding title. Elementary school social studies was so much fun, at least in Mrs. Dunbar's room. As a matter of fact, she has had more influence on my wanting to become a teacher than any other single factor."

Mrs. Dunbar was an exceptional fifth-grade teacher with many years of experience when I met her during my first year of teaching. A model professional, Mrs. Dunbar had a tremendous influence on my career. This remarkable woman motivated children to do good things. She did all she could to encourage them to be curious, active, and persistent in finding answers to life's mysteries. If you were lucky, somewhere along the line you would have had a Mrs. Dunbar: Someone who made the sun come up. Someone who smiled and was patient. Someone who wouldn't let you quit. Someone who told you that all things really are possible. Someone who motivated you to do good things.

The "Mrs. Dunbars" of our world realize that the goal of dynamic social studies education is to develop informed, caring citizens who will one day make a difference in our country. To help achieve this goal, our Mrs. Dunbars make social studies a subject that children can sink their teeth into by reading fascinating stories of Alaska's untamed frontiers, helping the class write with Chinese characters called *ideograms*, making special Pueblo paper-thin bread called *Piki*, creating a salt-and-flour relief map of the United States, or searching through the Internet to uncover fascinating information about coal mining.

William Webster (1993) had his "Mrs. Dunbar," too—a special teacher who evoked similar memories (his "Mrs. Dunbar" was named Miss Monroe). While a little boy in Miss Monroe's room, Webster first learned what an engrossing place our world was as Miss Monroe stimulated real interest and curiosity.

> When we studied the Colonial era, we visited Philiplse Manor, the home of an early Dutch patroon. We danced the minuet and made candles. We also wrote stories, created plays, and learned our arithmetic by preparing Colonial food and building sets for our dramas.
>
> That year we also studied Brazil and the great Amazon River, and it sparked in me a lifelong desire to go see the river, the rubber trees, and how tapioca grew. . . . My mother came to school and helped my teacher, Miss Monroe, cook the tapioca, something we now call parental involvement. My classmates and I constructed a native village using papier-mâché for the rubber trees and weeds for thatch-roofed houses. The richest girl in the class was the daughter of a ship captain who had been to Belem. Her father came and told us about the city, which we now label "using community resources." (p. 45)

As a result of his involvement in these splendid activities, Webster developed a lifelong infatuation with Brazil and the Amazon River—he had a powerful dream to travel there. Decades later, Webster's little-boy dream came true; he visited Brazil and traveled the Amazon, living everything he had learned about in Miss Monroe's class. Webster was fortunate to have a teacher who brought such life to her social studies program that she was able to inspire lifelong dreams within her students. However, Webster laments that an overwhelming emphasis on meeting standards and achieving high scores on standardized tests are cheating many of today's social studies teachers of their dream-building time. Has the quest for inspiration been squelched by a sanitized and bland school curriculum? Have we seen the last of our Mrs. Dunbars and Miss Monroes?

When your first class comes eagerly marching into your classroom, whom will they face as their teacher? Will you take your place among the Mrs. Dunbars and Miss Monroes of educational history? Or, will you be satisfied to be counted among the ordinary? Some of your children, like Webster, will only need to have their natural enthusiasm for learning kept alive, while a number of unlucky others will need to have it revitalized. Fortunately, you can reach both groups by offering learning materials that fascinate children and capture their interest. Sometimes you will use textbooks and workbooks, but not all the time; these will be but one kind of teaching tool. A true test of your skill lies in how well you choose all other materials for instruction. What will work best today: trade book, textbook, video, computer, slides, filmstrip, pictures, overhead transparency, cooking, singing, dancing, field trip, or guest speaker? *Sometimes* the children will need to sit, but most of the time they will be up and involved, performing tasks and investigating mysteries.

DO SOMETHING REAL

Because children of all ages must be offered experiences in which they can actively participate, the first job of all dynamic social studies teachers is to challenge them with fascinating materials and abundant stimulation. Direct involvement in real classroom experiences such as making fabric dyes from berries or designing *petroglyphs* (early Native American cave drawings) not only deepens their understandings, but also widens their interests. For example, Elizabeth Susko, a fifth-grade teacher, wanted to take the study of the Pueblo Indians of the Southwest beyond the textbook to where the action is. Her students had learned that early Pueblo Indians, the Anasazi, built adobe homes by using the "puddling method." They built walls by placing handfuls of wet adobe on a mound, letting it dry, and adding another layer. They would add layer after layer rather than stacking bricks. The Spanish introduced adobe bricks to the Pueblo Indians, and, because of its many favorable properties, adobe remains a building material of preference today.

Classroom Vignette

To help her students understand how to build with the adobe material, Ms. Susko adapted the adobe brick-making process accurately described by Byron Augustin and Michael Bailey (2001). Before actually involving students in the activity, Ms. Susko built wooden frames from common lumber, one frame for each team of four students (see Figure 10–1), to be filled with wet adobe to make bricks. The inside dimensions of the frame were $8 \times 10 \times 4$" (although the traditional size of adobe bricks is $10 \times 16 \times 4$"). Ms. Susko left an additional 6 inches of wood on each corner of the long sides to serve as handles for removing the frame once the adobe had set.

Real experiences help children view the world from an expanded horizon free of traditional stereotypes.

FIGURE 10–1
Form for Shaping
Adobe Brick

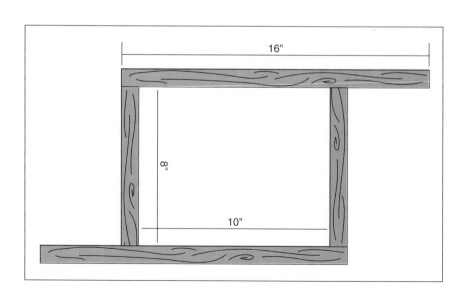

Ms. Susko set up work stations outside on a hard asphalt surface and placed at each station a wooden form, a sheet of wax paper that extended beyond the sides of the form, two large plastic bags, a putty knife, a one-gallon plastic water jug, and three or four handfuls of hay. She then directed the students through the following course of action:

1. Two students in each team went to a caliche pile and filled their plastic bag approximately two-thirds full of caliche. (Caliche has a high clay content that serves as an adhesive for the adobe brick. Ms. Susko's source of caliche was a local building supply company.) The two other students went to a dirt pile and returned with a plastic bag two-thirds full of dirt. Because caliche absorbs moisture, some students' hands got very dry. (It might be best to wear rubber gloves if your students will be handling the caliche at any time during this activity.)

2. The first two students thoroughly hand mixed the caliche and dirt (in equal portions). They then slowly began to add water to the mixture. Ms. Susko stressed that the water needed to be added slowly.

3. After the first two ingredients were mixed, the other two students mixed in the hay, making sure that not too much was added. Some groups needed to add water to acquire the desired texture and consistency (it is best to err on the dry side). Ms. Susko showed the students how to test the consistency by forming a baseball-size patty from the material, hold it waist high, and let it drop to a hard surface. If the adobe splattered everywhere, it was too wet, and more caliche or dirt was needed. If the material broke apart into pieces, the material was too dry, and a little more water was needed. If the patty stayed together and made a solid "thump," the material was ready to shape into an adobe brick.

4. Students spread out the sheet of wax paper and placed the form on top, making sure that the wax paper extended beyond the outside edges of the form. They placed the adobe mixture in the form and packed it tightly to force out all air pockets.

5. The adobe set in about 10 to 15 minutes. During that time, the students cleaned their hands with a hose and soap and water. Children handling the caliche without rubber gloves applied some hand lotion.

6. The students removed the bricks from the forms. They took the putty knife and cut around the interior edges of the form, loosening the brick. When they took the form by the handles and lifted it from the brick, they viewed their very own adobe brick. The bricks were allowed to dry in the sun for several days.

"Congratulations, class," smiled Ms. Susko wryly. "Only 6,000 more and we'll have enough to construct our own little house!"

Like Ms. Susko, social studies teachers around the country have developed significant learning situations in which their children are actively and directly involved in doing something real:

- Designing an Egyptian-style calendar, creating numbers the ancient Egyptian way.
- Decorating Ukrainian Easter eggs, called *pysanky.*
- Preparing and eating Russian pancakes, called *bliny* (blee-NEE).
- Celebrating "Juneteenth," a slang combination of "June" and "nineteenth." Food, music, and dance commemorate the day in 1865 when the slaves discovered the Civil War had ended.
- Reciting this prayer chant, to the slow beat of a drum, that was used by the Navajo people:

> *May I walk in beauty before me.*
> *May I walk in beauty behind me.*
> *May I walk in beauty below me.*
> *May I walk in beauty above me.*
> *With beauty all around me, may I walk.*

- Folding paper in the Japanese origami tradition.
- Playing "Marble Bridge," a game German children like to play.
- Constructing a piñata to hang from the ceiling during the classroom celebration of *Los Pasados,* a Mexican holiday recalling the Holy Family's journey from Nazareth to Bethlehem.

REALIA

In addition to providing children opportunities to *do* real things, dynamic social studies teachers must also arrange opportunities for them to *explore* real things, too. Children are spellbound as they handle a real Akua-ba doll from Ghana while you tell the story of how it is tucked into a skirt at the waist and carried by girls who hope to have children in the future. Opening up a matryoshka (mah-tree-OSH-ka) doll from Russia to reveal smaller and smaller dolls helps the children conceptualize this favorite nesting toy in ways that words alone can never equal. A set of worry beads from Greece is the only way to meaningfully explain this great relaxation technique. And what child wouldn't enjoy twirling a cowboy lariat while studying America's Old West? Real items inspire fascination and awe in any social studies topic.

This is not meant to imply that teachers should never talk, but only to stress that they enhance learning by allowing children to interact with something real. Think about all the possibilities for bringing in realia during any unit of study:

- Clothing (cowboy chaps, Japanese happi coat, chef hat and apron).
- Money (ruble, yen, mark, peso).
- Documents (wills, letters, newspapers, court records).
- Household items (colonial butter churn, Asian wok, African calabash).
- Musical instruments (Mexican guiro, Japanese den den, Zulu marimba).
- Tools (stethoscope, mortar and pestle, fishing net).
- Food (Pueblo Feast Day cookies, Mexican wedding cakes, Nigerian peanut soup).
- Toys (Chinese kites, Colonial "buzz saw," Jewish dreidel).

Oftentimes, teachers make collections of objects related to a theme and store them in boxes so they are kept well organized from year to year. These "prop boxes" can represent countries, cultures, community helpers, or most any other topic. In selecting items for cultural boxes, for instance, be sure that you incorporate "present-day" samples whenever you include traditional items such as kimonos, kilts, or sombreros. If you want to put together a box of serapes, sombreros, and other traditional Mexican attire, for example, you will give an unfair picture of what people dress like in Mexico today unless you also include examples or photos of contemporary Mexican clothing.

To make a prop box, get a large, sturdy cardboard carton that can be painted and easily decorated (preferably by your students). Place real objects inside; some can be bought, others might be donated by businesses, and parents are always willing to contribute items (if they're returned in the condition you received them). Below, you will find two suggestions for prop boxes: One is specific to community helpers, the other to a culture.

Physician's Box
  Doctor's bag
  Stethoscope
  Bandages
  Empty pill bottles
  White shirt to use as a uniform
  Nurse's hat
  Plastic digital thermometers
  Eye chart
  Pad and pencil
  Telephone

China Box
  Chinese tops
  Kite

Wok

Chopsticks

Tea cup

Dried lentils (bean sprouts)

Noodles

Silk cloth

Bamboo mat

Pleated fan

Lantern

Calligraphy brush

Abacus

Tangram puzzle

Game materials for *Ti Jian Zi* (similar to *Hackey Sack* in the United States)

Shirts with mandarin collars

Farmer hats

Silk dresses

Pictures of Chinese people in modern apparel

Classroom Vignette

Orpha Diller, a student in my social studies methods class, was assigned to teach a short, 1-week mini-unit of her choice during the field experience component. Because we were addressing multicultural studies and the global community at the time (experiencing our rich diversity of traditions and customs), Orpha chose to use the cultures and traditions of Kenya as the topic for her unit. Being convinced that the best way to learn about a culture is to examine ordinary objects used in daily life, Orpha obtained from fellow students, parents, relatives, friends, and local ethnic stores objects that would help her students learn more about Kenya and its people.

On the opening day of the unit, Orpha hauled in an antique-looking trunk that she called "Grandma's Trunk." She told the students that hers is a very close family and everyone pitched in to help her get this important unit ready for her first big teaching responsibility. She went on to explain that the trunk was a gift from her grandmother who was so very proud of Orpha becoming a teacher. It was a special trunk, her grandmother explained, because whenever Orpha needed something special to use in her classroom, all she had to do was open it up and look inside. In anticipation, the students and Orpha slowly opened up grandma's trunk and found the following objects inside: *shanga* (beaded jewelry), *batik* fabric, a *skafu* (head scarf), a *mkeka* (straw mat), a *kikapu* (straw basket), *pesa* (Kenya shilling—money), *stempu* (stamps), and *kinu na mchi* (mortar and pestle). Orpha introduced the objects one at a time, inviting the students to examine each as she told about them. Orpha arranged the objects on a table display and added large-size index cards labeling the objects and briefly describing their use. As students made their own masks, baskets, and beaded jewelry, their items were added to the "mini-cultural museum."

FIELD TRIPS

Field trips are first-rate learning adventures for any age—kindergartners, upper-graders, high school or college students, and adults. Who doesn't learn best when taken to a firsthand source of information?

Good field trips for elementary school youngsters involve students as active participants. A trip to the automobile museum during which children are lectured at by a guide or required to be silent onlookers is not as good as a trip that allows them to get into an automobile, sit in the driver's seat, talk to the owner, listen to the engine's clatter, and possibly even be taken on a short ride. A trip to the bakery where the children can only watch something being made is not as good as a trip that allows them to measure and mix the ingredients for a batch of muffins that they will devour later. A good field trip must envelop elementary school children in opportunities for direct, meaningful involvement.

Classroom Vignette

At no time was this point clearer than when Erika Ziegler heard a discussion about firefighters among a small group of her first graders as they were looking at a book in the library corner. Because she was planning to introduce firefighters as part of their year-long study of community workers, she listened as the children talked excitedly about the big rubber boots and hard red hats the firefighters wore in the pictures. Seizing the moment, she suggested that it might soon be possible to visit firefighters at the fire station near the school. "We could really go there?" asked Chin-Jen. "Wow, that would be fun!" exclaimed Sarah. Ms. Ziegler had no doubt that interest was high, so she called the fire department to ask if they could visit. She was informed that not only was such a visit possible, but that arrangements could be made for two firefighters to take them for a short ride on a truck.

On the day of the trip, the children eagerly disembarked the school bus after it arrived at the fire station. Their anticipation grew into excitement as they peered in at the bright red and silver truck parked in the garage. Keeping the children well in control, Ms. Ziegler reminded them to stay behind her until the firefighters explained what to do. When two firefighters invited them to enter, the children gingerly approached the vehicle. The firefighters showed the children all their paraphernalia and explained their jobs. "The truck is so shiny," remarked Rebecca. "Yeah, and look at all those big hoses!" shouted Ben. "That fireman is a lady. Wow, a lady fireman!" shouted Denise, all agog. "Do you have to go to college to learn how to be a fireman?" asked Oren. The children watched, listened, commented, and asked questions as they tried on the hard hats and floppy boots, watched the brilliant lights, and listened to the firefighters talk. The most adventurous youngsters even accepted an invitation to climb up into the cab and sit in the fire truck. All this wonderful activity culminated in a well-supervised short trip around the parking lot on the back of the truck. Ms. Ziegler was rewarded by the fact that this learning experience was thoroughly enjoyed by the children, and that they had learned a great deal about a valuable community service.

When they returned to school, Ms. Ziegler brought the children together as a group to talk about their experiences at the fire station. They then drew pictures and wrote about their favorite experiences (see Figure 10–2). Later that day, the children dictated a thank-you note that was mailed to the fire-house.

Ms. Ziegler's field trip was successful because she always followed a set of specific guidelines whenever she planned field trips for her students. She learned that appropriate field trips include the responsibilities as shown in Figure 10–3.

It is a fact of life, though, that regardless of how well you plan the trip, "surprises" will be sure to pop up: for example, Ethan yanked on Jamielle's homemade necklace and sent beads rolling in every direction; midway through the visit, Blaine got frightened by a goat's loud bellow and wailed, "I want my mama. I wanna go home!" After you resolved those problems, Joseph wet his pants and Carrie brought up her snack. And when you returned to school, individual letters of thanks revealed some interesting "confessions":

FIGURE 10–2
Drawing Made After a Field Trip

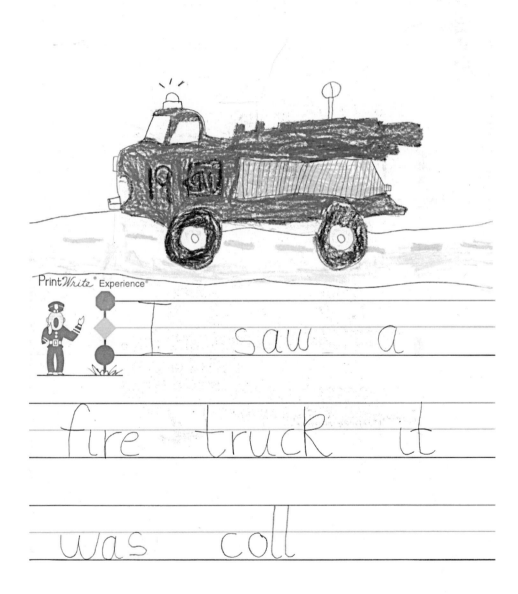

Jennifer   10-10-97

PrintWrite® Experience

I saw a fire truck it was coll

FIGURE 10–3
Field Trip Checklist

1. Address logistical concerns associated with the trip:
   - Have you taken the trip beforehand?
   - Have you made all necessary arrangements at the field trip site—what time you will arrive, where the restrooms are located, accommodations for children with special needs, places for lunch and snack?
   - Are there any special rules or regulations that must be followed at the site?
   - Have parental permission slips been signed (even if the school uses a blanket permission form)? Do not take children who haven't returned them.
   - Has transportation been arranged? (For liability reasons, a school bus is much better than a private car.)
   - If private cars are to be used, have you verified that each driver has adequate insurance and a valid driver's license? Have you provided each driver with a map and precise directions?
   - Have you planned proper supervision? (I always felt a 4:1 child-adult ratio was maximum.)
   - Are chaperones informed about what you expect of them? Do they know the behaviors expected of the children?
   - Do parents understand what clothing is appropriate for their children?
   - Do the students understand the expected standards of conduct? (Children respond better when they have a voice in determining these.)

2. Establish a clear purpose for the trip; be sure the children understand clearly why they are going ("What are we looking for?" "What do we want to find out?"):
   - Read books, share brochures or posters, and talk about what to expect at the site. Give the children an idea of what to expect.
   - Involve children in planning the trip.

3. Prepare for the trip:
   - Use name tags (including the school name and teacher's name). This helps if a child gets lost, as well as assisting volunteers when they need to call a child by name.
   - Assign a partner, or buddy, for each child. Explain why it is important to stick together.
   - Divide the class into groups. Give each student a specific responsibility (recorder, photographer or illustrator, organizer, and so on).
   - Show the students how they will record information from the trip (worksheet or guidesheet).

4. Take the trip:
   - Be sure to take roll when you leave and each time you depart or return to the bus.
   - Take along a basic first-aid kit (or nurse). Several wash cloths or paper towels will be needed if a child gets sick on the bus.
   - Arrive on time.
   - Keep the children who will need your attention close to you.
   - Introduce your class (but not each individual child) to your guide, if there is one.
   - Enjoy the experience!

5. After the trip:
   - Have the children write (or dictate) thank-you notes, or draw pictures expressing their appreciation to volunteers and field site personnel.
   - Talk with the class about what they liked best.
   - Provide enrichment activities—draw pictures, write stories or poems, create a dramatic skit, make a map, or conduct any related activity to help the children deepen their understanding of the trip.
   - Have the students evaluate the trip: Did they accomplish the purposes identified at the beginning of the trip?

Classroom Vignette

Dear Sir:
I am sorry about the way our class acted on our trip to your park, but I did not personally curse or anything. I hope you let us come back. Our friend who pulled down his pants is sorry!

Dear Sir:
I am sorry that our class was bad on its trip to the park. I apologize for exposing myself. I hope you let us come back. (P.S.—I am a boy.)

These are the moments that try teachers' souls. Perplexed, these disillusioned few often wonder, "What in the world am I doing here? I should have stayed back in the classroom and just read a book!" It's quite normal to have such reactions, but don't give up on field trips. Experience and good planning will help you avoid most of these disasters.

RESOURCE PERSONS

Resource persons are individuals from the world outside who come into your classroom to share some expertise or knowledge with your students. They serve the same purpose as field trips, but offer opportunities for direct action when the children cannot go on field trips. As with field trips, the resource person should not be a talker putting on a demonstration while the children serve as an attentive audience. The children must be involved in the action.

Although not as intrinsically motivating as going somewhere, children nevertheless enjoy contact with outside visitors and the interesting ideas and materials they share. You might invite people who provide goods or services in your community: police officers, firefighters, farmers, delivery people, construction workers, doctors, nurses, newspersons, bakers, industrial workers, store workers, craftspersons, lawyers, bankers, clergy, government officials, and so on. When introducing children to different cultures or ethnic groups, you can ask people with unique backgrounds to provide information and carry out demonstrations. Visitors can help break down stereotypes, such as senior citizens with special skills or hobbies, women carpenters, or male nurses. Care must be exercised in the way guests are selected, however. The safest approach seems to be requesting recommendations from other teachers, involved parents, and other school personnel. In this way, you will be able to select speakers who will appropriately inform and motivate your children. You and your

Carefully selected and planned field trips or classroom visitors are core experiences in dynamic social studies programs.

students will benefit the most if you send your visitor a list of helpful suggestions similar to the following:

1. *Think young.* Remember that you are speaking to young children, so be especially careful of your vocabulary. Also, limit your talk to the attention level of the children—about 20 minutes for K–2 children and 40 minutes for grades 3 and up.

2. *Bring something.* Real, touchable objects that relate to your area of expertise help children understand your life and work. For example, a construction worker recently visited our class and brought along a hard hat, small tools, and a lunch box.

3. *Move around.* As you speak, try to move to various sections of the classroom so the children will feel personally involved. It helps to ask them questions once in a while, too.

4. *Share personal stories.* Is there something that happened during your childhood that motivated you to do what you're now doing?
5. *Allow time for student questions.* When students do ask you questions, repeat them back to the class so all can hear.
6. *If possible, leave the children a memento.* You might bring something small to leave with the children (a florist brought a small flower for each child, an artist left an autographed print, and a flight attendant gave out some "wings" pins). If you can't leave a small souvenir, you might want to create something (the construction worker helped the children build a structure with blocks).

VIDEO, FILMSTRIPS, AND SLIDES

When it is not possible to provide either a field trip or resource person, it is wise to use various audiovisual materials including videotapes, filmstrips, slides, models, posters, and other displays. In most classrooms, the television set has replaced the movie projector and screen as the most popular piece of audiovisual equipment. Videos on virtually every social studies topic can be found in school libraries.

What are the benefits of videotapes? First, they involve action. When action is essential to the concept being taught, a good video will get the idea across. And if stop-frame, time-lapse, or slow motion can help clarify a point, the VCR is ready to respond. Second, the video can transport students to other times and places too difficult to reach in any other way. A trip back to the Battle of Gettysburg or a voyage on the *Mayflower* is as easy as the click of a few controls. Third, the video adds interest and variety to teaching. Through music and excellent camera work, the viewer becomes immersed in the mood of educationally significant times and places.

Some videos are interesting when they are viewed a second time. The second viewing can occur the same day if the program is short, but if it is lengthy, the second viewing should be done the next day. As the video is shown the second time, you could ask the students to interpret the action as the sound is turned all the way down. The students might also narrate each scene.

Portable video cameras make it possible to record a variety of student activities. Recording a field trip, resource person, or student project offers endless opportunities to supplement instruction in the social studies classroom. Use audiotapes and videotapes to prepare a student program each week. The format of the show can be anything you like, but one that works especially well is the music/news/talk show format: "My name is Lisa Forrest, and this is television (or radio) station WOW right here in Cedar Elementary School." (Changing the host weekly generates more interest.) "Hope you're having a good day and are ready to listen to the top tune of the week as voted on by the fifth graders of Cedar Elementary. Here it goes!" (Once the song is over, you may wish to provide a teletype background for the news—an electric typewriter does fine.) News items may range from actual current events to announcements of school events, special student accomplishments, or other local happenings. Commercials may be interjected to promote items such as a school band concert, the school store, and so on. Some children enjoy creating parodies of actual commercials. The talk show format may conclude the broadcast as the host interviews a teacher who has taken an interesting trip, a student who has won a special prize, and so on.

Suggestions for using videotapes can be directly applied to slides and filmstrips. Filmstrips are a connected series of images on a roll of film. Audiotapes provide narration and background music for filmstrips. Filmstrips are relatively inexpensive; they can be purchased from many commercial instructional supply firms for almost every common social studies topic. The major benefit of filmstrips is that they offer teachers flexibility; the filmstrip can be stopped at individual frames and then discussed or analyzed. Slides are quite similar to filmstrips, except that the individual frames are not connected. Despite the ease and convenience of videotaping, slides remain quite popular among elementary school teachers. Taking along a still camera on a field trip and making slides from the film is more convenient than hauling along a video camera. And some schools have equipment to make slides from computer programs—a valuable source of in-class experience.

PICTURES AND STUDY PRINTS

A rich storehouse of social studies content can be found in the countless pictures found in such sources as posters, literature books, textbooks, prints, newspapers, travel brochures, calendars, and magazines. Pictures help children envision people, places, events, or feelings that are difficult to perceive in other ways, proving the truth of the old expression, "One picture is worth a thousand words."

As you display and discuss pictures, remember that children vary in their ability to read and interpret them, much as they vary in their ability to read and interpret words. Some children may function only at the *literal* level, being able to simply name, list, and describe specific details about items being observed. Most, however, extend their skills to the *inferential* level, where they are able to speculate about such things as character traits, missing details or elements, or cause-and-effect relationships, and to the *critical* level, where they are able to share feelings and values. You should attempt to structure discussions with questions of varying degree of difficulty, giving children of different abilities the chance for success at their individual levels and exposing less-mature students to the higher-level thoughts expressed by their more mature peers. Sample questions follow.

- *Literal Questions*

    "Tell me what you see."

    "How many _____ ?"

    "Describe the _____ ."

    "What color (size, distance, and so on) is the _____ ?"

    "What is this person wearing?"

- *Inferential Questions*

    "Where (or when) does this take place?"

    "What will happen next?"

    "What are the _____ doing?"

    "How is this the same as (different from) _____ ?"

    "What kind of a person do you think _____ is?"

- *Critical Questions*

    "Why did _____ happen?"

    "Do the people like one another? How do you know?"

    "What do (don't) you like about the picture?"

    "What conclusions could you make about _____ ?"

    "How might the information in this picture be used?"

SIMULATION GAMES

As students progress through the elementary grades, they should continually experience developmentally appropriate activities that require them to make decisions similar to those people actually face in life. *Simulation games* place students in these kinds of situations; they involve students personally by requiring them to assume roles and seek solutions to problems. The degree to which actions and decisions meet with positive results determines the winner of a simulation game. The "Stores and Shoppers" game illustrates the major features of a simulation game designed to deepen economic concepts (see Figure 10–4). The objectives for the game follow:

1. Helping children discover that people may prefer to buy for any of several reasons—lower prices, better goods or services, convenient location, or customer confidence.

2. Helping children understand that owners of stores earn income from the production and sales of services and goods, and that from income they must pay for goods and materials to replace what they have sold, wages for their workers and selves, rent and utilities, repair and replacement of tools and equipment, and taxes.

FIGURE 10–4
Stores and Shoppers

The Situation

The players are divided into two groups: shoppers and store owners. In a class of 30, there might be four stores with three owners per store. Each store selects an owner to be the treasurer. Pupils serve as scorekeeper, resource keeper, and card dealer. Others are shoppers.

Resources for Players

Each shopper has some sort of medium of exchange, which can be play money or a simulated medium such as red paper circles. Each shopper receives an equal amount of "money," and all shoppers receive identical shopping lists of items to be obtained at the stores. All the stores have equal amounts of the exchange medium, but their amounts are not equal to that of the shoppers. The stores are also provided with goods for the shoppers to buy, but the quantity and prices vary among the stores. Goods are represented by different colored paper squares, triangles, and rectangles. Prices are set by the teacher; for example:

> *Store 1:* Two green triangles sell for three circles.
>
> *Store 2:* Six green triangles sell for one circle.
>
> *Store 3:* One green triangle sells for five circles.
>
> *Store 4:* One green triangle sells for five circles.

The card dealer has small cards that designate amounts of exchange medium that must be paid by store owners at different intervals. For example, "Pay rent—five circles."

The resource keeper is only used in a more complex game for a middle grade. He or she sells goods to stores when they want to use their profit to buy more goods.

Goals for Actors

The shoppers try to buy all the things on their shopping lists. The shopper who completes or comes closest to completing his or her list in the given time is the winner. If two shoppers tie, the shopper with the most exchange medium left is the winner.

The store owners try to sell all their goods at the prices the teacher has set. The store with the most profit is the winner.

The scorekeeper is in charge of counting the stores' profits and determining the winner among the shoppers.

Special Rules and Limits

1. When the card dealer blows the whistle, each store treasurer must draw a card and pay what the card says to the dealer.
2. All sales are final.
3. Shoppers cannot resell goods.
4. Stores cannot trade goods; shoppers cannot trade.
5. Playing time is set by the teacher—approximately 20 to 30 minutes.

Follow-Up

The most important part of a simulation game is the follow-up. Leading questions asked by the teacher help children verbalize the objective of the game and the meaning of the game symbols. Such a question for this game might be, "Why did you buy green triangles at Store 2 instead of Store 1?" Hopefully the answer would be that the price was lower at Store 2.

Source: Tivoler, J., Montgomery, L., & Waid, J. (1970). Simulation games: How to use. *Instructor 68*. Reprinted from *Instructor*, copyright © March 1970 by the Instructor Publications, Inc. Used by permission.

3. Helping children understand that the income left after the business owner has paid all expenses is profit. The owner earns this profit by taking risks, since she can't be sure that her customers will buy the goods and services she sells.

4. Helping children discover that business owners compete to attract customers with better goods and services or lower prices.

5. Helping children understand how stores use advertising to tell customers what they are selling and what their prices are. This helps children understand how advertising assists people in making choices.

Simulation games are enjoyable and easy to play. They actively involve children in social studies learning; because the game situations represent real-life circumstances, the students view them as relevant to their lives. Another advantage of simulation games is that children work cooperatively through sharing ideas and helping one another with special strategies. Despite the fact that simulation games may require up to an entire class day to play, teachers have found that students' motivation remains high, and that the positive outcomes far outweigh any sacrifices made in class time.

INTEGRATING THE ARTS

Teachers around the country are being encouraged to view the world from a perspective that integrates social studies with the expressive arts—*music, drama, dance, literature,* and *visual arts.* (Literature was discussed separately in other chapters.) This larger perspective calls for joining together history and music, geography and art, creative dramatics and political science. The overall aim is to help students gain a sense of a culture's human spirit by examining the great works of art created by its people. That idea is simple, but the harsh truth is that arts education is denied to most children in our elementary school classrooms today. It seems that many people think of the arts (for young children at least) as having little value other than being something to put on the refrigerator door. That conception of art can never compete adequately for time in the school program. Fowler (1989) offers this forceful reaction to our failure to tender a role for the expressive arts in our schools, particularly as it relates to the social studies:

> By denying children the arts, we starve our civilization. We produce children who are more fitted for an age of barbarism than the advanced civilization of the information age. . . . But equally important, we fail these children educationally by depriving them of the insights that the arts afford. The arts provide windows to other worlds. . . . The arts illuminate life in all its mystery, misery, delight, pity, and wonder. Encounters with the arts invite us to

explore realms of meaning that, according to an old Persian proverb, lie next to the curtain that has never been drawn aside. (p. 62)

Dynamic social studies programs do not exist simply to create a data package that often results in a partial or shallow view of the world. Instead, they offer comprehensive and interconnected experiences that enlighten students' understandings of humanity past and present. They pull from many areas of the curriculum, including the social sciences and the arts (creative writing, dance, music, drama, and visual arts) to present a cohesive, compelling impression of the world. The social sciences convey the content; the arts give us the emotion. The arts tell us about people—how they feel and what they value. Isadora Duncan, a renowned dancer, was said to have once commented, "If I could tell you what I meant, there would be no point in dancing." Through dance, Duncan was able to communicate emotion as few dancers had done before her, as well as instruct about history and culture. She accomplished all of this using a special "language" comprised of a series of graceful movements. If Duncan were not able to express her message through dance, whatever she wanted to convey would have been lost forever. (The story of this talented dancer is portrayed in Rachel Isadora's splendid book, *Isadora Dances* [Viking].) Thus, the social sciences *inform* us while the arts *impassion* us. Fowler (1994) argues in support of connecting these two areas in dynamic social studies instruction:

> We need every possible way to represent, interpret, and convey our world for a very simple but powerful reason: No one of these ways offers a full picture. Individually . . . history [and the social sciences] convey only part of the reality of the world. Nor do the arts alone suffice. A multiplicity of symbol systems are [*sic*] required to provide a more complete picture and a more comprehensive education. . . . Both views are valid. Both contribute to understanding. . . . The British aesthetician and critic Herbert Read once said, "Art is the representation, science the explanation—of the same reality." (pp. 5–8)

The social sciences supply insight and wisdom; the arts reflect the spirit of the people who created them. Every child should have the opportunity to explore the expressive arts as "language of civilization," for the arts may well be the most telling feature of any society.

Involvement in the Arts

Children need access to the arts as a fundamental part of culture, and some might not find success in school without the arts. Because they address many of the multiple intelligences, the arts provide an avenue into the academic areas that may have otherwise been inaccessible. Teachers who value the arts often integrate them with other subjects, particularly as they teach certain topics in social studies through the "arts lenses." This means that you have the responsibility to consistently search for and bring your students into contact with the arts of cultures under study and, while interest is high, help them become involved in related arts activities.

Bringing Students Into Contact With the Arts

Students must be exposed to a wide range of artistic encounters. Elementary school children must experience original art (concerts, museums, and performances) and be presented opportunities to enjoy reproduced art (records or CDs, CD-ROMs, videos, posters, and the like). These valuable firsthand encounters help students understand the place of the arts in people's lives. The arts that are meaningful to children are not the same as "great art," but children must gain a sense of art in their own lives before they can be expected to value artistic achievements from the past or from other cultures.

Involving Students in the Arts

Exposure to the arts must not stop at examining and responding. You should give children opportunities to recreate what they observe as an extension or enrichment of the topic under study. Because the process of recreating an expressive arts product is intended to reinforce specific understandings, authenticity and accuracy of the representation are primary considerations. Few valid understandings about early dances in Colonial America could be expected, for example, if students were allowed to move freely to the music instead of duplicating the specific steps of a minuet. The designated purpose of the activity dictates the amount of conformity you must demand. If the purpose of an activity is to "deepen understandings of Kachina dolls as religious educational objects of the Hopis," then you must use accurate details, colors, and distinctive symbols. For example, a rain-cloud design on each cheek of the Kachina is a symbol with a specific meaning. The children should use this symbol only if they intend to communicate that meaning. Likewise, other special symbols, shapes, and colors convey specific messages to the Hopis. Experiences with Kachinas, then, must begin with research. By examining a Kachina doll (or its representation), you must determine where, why, how, and by whom it was made. Give students the opportunity to apply what they learn to the accurate reproduction of Kachinas. Experiences in recreating a culture's art, then, must be authentic; these projects lead children to deepened interest in academic pursuits and strengthen their understanding of culture.

A primary aim of the arts is to maximize children's creative potential, so social studies should move beyond opportunities to examine and reproduce art toward providing opportunities for original self-expression. These experiences are recognized for their value in developing divergent thinking and creativity as well as cognitive enrichment. To return to the Kachina doll example, suppose that you have approached the project in a way similar to our recommendations—learning about real Kachinas followed by having students create an accurate reproduction, or model. Now you wish to encourage children to personalize the information to a greater degree by encouraging them to offer original symbols for Kachina dolls. Therefore, you inform your students that there are over 200 actual Kachinas, and that new ones are often invented. You might approach the creative phase of the experience by encouraging the children to enhance their Kachinas by creating new symbols to represent any spirit they wish.

John Dewey once made a clear connection between these complementary aspects of the arts when he said, "A beholder must create his own experience." Children must be given opportunities to enrich and express what they learn about the arts through as many expressive experiences as possible, for students who study the arts are also more likely to display originality and creativity in all of their social studies work.

Select activities not because they appear to be "fun" or "cute," but because they contribute directly to students' understanding and appreciation of a culture. What rationale, for example, could you use to support making log cabins with pretzel sticks and peanut butter or constructing tipis with painted construction paper cones? These activities commonly appear in idea resource books, but they create a weak connection between the experience and an understanding of early colonial or Native American cultures. This word of caution is not meant to stifle your eagerness for a "hands-on" approach to social studies instruction, but rather to point out your responsibility to organize and coordinate meaningful knowledge through the visual arts to channel students' imagination and enthusiasm into creations that reflect *accurate* understandings of a culture.

The Visual Arts

Dynamic social studies teachers believe that immersing their students in any culture they are studying means the visual arts must play a significant role. This can be done by bringing to class examples of the arts and crafts of people under study to help clarify their cultural distinctiveness. These materials not only arouse a great deal of interest in a topic and convey meaningful information, but they stimulate many questions, too.

In elementary school classrooms where children's self-expression is valued and nurtured, you can carefully follow-up exposure to the visual arts with appropriate materials that encourage students to express themselves with individuality. When elementary school youngsters are free to communicate their observations and feelings, they develop pride in what they produce and realize that their thoughts and actions have value.

As students explore the art of Africa, for example, they will be eager to try their hands at tie-dye, weaving, mask construction, or sculpture. You could enhance the study of Greece by having students create plaster reliefs or signature seals. Enliven a lesson on medieval Europe by having children study and recreate coats of arms, stained glass, and castle designs. You can use the arts of *Gyokatu* (fish printing), block printing, kite making, batiks, and folded-paper design to introduce students to Asian cultures. Native American sand paintings, blankets, Kachina dolls, and totem poles help students learn about the earliest cultures in North America. *Pysanky* (intricately designed Easter eggs) and flax (straw) dolls bring to life important aspects of Ukrainian culture. The potential is unlimited for integrating the visual arts and social studies:

Classroom Vignette

Mildred Spriggs, for example, enriched her fifth graders' study of prehistoric life by engaging them in a study of ancient cave paintings. Displaying large study prints of actual cave paintings and those found in art books available from the library, Ms. Spriggs directed the students through a discussion of what was important in the life of prehistoric humans, as depicted by their art. Then, making paint from grass and berries, and using homemade brushes of sticks and reeds, students created designs on large rocks illustrating subjects important to them.

When children become aware of art as something valuable that people do in real life to communicate ideas and feelings, they are inspired to create their own.

Illustrations

Although dynamic social studies programs offer countless opportunities to involve children in standard art experiences such as *dioramas, murals, mosaics,* and *collages,* the most popular art technique used in elementary social studies classrooms seems to be *illustrations.* You may have children draw the details of the process involved in making linen from flax, to use crayon for indicating various areas on a

Children enjoy expressing what they learn in social studies through the arts.

state map, or to paint a picture of a landscape in the high Andes Mountains. Children like to illustrate group notebooks and decorative charts. There are literally hundreds of possibilities.

Classroom Vignette

For example, while studying the Native American Indians of the Great Plains, Maurice Tatum organized daily learning experiences around such motivating questions as "What would it be like to eat a meal with a Cheyenne friend of the past?" (food gathering and consumption) and "What games would you play if you visited the home of a Sioux friend of long ago?" (leisure, recreation, and education). Today's question of intrigue was, "How did the Plains Indians teach history to their children?"

Mr. Tatum selected the Sioux as the focus of exploration to answer that question. He informed the students that, like all native North Americans, the Sioux of the Great Plains had no written language. Mr. Tatum explained that stories and history were passed down from one generation to the next in the oral tradition. To set the scene, Mr. Tatum sat together with his students on the floor and told them a story that he had found in the book, *Why the Possum's Tail Is Bare: And Other North American Nature Tales* by James E. Connolly (Stemmer). After the storytelling was completed, Mr. Tatum described one way that the Sioux wise men recalled the major events of passing years for their tales—the "winter count." Each year a tribal artist would depict a special event, drawing it on bison hides that the women had stretched and dried in the sun on drying racks. Mr. Tatum showed an example of one event, the story of which was centered on a drawing of a horse. He explained that about 250 years ago the Sioux got horses from the Spanish and that these horses had both positive and negative influences on the lives of the Sioux. They allowed the Sioux to hunt bison more skillfully, but they also caused them to stop growing their own food. By drawing a single picture to remember an important event, the wise storyteller could remember enough to retell the history of the tribe to its youngsters, such as the introduction of horses to the Sioux.

Mr. Tatum encouraged his students to retell some of the important events they had been learning about in Sioux history by making their own "winter counts" following these steps:

1. Cut a long sheet of brown butcher paper (8 × 15") or a section of a grocery bag. Tear it into the shape of a bison hide.
2. Crumple the sheet into a ball and dip it into a container of water until it is thoroughly soaked.
3. Press out the sheets on newspaper and let them dry.
4. Recount important events in the lives of the Sioux.

5. As the students select events to depict, draw them on the sheets with crayon (press hard with the crayon).
6. Place the sheets between two pages of newsprint and press with a warm iron. The colors will darken and the "winter count" will have a weathered appearance.

Music

The folk songs of cultures around the world vividly reflect a group's personality, expressing cultural characteristics that cannot otherwise be effectively communicated. People have sent messages across distances by pounding on drums. People have marched off to war to stirring music and walked down the aisle to the joyous strains of a wedding march. People have raised their voices in national anthems. People have sung work songs or chants to keep their work rhythm straight as they pulled in nets or pried rails back into line. People have sung songs of worship in camp meetings and cathedrals, and they have celebrated harvests with festive dances. People have expressed their grief in tender ballads and their love in songs. Wherever and whenever possible, people have gained satisfaction and joy in making music for themselves and for others.

How can you help children to understand all of this? Certainly, they must have experiences with many kinds of music. Different people and faraway places may be abstract notions for some children, but they can be made more concrete by forging an integral relationship with music and social studies. Sharing the music they create may effectively develop important concepts about people around the world. Exposure to folk music is, perhaps, the most fitting vehicle for teaching about the lifestyles and beliefs of various groups. *Folk music* is, generally, traditional music created by people as part of their everyday existence and handed down from one generation to another, and it reflects the cultural heritage and traditions of a group.

Classroom Vignette

Leona Boreski sought to enrich her unit on the state of Hawaii by including a study of its folk music. This segment of the unit was to emphasize the idea that groups create special music to reflect the way they live. She realized that for the music to have a valuable role in the unit, she must have accurate background information. After hours of informative library research, Ms. Boreski discovered that the early Hawaiians created their musical instruments from objects found in the island's natural environment—the low "wail" of the conch shell, the resonating "rattle" of hollowed-out gourds filled with pebbles, the "clack" of split bamboo, the hollow "tock" of hardwood sticks, and the "click" of smooth stones tapped lightly together. From such sounds created by common objects found in the sea, forest, garden, and river, Hawaiian musical instruments first accompanied chants called *olis* and dances called *hulas*. Since there were no written records in ancient Hawaii, the history of its culture was often told through and passed on by olis and hulas. Some olis and hulas were used for entertainment, and others only for religious ceremonies. Ms. Boreski was able to display several native instruments for the children to observe as she played recorded songs and chants for them. She was careful to use appropriate language labels as she introduced the instruments—*ili-ili* (smooth oval pebbles), *kalaau* (hardwood sticks), and *ipu* (large hollow gourds filled with pebbles). Ms. Boreski emphasized that times have changed in Hawaii, as they have in any location that has been modernized. But even today, original instruments still accompany programs or other special pageantry of the islands. Ms. Boreski then shared a brief video clip of a performance of the "Va Nani O Nuuanu" ("Pebble Dance").

Listening to the unique and fascinating music of the folk song tradition is an important activity while children learn social studies content, but, as we found with the visual arts, a learner's experiences in music should include opportunities to reproduce the music accurately through singing songs, moving to rhythm, or trying out musical instruments. Ms. Boreski realized the value of this dimension of music integration and capitalized on it during her unit on Hawaiian customs and traditions. After her students learned of the early music of the Hawaiian people, she introduced them to "Kauiki," an old song about a dormant volcano that juts out into the sea. Background music for "Kauiki" is ukulele accompanied by *ipu* beats. Ms. Boreski was able to obtain a few ukuleles on loan from parents, and brought in her own *ipu*. Volunteers strummed the ukuleles (the strings were tuned to the proper chords) and beat the *ipu* while the rest of the class sang the lyrics in both Hawaiian and English.

Like the visual arts, music establishes a cultural link among people, past and present.

Creative Movement and Dance

Music, creative movement, and dance are integrally related. People in all cultures respond to music with their bodies. Use songs to develop a repertoire of associated dance movements, both to enhance understandings of people and for creative self-expression. Folk songs, as we saw in the discussion of Hawaii, can be used for these purposes.

Classroom Vignette

In Ms. Boreski's classroom, for example, the students made grass skirts (to represent the authentic ti-leaf skirts used on the island for performances) and leis. The girls wore the grass skirts while the boys dressed in clam diggers or shorts and Hawaiian pattern shirts; both boys and girls were bedecked in flowers and leis. Ms. Boreski, dressed in a unique muumuu with flowers in her hair, played a recording of the traditional Hawaiian chant "Ka Hana Kamalii" ("What Children Should Do"). The chant was composed for Queen Emma of Hawaii and recited by children to demonstrate to the queen that they had learned the lessons in personal hygiene that the missionaries had taught them—to wash with soap and to use mirrors to see that they were clean. Their words were accompanied by a hula dance to show what they should do and how they should do it. Ms. Boreski taught the chant to the children and demonstrated the hula routine. Throughout the entire experience, one child beat an *ipu* (hollow gourd) in a continuous rhythm. After the class finished the hula, they proudly demonstrated a *pau* (bow).

To be sure that children acquire accurate concepts of the culture, it would be helpful to invite resource people into the classroom to share their talents with the children. Capitalizing on the value of dance and movement in the social studies program demonstrates to children that all forms of music are integral parts of people's lives rather than a form of expression for only a talented few.

Drama

Our final arts-based strategy is helping students acquire social studies understandings through drama and plays. Since the dawn of civilization, people have used dramatic performance to reveal their innermost feelings of conflict, despair, truth, beauty, hope, and faith. Drama brings out the essence of a culture clearly before one's eyes and is an excellent vehicle to challenge students to think about and appreciate societies from generation to generation.

The dynamic nature of social studies is a perfect match for creative drama. Drama is a vital aspect of a society; by its very nature, drama presupposes an important kind of social communication. For primitive humans it was a ritualistic attempt at communicating with a spirit in a magical sense; through ritualistic half-acting episodes, while dressed in masks and skins, groups would try to seek cooperation with special spirits to control certain events (a good hunt, the power of thunder, fertility, and the like). As humans began to settle in regular communities and rely on agriculture, their greatest fears became bitter weather and a failed harvest. Their simple half-acting gradually evolved into a ritual of more formalized movement (e.g., a king overcomes death (winter) to bring life (spring)). From these early patterns, drama developed into a stylized form of communication with intricate dialogue, disguises, and symbolism. Most civilizations throughout history have used some form of drama to express their important thoughts.

To fully understand any group or civilization, students must receive and respond to its dramatic creations. Therefore, students will often assume the role of "audience" as others communicate drama to them. Whether witnessing an example of Japanese *Kabuki* theater or a puppet play about North American cowboy life, you must guide students through any quality dramatic episode as effective audience members if you wish them to achieve specific learning outcomes.

Provide students with opportunities to observe quality dramatic experiences throughout the school years. These can range from professional theater groups to students in drama classes in middle or high school. Regardless of the source, however, children must have opportunities to discover for themselves the power of drama as it is used to communicate representative ideas and feelings of people around the world.

By nature, children are born to create. As it applies to drama, they create in a joyfully daring way. They are confident and expressive in their actions and improvise freely (make up dialogue and actions). Such informal dramatic experience is important to the social studies curriculum, for, as the Joint Committee of the National Council of Teachers of English and the Children's Theatre Association (1983) explains:

> Informal classroom drama is an activity in which students invent and enact dramatic situations for themselves, rather than for an outside audience. This activity, perhaps most widely known as creative drama, . . . is spontaneously generated by the participants who perform the dual tasks of composing and enacting their parts as the drama progresses. This form of unrehearsed drama is a process of guided discovery led by the teacher for the benefit of the participants. (pp. 370–372)

Creative drama in the social studies classroom often utilizes these major forms of creative dramatics: *pantomime, improvised skits,* and *dramatizations.*

Pantomime

Children of all ages enjoy using body movements to convey ideas without words, and this is a good technique for introducing creative dramatics to your classroom. Use your imagination to concoct stimulating situations for pantomime in the social studies classroom. One possibility is the popular "What Am I Doing?" or "Who Am I?" game. Many social studies concepts have to do with people making or doing something; and children enjoy imitating their distinctive movements. They like imitating activities of people in a topic under study and challenging their classmates to guess who they are, such as pilgrims landing at Plymouth Rock, the first landing on the moon, Ben Franklin's kite-flying experience, Barry Bonds' record-breaking 70th home run, or the type of work people do at home or in the community.

As each child or small group shares its pantomime with the class, it is important to lead a follow-up discussion: "Who is represented in the action? What helped you recognize this person? Where do you think the action took place? What helped you think this? What do you think the people were talking about? What are some of the feelings you think the characters experienced? From what you know about the actual situation, do you think the actors accurately portrayed what really happened? What feelings did you have as you watched the action? What would you have done if you were any one of the characters in the same situation? Can you predict what might happen to the people after what was shown in the pantomime?"

Improvised Skits

A charming characteristic of young children is their ability to use their creative imaginations in *improvisations*—unplanned situations where dialogue is necessary but there are *no* learned lines, *no* costumes, and *no* sophisticated scenery. With a thorough knowledge of the story situation, however, dialogue begins to flow, and children further their understanding of a social studies condition by describing its characters more completely.

Many of the classroom situations described as pantomiming activities are also suitable for improvisation experiences. Begin the initial improvisation experiences with simple situations. As the children gain confidence, they can attempt more challenging material.

Classroom Vignette

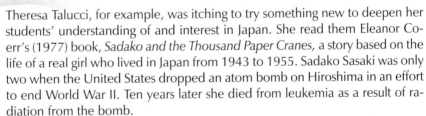

Theresa Talucci, for example, was itching to try something new to deepen her students' understanding of and interest in Japan. She read them Eleanor Coerr's (1977) book, *Sadako and the Thousand Paper Cranes,* a story based on the life of a real girl who lived in Japan from 1943 to 1955. Sadako Sasaki was only two when the United States dropped an atom bomb on Hiroshima in an effort to end World War II. Ten years later she died from leukemia as a result of radiation from the bomb.

The morning after she was diagnosed with leukemia, Sadako woke up in the hospital and hoped the day before was a bad dream. It wasn't long before she was jolted with the realization that everything was indeed real. That afternoon, Chizuko, Sadako's friend, was her first visitor. Chizuko brought a square piece of gold paper that she folded over and over until it turned into a beautiful crane. "If a sick person folds one thousand paper cranes, the gods will grant her wish and make her healthy again," She handed the crane to Sadako. "Here's your first one." (pp. 35–36)

Sadako folded 644 cranes before she died. They hung above her bed on strings. Her classmates folded the rest. Today, Sadako is a hero to the children of Japan, who visit her memorial in the Hiroshima Peace Park to leave paper cranes they make in her honor.

Ms. Talucci's students found Sadako's story to be unforgettable and were touched by her life. Because the story was written tenderly, it was neither morbid nor weepy. The children were touched, but not crushed or brokenhearted. Therefore, Ms. Talucci chose to review the events of the story as well as promote creative thinking by using a role playing strategy called *character interview*.

To prep the class, Ms. Talucci announced, "Today we are going to pretend to be somebody else, and we are going to do that by using our imaginations. We have just finished reading a tender story of Sadako Sasaki, and I would like for some of you to share with us what you learned about her life. You will do that by pretending to be Sadako."

Ms. Talucci continued, "I've made this cardboard name sign with a string loop that will fit over your head and hang around your neck. The name sign has the words *Sadako Sasaki* on it. One of you will come up here and put the sign around your neck. You will then become Sadako. Of course, we know that Sadako is no longer alive, but through the use of our imaginations, we will have an opportunity to hear her innermost thoughts. The rest of the class will ask questions about Sadako's life."

Ms. Talucci's final directions were, "We are going to use our imaginations. When she puts on the sign, we won't see Tara anymore; instead we will meet her as Sadako Sasaki." Once Tara placed the sign around her neck and the transformation occurred, the class interviewed the character. Ms. Talucci took the lead: "Welcome, Sadako. We are very honored to have you here today. What questions do we have for Sadako, class? What things might she share that she may have never said before? What dreams did she have? What are we curious to know more about?"

After several questions, Ms. Talucci checked to see if the student playing Sadako seemed uncomfortable. When she sensed that the student was struggling with answers, she simply thanked the character, removed the sign, and had the student face the class for applause. Ms. Talucci then went on, "Now, who else would like to be Sadako Sasaki. . . ?"

Dramatization

Many times, children wish to extend informal dramatics by putting on a formal play with script, costumes, and scenery. The most effective type of formal play is one in which the children are responsible for most of the planning. Your role in the process is to guide the students to write the different scenes. As the children plan their script, ask them, "What are the most important things to tell? What are the people really like? What kind of place do they live in? What do the people do to make their lives interesting? How will you stage your play?"

Dramatizations should emerge from the social studies content in which the students are immersed. The students' concern for elaborate scenery and costumes should not take precedence over the concepts or ideas the play is designed to convey, however. Simple objects can effectively represent more intricate objects; for example, a mural or bulletin board design can serve as a backdrop; a branch in a big can filled with dirt makes an excellent tree or bush; your desk becomes a cave; chairs placed in a straight line can be seats on a train or airplane; a pencil can become a hand-held microphone. The following example describes a highly appealing and successful dramatic skit composed by a group of fifth graders following a reading of *Mufaro's Beautiful Daughters* (Lothrop). At the end of the book, Nyoka and Nyasha are married. The children continued the story by creating an authentic naming ceremony for the couple's first baby.

Classroom Vignette

African villagers celebrate the birth of babies with great enjoyment. Babies are an important part of African life; the African baby belongs not only to its mother and father, but rather to everyone in the village. Everyone in the village loves the baby and is responsible for giving it protection and direction in life.

SCENE: People are milling about the village. Nyoka and Nyasha are standing and holding their baby. Traditional African music is playing in the background.

Nyoka: With pride and gladness, we announce that on this special day our child is to be named.

Nyasha: Come be our guests at this joyous celebration.

SCENE: Everyone comes together. All guests sit in a circle with Nyoka and Nyasha. The ceremony begins when the oldest living member of the family recites the family history.

Oldest: Welcome to all who join us to bless and honor the child of Nyoka and Nyasha. Out of deep respect for their ancestors, we begin the ceremony with a short family history.

The baby's mother is Nyasha, who is the most worthy and most beautiful daughter in the land. Nyasha's father is Mufaro. Mufaro is very rich because he has two most beautiful daughters, Nyasha and Manyara. Mufaro is known for his great wisdom and patience. Dayo (DAH-YO) is Nyasha's mother. She can sing so sweetly about the land! It is said that she composed a song for each cow in a herd of a thousand. This is Nyasha's gift from Dayo—a voice as sweet as honey.

The baby's father is Nyoka. He has the special power to appear as all living things. Nyoka's father is Olu (OH-LOO). Olu had a large family, all of whom were fearless warriors. Olu is the one who built the great city. As an adventurous young man, Olu spent much time in the tropical forest by himself. Olu learned many new things from the beasts and serpents he befriended. Olu was a great king. He was handsome, tall, and strong. Olu had a son and named him Nyoka. Olu taught Nyoka all that he had learned in the tropical forest. That is the reason Nyoka now has the power to appear as all living things.

Now, everyone will join together to introduce this child to the nature of life.

> ***Friend 1:*** I offer a few drops of wine to the baby because the wine ensures the child a full and fruitful life.
>
> ***Friend 2:*** I splash a few drops of water on the baby's forehead and put a drop or two into its mouth when it cries. This is a way of showing that water is important for all living things. It also tells me if the baby is alert. This looks like a very fine baby—alert and responsive. I now wish it a smooth sail through the sea of life.
>
> ***Friend 3:*** I drop a bit of honey onto the baby's tongue. This is to show the child that life is sweet.
>
> ***Friend 4:*** I place a pinch of pepper on the baby's lips to represent the spice of life. Although we have learned that life is sweet, we must also know that life is exciting.
>
> ***Friend 5:*** I lay a dab of salt on the baby's lips. The salt stands for the liveliness and zest of life.
>
> ***Oldest:*** The baby of Nyoka and Nyasha will be rich in its pursuit of life.
>
> ***SCENE:*** Everyone, including Nyoka and Nyasha, must give the baby a name. Girls are named on the seventh day after birth; boys are named on the eighth.
>
> ***Oldest:*** On this joyful _____ th day after birth, all who are present are invited to offer a chosen name for the baby of Nyoka and Nyasha.

The rest of the group uses the sentence pattern shown below to offer a name for the child.

> ***Child:*** I have picked the name NAYO because it means *we have joy.* This child brings much happiness to Nyoka and Nyasha.
>
> ***Oldest:*** The ceremony is now complete. May Nyoka and Nyasha have many beautiful days with their lovely new child.

Print Materials

Textbooks and Trade Books

Oftentimes, when teachers are asked to describe their social studies curriculum, they will tell you all about their textbooks. Principals regularly introduce new teachers to the school's social studies curriculum by informing them, "Here's the textbook. You should become familiar with it." Starr (1989) estimates that "ninety percent of all classroom activity is regulated by textbooks" (p. 106). In recent years there has been a great deal of debate about whether textbooks or trade books (stories, informational books, poetry) should be used as the major source for teaching social studies. Most educators agree that textbooks and trade books can coexist, and that students should read both types, suggesting that textbooks should not dominate instructional experiences; they should be considered only one of many teaching tools, not the ultimate word.

It is my contention that using social studies textbooks exclusively is not a good idea. Children need to experience a wide variety of print sources in the social studies program. Textbooks can be used to survey a topic; trade books can add depth and passion. Beginning teachers often rely on textbooks and their accompanying teacher's guides to gain confidence in the classroom; as their teaching abilities grow, they rapidly move toward using trade books. Many experienced teachers, especially those using an integrated approach to social studies instruction, incorporate reading from a variety of sources, including textbooks and the rich treasures of children's literature. In support of this idea, the authors of *Becoming a Nation of Readers* discuss the strength of the literacy–social studies connection: "[T]he most logical place for instruction in most reading and thinking strategies is in social studies . . . rather than in separate lessons about reading" (Anderson, Hiebert, Scott, & Wilkinson, 1985, p. 73). There are many opportunities to use literacy strategies in the social studies program. Teachers familiar with the concept of "literacy across the curriculum" know that they must supplement the textbook with some of the many superb trade books available today. Information books, historical fiction, folklore, biographies, and poetry collections, as well as newspapers, add depth and substance to any topic under study.

Newspapers

Amendment 1—United States Constitution

Congress shall make no law respecting an establishment of religion, or prohibiting the free exercise thereof, or abridging freedom of speech, or of the press; or the right of a free people to peaceably assemble, and to petition the Government for a redress of grievances.

People over the centuries have argued that a *free press* (not only newspapers, but all broadcast media, too) is the keystone of a democracy, and predict that without it our democratic form of government would fail. Likewise, it has been argued that a *free education* is the keystone of a democracy. Combining these two popular sentiments, we find that neither a free press nor an informed citizenry could exist without each other. The free press ensures the public's right to know; our schools have a responsibility to prepare able and interested citizens to understand the issues involved in their lives. With important issues changing as rapidly as they do in contemporary society, the matter of being well informed is now more important than ever. The foundation for developing informed citizens should be established in the elementary grades as you establish an effective newspaper-based current affairs program.

Perhaps the best place to begin an effective current affairs program is to teach children how to read a newspaper. Children should learn to examine the various sections of the newspaper (world news, sports, community events, and so on), with special attention directed toward the sources of material for these sections (wire services, local writers, and syndicated columns). Children can understand that a good news article answers four basic questions (*who, what, where,* and *when*) in the first paragraph or two, and then should go on to state *why* and *how.* Lead the children through the literal interpretation of news articles by making copies of a news article, distributing it to each child, allowing him or her sufficient time to read it (for a *purpose,* of course), and then guiding the interpretation:

- "Who is the story about?"
- "What did the person do?"
- "When did the event take place?"
- "Where did the event take place?"
- "Why did the event come about?"
- "How can this event be extended (be prevented from happening again, teach us a lesson, and so on)?"

News stories are objective; they present the facts with an absence of personal reaction. Encourage the children to share their feelings regarding any particular news article with questions such as the following:

- "How did you feel about the situation?"
- "What would you have done?"
- "Would you be willing to do the same thing?"
- "Do you agree with the central character?"
- "Is there any information you can add to the article?"
- "Do you think the story was written fairly (accurately)?"

Teachers can and should use newspapers in creative ways. The following are some guidelines you might find helpful with your students:

- Fictionalize a story about what happened just before and just after the moment captured in a selected newspaper picture.
- Give students a newspaper headline and ask them to write their own news story based on it.

- Take a student survey to find out what part of the newspaper interests them most. Divide the class into groups, each with a variety of interests. Each week, assign a group to report to the class. Urge "reporters" to read all they can find in the newspaper each day concerning their field. Each "reporter" then reads to the class one especially interesting item each week.

- Read only the first paragraph of a news article, and then make up the rest. Be original! Compare your story with the original article if you want.

- Have each student study newspaper articles about different well-known personalities and pantomime personality sketches as other students try to guess who is being portrayed. As students become skilled in mime techniques, have them scan the newspaper for simple situations to play out in pairs or small groups. Other students will enjoy guessing the actions.

- Have a student assume the personality of a person in the news and have classmates interview him or her. Remember, the student who assumes the character's personality must answer questions using the character's observed speech and other mannerisms.

- Have students take a favorite nursery rhyme and write it up as a news story, giving it an appropriate headline; for example, *Mr. Egg Fractures Skull.*

- Let your students be advice columnists. Have them read letters to a popular advice columnist. Pupils then write their own advice on real problems affecting playground behaviors, classroom problems, and other sources of conflict.

- Clip headlines from stories, but keep one newspaper intact. Have students read news stories and write their own headlines. Did students discover the main idea of the story expressed in the headline? Compare student headlines with the headlines in the uncut issue.

- After identifying the main topic of an editorial, scan the paper to locate stories related to the topic and read them. Study how the editorial was developed and then have students write their own editorials on the same issue.

Electronic Newspapers

Knowing what is going on in the world creates better citizens. One of the major goals of elementary school social studies programs is to help children acquire a life-long interest in newspapers so they can keep in touch with what is going on in the world outside. However, when teachers try to interest children in current events, they often assign a child to bring in a newspaper article the next day. What normally happens is that, just at bed time, the child realizes that she needs a current events article for tomorrow and a parent frantically searches the front page to clip out something the child can take to school. Or, even worse, the child informs a parent just before he goes to sleep that he needs a specific current events article on Tasmania. The newspaper has probably been discarded, but even if it hasn't, did anything newsworthy happen in Tasmania that day? The value of such hit-or-miss newspaper programs is questionable.

An option currently being carried out by many elementary school teachers is online newspapers. They have selected this option for several reasons. One that stands out is immediacy—the newspaper that is discussed in school at 9:00 A.M. does not include the important stories that broke at 8:00 A.M. Online news fills the gap; keeping the class informed creates links between what is in the newspaper and what is happening right now. In addition to immediacy, the multimedia presentation of online newspapers is impressive. Video clips, sound, and full-color images accompany online news stories. In addition, should a particular story capture the interest of the class, online newspapers allow them to search the story by key word and access a rich storehouse of related news. Students can incorporate all of these sources into their own multimedia packages, properly cited, creating a customized newspaper.

Several online sources offer colorful, easy-to-read newspapers. *The Christian Science Monitor* is one of the most useful (*http://csmonitor.com*). Others include *CNN Interactive* (*http://www.cnn.com*), *MSNBC* (*http://msnbc.com*), and *USA Today Online* (*http://usatoday.com*). Several local newspapers have online versions, too, so check to see what is available.

None of these online newspaper sources is meant to replace the daily newspaper; newspapers are here to stay. However, online newspapers can augment the use of classroom newspapers by serving as an alternative that many people believe will be the dominant source for delivering information in the future.

Current Affairs Periodicals

Teachers often rely on commercially produced current events magazines or newspapers such as *Weekly Reader* for most of their program in current affairs. This approach has both advantages and disadvantages.

Some teachers become too dependent on a formal reading–reciting technique when using periodicals, but they can be an important educational tool for several reasons. First, these periodicals can motivate. They are written especially for children, and even reluctant readers respond positively to the nontext format and the comfortable reading levels. Second, the periodicals select interesting, contemporary topics. This week's big news, a special TV presentation, or the latest technological breakthrough helps to make the classroom more current. Finally, comprehensive teacher's guides that describe creative teaching strategies accompany periodicals. Periodicals are used in thousands of schools around the country and are a valuable resource material.

Computers

When computers were first introduced into elementary schools during the late 1970s and early 1980s, they were used primarily as "electronic workbooks," providing drill-and-practice exercises much like conventional workbooks. Students were presented problems and entered their responses; the computer would send a graphic reward (smiley face or explosion, for example) for the right answer. This usage clearly concerned many educators, for they envisioned computer terminals as centers of "busy work" with the same potential for abuse as regular worksheets and workbooks.

Although software companies continue to produce "drill-and-practice" programs, educators everywhere are convinced that the singular "electronic workbook" role of computers is long gone. The evolution of tutorials and simulations quickly joined the drill-and-practice programs as powerful tools for teaching content and skills. Some, like *Oregon Trail* (MECC), remain among the most popular programs in elementary school social studies classrooms. Now educators everywhere are trying to figure out ways to enhance learning in elementary school classrooms with computers. "I see the computer as an irreplaceable tool in my classroom," comments Ms. Kathy Nell, a fourth-grade teacher from Philadelphia. "I don't see it as a workbook, but a skylight—a skylight that opens up so many new opportunities for my students. The computer is pretty much my ally."

Not all teachers share Ms. Nell's vision, however. Brandt (1995) reports that "few teachers in a relatively small number of schools possess the equipment and knowledge to have their students . . . send e-mail to people in other countries, consult CD-ROMs for information, and prepare multimedia reports. The vast majority do not have access to the newest technologies—and even those who do tend to use them in conventional ways" (p. 5).

Indeed, merely putting a computer into a classroom and connecting it to a power source will not guarantee effective use. If teachers are going to use technology, they will need to be as skilled in using computers as they are in reading and writing. Fortunately, "the cavalry is on the way" to help teachers get in touch with technology's promise and power; a blaring bugle call signals the "charge" of schools of education to include technology training for their undergraduate students.

It is true that drill-and-practice computer programs remain, but an expanded variety of increasingly sophisticated educational software gives more attention to higher-level thinking processes. These programs can be classified within the following functions: (1) tutorial, (2) problem-solving, (3) simulation, (4) word processing and database, (5) telecommunications, and (6) hypermedia presentations.

Tutorial Software

Tutorial programs focus on the presentation of new knowledge and skills, as opposed to the review and reinforcement of previously acquired skills and knowledge found in the outdated drill-and-practice programs. One example of a tutorial program is *Nigel's World* (Lawrence Productions), a fascinating world of adventures and learning experiences for children of all ages. Nigel is a fearless Scot who takes on the world, camera in hand, in search of a photograph that will earn him first place in a photography contest. As he journeys to the ends of the Earth snapping pictures, Nigel calls for the help of the students to make decisions involving geography and basic map skills.

Problem-Solving Software

Problem-solving programs often present highly complex situations where students face a dilemma, choose from a number of possible alternatives, and arrive at a solution. The computer encourages active exploration and discovery. One extremely popular problem-solving program for children in grades 3 to 8 is *Where in the World Is Carmen Sandiego?* (available from Broderbund Software). Students become crime fighters as they search the world for Carmen Sandiego (an ex-secret agent turned thief) and her gang of criminals, who are out to steal our most precious natural treasures. The venture starts with 1 of the 16 thieves hiding in a city and leaving a trail of clues for student "detectives" to decipher. The students read descriptions of various cities, visit clue locations, and check possible destinations, all the while using their problem-solving and analytical skills. Along the same principle, the program developer has added other adventures.

Simulation Software

Computerized instruction may take the form of *simulations,* or imitations of something real. Simulation programs place students into situations that are as authentic as possible. Social studies teachers employ simulations because they invite many modes of thought: conceptualization, problem solving, and varied opportunities to apply knowledge and skills. One of the most noteworthy simulations available for upper-grade elementary school classrooms is *The New Oregon Trail* (available from Minnesota Educational Computer Consortium). It presents a series of decisions that pioneers faced in 1847 as they set out in wagon trains to find new homes in the Oregon Territory. They stock up with provisions at the beginning of the 5-to-6-month journey, but heavy rains, wagon breakdowns, illness, and robberies eventually deplete their supplies. Children must make decisions along the way, but if they choose to hunt for food or stop at a fort, for example, they lose precious time and could suffer starvation or illness, or fail to pass the western mountains before the freeze and the blizzards. The computer mathematically determines the outcome of the children's decisions and gives the decision makers immediate feedback about the consequences of their choices. Although most simulations involve only one user per computer, *The New Oregon Trail* may involve a small-group approach by engaging a student family of four or five children to make the difficult decisions.

Another useful simulation for students in grades 4 to 8 is *Galleons of Glory: The Secret Voyage of Magellan* (available from Broderbund Software). The students imagine the world of 1519 by taking the place of Magellan and trying to sail a galleon around the globe. They are confronted with problems ranging from nature's fury to shipboard politics as they react to changing circumstances that arise during the voyage. Calm seas give way to vicious storms, and a frightened crew comes close to mutiny, challenging students to make decisions in situations representative of actual historical events.

Word Processing

Word processing systems are computer software programs designed to facilitate the efficient collecting, revising, storing, and printing of text. Word processing software has made it possible for elementary school students to type a first draft of a composition or report on a keyboard and simultaneously display it on a computer screen. By using a few quickly learned computer commands, the author can insert or delete text; combine sentences; "cut-and-paste" words, sentences, and paragraphs, moving them to new locations; change grammar and spelling errors; and print out the completed copy.

One type of word processing system helps students prepare neat, well-edited written pieces; another, called desktop publishing, combines graphics and text and allows users to produce other products such as signs, banners, greeting cards, newsletters, letterheads, travel brochures, and newspapers.

Telecommunications

In a small but growing number of classrooms, connections to the Internet are providing access to vast outside sources of information and creating new opportunities for collaborative learning. The Internet is often designated as a *telecommunications* service because it involves two-way communication via telephone lines. You are using a form of telecommunication when you chat with someone on the phone; a person who faxes a document from one fax machine to another transmits over telephone lines, too. In much the same way, we now find it possible to transfer data from one computer to another via telephone lines with the use of *modems.* If your classroom and others have computers, it is possible to connect to one another through a modem and telephone lines and go *online.* This means that you are connected to a network of users who can exchange files, send and receive electronic mail (*e-mail* for short), conduct joint projects, share data, and even play educational games with one another.

As fascinating as the Internet world can be, it can also confront you with a few problems. Perhaps the greatest is the availability of material not suitable for children. The Telecommunications Act requires Internet services to keep such material away from children, but nothing substitutes for careful supervision. Online services provide controls that allow you to limit what your children can access. Software such as *Cybersitter* or *Surfwatch* lets you supervise visits to the Internet.

Some teachers hail the Internet as a new era in learning. Ms. Nell maintains that her online connection has revitalized her teaching and motivated and excited her students, who are learning much of the same information as in their textbooks, but in a more engaging way. According to Ms. Nell, this added motivation comes from the fact that her students have an immediate, authentic audience for their work; through planning and sharing with their online peers, they find that the information they exchange has meaning—an instantaneous and utilitarian motive. One of the greatest benefits cited by Ms. Nell is the capacity for collaborative research. She and her students have joined other classrooms around the country to investigate everything from the weather to prices for consumer goods. The major complaint voiced by Ms. Nell is that there just are not enough online classrooms to make a comprehensive exchange of projects and ideas possible.

For many teachers, it's a challenge to bridge the gap between traditional instruction and technology. However, as many teachers have learned, tapping into the Internet helps children benefit from a rich source of new ideas, friends, and experts. Blagojevic (1997) describes several categories of project ideas that were carried out in classrooms using the Internet:

- *To make new friends.* Using electronic mail, children from Maine exchanged letters with Icelandic children.
- *To extend the curriculum.* After listing favorite storybook characters, children from Sacramento and Baltimore initiated a "story swap." Each group created and e-mailed original stories to each other.
- *To build cross-cultural comparisons.* Children from Oregon worked with children from Florida, Arizona, Japan, Russia, and South Africa on a collaborative book. Each location responded to the question "How do you like to play?" The children assembled a *How We Play* book of stories and drawings.
- *To produce information.* After a bee stung young Ted on the leg one day, his classmates suggested they build a Web page that would educate people about bees. They researched bees and added their own drawings to the informational pages.
- *To learn more.* A week after creating their Web page, the class received an e-mail message from a bee expert. He pointed out that their bee was actually a yellowjacket wasp and explained the difference between bees and wasps. The children used this information to correct their page.
- *To meet new people.* While they were studying penguins, a group of children electronically "met" a class of second graders from Dunedin, New Zealand. They were awed to learn that their New Zealand friends could walk a half mile down the road to see penguins in the bay.
- *To explore the world.* The Internet allows children to break out of the walls of their classroom to interact with people all over the world.

Hypermedia (Presentation Software)

A special area of interest that has contributed greatly to social studies programs today has been the use of *hypermedia,* a communications tool that combines video, graphics, animation, and text. Known as "presentation software," hypermedia authoring programs enable students to organize and communicate information in innovative and thought-provoking ways, accessing and integrating information from such diverse sources as the Internet, sounds or clip art pulled from public domain software, photographs from a digital camera or scanner, and clips from a video camera or CD-ROM. Three widely used presentation software programs are *HyperStudio*, *ClarisWorks*, and, especially for the younger set, the *SlideShow* portion of *KidPix*. These are not the only hypermedia tools available to teachers and students, but they are excellent examples of how such programs work.

Classroom Vignette

I once heard a computer expert say, "To be truly literate, one must learn to communicate in the dominant system of a culture." At no other time was this truism clearer to me than when I visited an elementary school "History Fair" and examined all of the wonderful projects. Every display attracted a great deal of attention, but one stood out above all the others. I was made aware of this special display by a neighbor who turned to me and asked, "Can you believe what Raymond did on the computer?"

I walked over to Raymond's space and was met with eye-popping graphics, clear text, dazzling animation, and breathtaking audio. Telling about the life of Harriet Tubman, Raymond's project gave every impression of a professional presentation. Scanned photos from literature sources, spoken text that highlighted the key events in Harriet Tubman's life, recorded spirituals, and the culminating video clip of Martin Luther King, Jr.'s "I Have a Dream" speech made the presentation a special occasion. The "flash" was not the priority in Raymond's presentation; the message was certainly the important element. However, the presentation software added much more to Raymond's research than if he had simply put together an oral or written report.

Some presentation software systems contain special "buttons" that allow the user to immediately access another part of the presentation by clicking on a prompt. For example, Raymond's presentation included a button at the point in Harriet's life when she gave slaves secret directions on how to flee to the North. By using the mouse to click the button, the user could listen to the song "Follow the Drinking Gourd" while looking at an illustration of the Big Dipper.

Speaking of presentation software, Muir (1994) comments that if students find learning more interesting and engaging as a result of creating an interactive project (perhaps one that makes information pop up on the screen when you click a button), then computers have served their purpose. If students become more enthusiastic about research—because they know that their final report is going to look good and be fully interactive—then computers have made a valuable contribution to the educational process (p. 32).

Presentation software is available, relatively inexpensive, and not difficult to use. Teachers should use this practical technology application and demonstrate modern, effective communication techniques to their students.

AFTERWORD

The heart of dynamic social studies instruction is balance and proportion. These elements do not normally emerge as part of a teaching personality during a student's undergraduate certification program, during student teaching, or even after a year on the job. They often emerge after repeated successes with textbook-based instruction. You will not rely on textbooks to guide you throughout your entire teaching career; a feeling of unrest and a strong desire to "spread your wings" will begin to entice you to expand your repertoire and experiment with varied instructional materials and activities during your earliest years of teaching. Katz (1972) describes this professional evolution as a stage-related process:

- *Stage One:* You are preoccupied with survival. You ask yourself questions such as, "Can I get through the day in one piece? Without losing a child? Can I make it until the end of the week? Until the next vacation? Can I really do this kind of work day after day? Will I be accepted by my colleagues?" Textbooks are useful tools that help teachers gain the confidence necessary to manage the routines causing most of the anxiety during this stage. (*First year*)

- *Stage Two:* You decide you *can* survive. You begin to focus on individual children who pose problems and on troublesome situations, and you ask yourself these kinds of questions: "How can I help the shy child? How can I help a child who does not seem to be learning? What more can I do for children with special needs?" (*Second year*)

- *Stage Three:* You begin to tire of doing the same things with the children. You like to meet with other teachers, scan magazines, and search through other sources of information to discover new projects and activities for the children. You ask questions about new developments in the field: "Who is doing what? Where? What are some of the new materials, techniques, approaches, and ideas? How can I make social studies (or any other subject) more powerful?" (*Third and fourth years*)

According to Katz's developmental theory, then, new teachers should not expect to move away from a deliberate textbook-based routine until sometime during the third year of teaching. At first, you will feel more comfortable teaching with the help of textbooks and with ideas learned from others. The need to grow and learn will become evident as an inner drive gives you no other choice but to branch out. You should then begin to formulate and refine a personal philosophy of instruction that will serve as a foundation to undergird all professional decisions in the future. The difference between teachers who are good "technicians" and those who are educational leaders appears to be their willingness to constantly think about and work toward methods based on a sound personalized philosophy of teaching and learning.

You will use textbooks, then, and they will contribute immeasurably to your social studies program. But the emphasis of your instruction should be on variety. Continually confront youngsters with significant experiences so they get to know our world and build the qualities that help them become constructive, active citizens. The kinds of citizens our boys and girls grow up to be is determined to a great extent by the ways they live and grow in school.

References

Alliance for Arts Education. (1985). *Performing together: The arts in education.* Arlington, VA: Author.

Anderson, R. C., Hiebert, E. H., Scott, J. A., & Wilkinson, I. A. G. (1985). *Becoming a nation of readers.* Washington, DC: National Institute of Education.

Augustin, B., & Bailey, M. (2001). Adobe bricks: Building blocks of the Southwest. *Middle Level Learning, Issue 12,* 4–9.

Blagojevic, B. (1997). Internet interactions. *Scholastic Early Childhood Today, 11,* 47–48.

Brandt, R. (1995). Future shock is here. *Educational Leadership, 53,* 5.

Coerr, E. (1977). *Sadako and the thousand paper cranes.* New York: Dell.

Fowler, C. (1989). The arts are essential to education. *Educational Leadership, 47,* 62.

Fowler, C. (1994). Strong arts, strong schools. *Educational Leadership, 52,* 4–9.

Goodman, K. S. (1988). Look what they've done to Judy Blume!: The "bazalation" of children's literature. *The New Advocate, 1,* 29–41.

Joint Committee of the National Council of Teachers of English and the Children's Theatre Association. (1983). Forum: Informal classroom drama. *Language Arts, 60,* 370–372.

Katz, L. G. (1972). Developmental stages of preschool teachers. *Elementary School Journal, 73,* 50–54.

Lapp, D., Flood, J., & Farnan, N. (1992). Basal readers and literature: A tight fit or a mismatch? In K. D. Wood & A. Moss (Eds.), *Exploring literature in the classroom: Contents and methods* (pp. 35–57). Norwood, MA: Christopher Gordon.

Muir, M. (1994). Putting computer projects at the heart of the curriculum. *Educational Leadership, 51,* 30–32.

Starr, J. (1989). The great textbook war. In H. Holz, I. Marcus, J. Dougherty, J. Michaels, & R. Peduzzi (Eds.), *Education and the American dream; Conservatives, liberals, and radicals debate the future of education.* Grandy, MA: Bergin and Garvey.

Steptoe, J. (1987). *Mufaro's beautiful daughters: An African tale.* New York, NY: Lothrop, Lee & Shepard.

Webster, W. (1993). Thank you, Miss Monroe. *Educational Leadership, 50,* 45.

11

Equal Opportunity for All

St. Guadalupe: Knock, knock (pretends to knock on a door).
Rainbow: Who is it?
St. Guadalupe: St. Guadalupe.
Rainbow: What do you want?
St. Guadalupe: I want a color.
Rainbow: What color?
St. Guadalupe: Rojo!

In this game, one child is chosen to be St. Guadalupe and another is chosen to be Mother-of-Color, or Rainbow. All of the other children are assigned color names, each one given by Rainbow. [You could use English color names only, or create an opportunity to teach their Spanish names: *rojo* (red), *verde* (green), *azul* (blue), *amarillo* (yellow), *negro* (black), or *blanco* (white)]. All the colors belong to Rainbow, and they line up behind her—across the play space from St. Guadalupe, whose area is designated as "Home Base." St. Guadalupe then initiates the preceding dialogue. All the children of the color named by St. Guadalupe must attempt to reach "Home Base" without being tagged. If they reach it safely, they are named another color and get ready to go again. Those who are caught help St. Guadalupe tag other players until everyone (including Rainbow) is captured. Then the game is over.

The children in this second-grade classroom were playing a game called "Los Colores" ("Colors"), a group game well-known to Spanish-speaking children (Perez, 1993). Their teacher, Maria Quinones, selected games from various cultures as a vehicle to teach children that we are each different from each other in some ways, but we are also alike in many ways. Therefore, Ms. Quinones fashioned a unit of instruction around games. She believed that games demonstrate the idea that, despite many fundamental differences, cultures are alike in many ways. In fact, this game is much like the popular childhood game you may have played, "Red Rover." Ms. Quinones believes that if we focus on children's likenesses (their common bonds) then they will be able to celebrate their differences with pride.

Ms. Quinones includes multicultural experiences as an integral part of her classroom program, establishing an all-inclusive learning environment that nurtures mutual trust and respect for all people. She operates with a strong conviction that children who are equipped with a knowledge of and appreciation for the glorious diversity among people will more likely be wise citizens who respect our nation's rich variety of cultures, heritages, abilities, and interests. She understands that "tourist" approaches to learning about human diversity (Jones & Derman-Sparks, 1992), where classes occasionally visit other people and cultures before returning to the mainstream curriculum, although often carried out with the best of intentions, rarely broaden children's understandings of those people and cultures. "Tourist" approaches are characterized by "special" activities where students do isolated projects, usually associated with a holiday or special situation—such as performing a dragon dance during the Chinese New Year or running their fingers over a Braille card in an attempt to understand the conditions of sightlessness. Although these activities may be fun for the children, they have little to do with the interpersonal concerns of Yatao, who recently moved to the United States from Taiwan, or Carl, a young boy constantly teased about his thick eyeglasses. These disconnected experiences might make students a *bit* more aware of the world around them, but genuine understanding comes from *all* that children do during the day. Perez (1994) believes that a "tourist curriculum emphasizes the exotic differences [among people], while ignoring the actual experiences of diverse groups. This way of teaching misrepresents [diversity] and leads to stereotyping" (p. 153). Educators today recommend replacing the tourist approach with strategies like Ms. Quinones—infusing multicultural content into the broad curriculum. The primary goals of infusion are to extend equal recognition for all groups and to help all people achieve equal opportunity in society.

Elementary school programs employing an infusion approach to instruction are popularly referred to as *multicultural programs*. *Multicultural* was once a term that considered differences, especially cultural and language differences, as "handicaps." Resulting educational programs often pressured diverse students to conform to majority standards and mainstream cultural expectations. Now, however, multiculturalism accepts cultural and linguistic diversity as a *resource* rather than a handicap, and has been expanded to include groups based on exceptionality, gender, and class, as well as language and culture. Teachers who employ practices that affirm and respect the cultures of students, regard the students' cultures as strengths, and blend the students' cultures into the educational program are described as *culturally responsive teachers.* Cultural responsiveness is the core of multicultural education and a critical ingredient of dynamic social studies classrooms. Given the diversity of contemporary society, all dynamic social studies teachers must become culturally responsive and develop instructionally effective strategies to use with widely diverse groups of students.

A rich storehouse of information provides insight into what teachers can do with children in a culturally responsive classroom, but before we examine these recommendations, it must be noted that the cultural group categories used to organize this chapter are merely social constructions, which cannot possibly embrace the life of any single individual. For example, a person might be "Caucasian," but also French, Southerner, farmer, Roman Catholic, hearing impaired, female, and lower-middle class. This person's group memberships, as we find true for most people, include race, ethnic group, region, occupation, religion, disabled or nondisabled, gender, and social class. Many aspects of one's life are shaped by membership in many

It is important that we use the exceptional sincerity of the elementary school years to start building the best possible interpersonal relationships.

groups, making the person previously described a much different person than an Asian American male from a large Eastern city who grew up in a crowded high-rise apartment, or a Native American female whose family inhabits a pueblo in the desert Southwest. Groups can be defined along many different lines, but everyone is a member of assorted groups, each of which creates its own culture (knowledge, rules, values, and traditions that guide its members' behavior).

Membership in any specific group may reflect certain tendencies and probabilities of expected behavior; for example, Hmong families (from Laos and the mountains of Vietnam) tend to have many children, and low-socioeconomic status (SES) students tend to exhibit lower levels of school achievement than high-SES students. However, such tendencies do not tell you about *individual* students. Banks (1993) clarifies: "Although membership in a gender, racial, ethnic, social-class, or religious group can provide us with important clues about an individual's behavior, it cannot enable us to predict behavior. . . . *Membership in a particular group does not determine behavior but makes certain types of behavior more probable*" (pp. 13–14).

All of this is not meant to confuse you, but only to emphasize that the children you teach are not just Puerto Rican, male, or middle class; they are uniquely complex individuals who have become who they are through the interaction of many intricate factors. The labels we choose to describe the groups they belong to are not meant to stereotype, only to provide insight into the best practices for social studies instruction.

EXCEPTIONALITY

Nearly every day, you will come into contact with individuals who fall into one or more categories of exceptionality. Exceptional people include individuals with disabilities as well as gifted individuals. Anyone who has spent time with children knows that there are many ways they are all alike and some ways they stand apart from one another. It is important to keep this perspective in mind, for exceptional children are more like than unlike nonexceptional children. This perspective is important, for children who are exceptional are, most importantly, children. They therefore have the same basic needs as nonexceptional children.

Children With Disabilities

Children with disabilities are similar in many ways to children without disabilities. Wolery and Wilbers (1994) clarify:

> All children share needs for food and shelter, for love and affection, for affiliation with others, for opportunities to play and learn, and for protection from the harsh realities of their environments. All children deserve freedom from violence, abuse, neglect, and suffering. All children deserve interac-

tions and relationships with adults who are safe, predictable, responsive, and nurturing. All children deserve opportunities to interact with peers who are accepting, trustworthy, kind, and industrious. All children deserve . . . educational experiences that are stimulating, interesting, facilitative, and enjoyable. (p. 3)

Despite these commonly shared needs, children with disabilities are different from children without disabilities. Wolery, Strain, and Bailey (1992) explain: "They need environments that are specifically organized and adjusted to minimize the effects of their disabilities and to promote learning of a broad range of skills. They need professionals who are competent in meeting the general needs of . . . children *and* are competent in promoting learning and use of skills important to the specific needs of children with disabilities" (p. 95).

In addition to sharing a set of basic needs, all children go through similar *stages of development.* Although the timing will not be exactly the same for each, children all around the world move through predictable patterns of *motor development* (they will walk before they run), *language development* (they will babble before they speak in sentences), *cognitive development* (they will want to explore their surroundings with their hands and fingers before they will try to read a book for information), and *social-emotional development* (they will scream to get their own way before they ask permission for things). Chandler (1994) explains that children with special needs may develop at a rate different from that of more typical children, but the sequence of development remains the same:

> For example, we know that children learn to sit before they stand, stand before they walk. This sequence is the same whether a child is nine months old or three years old. If three-year-old Amanda is unable to walk, we consider her a child with special needs. However, our knowledge of child development still tells us that she first needs to sit, and then stand, before she can walk, even though her development of these skills is delayed. Again, understanding typical development provides the information needed to teach and care for the child with special needs. (p. 21)

Despite the fact that children are similar in many ways, some children with exceptionalities will exhibit characteristics not quite like most others. It is estimated that between 10 to 12 percent of all children in the United States fall into the *children with disabilities* category; they deviate far enough from the typical in at least one respect that an individualized school program is required to address their needs. Who are these *children with disabilities?* Public Law 101–476, the Americans With Disabilities Act of 1990, defines children with disabilities as those:

> A. With mental retardation, hearing impairments including deafness, speech or language impairments, visual impairments, including blindness, serious emotional disturbance, orthopedic impairments, autism, traumatic brain injury, other health impairments, or specific learning disabilities; and
> B. who, by reason thereof, need special education and related services.

The Concept of Inclusion

Much effort these days is being directed toward *inclusion*. Inclusive classrooms operate with a conviction that students with disabilities have a right to be brought together into regular classrooms with their nondisabled peers. An educational system for all students is a major focus of our nation's schools for a number of reasons. First, state and federal laws mandate, support, and encourage it. Second, some parents of children with special needs were troubled that their children were required to attend separate programs. They viewed these programs as a form of segregation. Third, educators, parents, and children have had rewarding experiences in inclusive environments. Despite these strong points, however, not everyone is sold on the idea of inclusion. First, not all parents want their children with disabilities taken from their special programs. They believe that their children are best served in separate special education facilities. Second, many teachers feel inadequately prepared to provide for the special disabilities brought to their classrooms. Third, some people think that the great cost of inclusion outweighs its benefits.

Federal Legislation

Despite these opposing positions, inclusion is currently an educational reality and a process in which elementary school teachers find themselves deeply involved. The trend was initiated with Public Law 94–142 (the Education for All Handicapped Children Act), signed into law in 1975 and implemented in the fall of 1978. It was a valuable outcome of the many social efforts during the early 1970s to prevent the segregation of any child from regular classrooms, whether because of special needs or race. Specifically, Public Law 94–142 made free public education mandatory for all children older than age 5 who were identified as having special needs. Such education was to take place within a "least restrictive environment." A least restrictive environment was defined as a place where the same opportunities as those available to any other child are offered to children with special needs (those who need special attention to overcome conditions that could delay normal growth and development, distort normal growth and development, or have a severe negative effect on normal growth and development and adjustment to life).

A comprehensive educational, medical, sociocultural, and psychological evaluation by a multidisciplinary team determined the extent of a child's disability. From there, possible remediation strategies were proposed. Schools did this by scheduling a meeting with the prospective teacher, the child's parents, a representative from the school district (usually a special educator), and a member of the assessment team. All information about the child was shared and a personalized education plan, the individualized education program (IEP), unfolded.

In 1990, Public Law 101–476 amended Public Law 94–142 in several very important ways. First, the legislation clarified what parents could demand for their children with disabilities. It reinforced the idea that all children with disabilities between the ages of 3 and 21 should receive a free and appropriate public education in a "least restrictive environment (LRE)" with their nondisabled peers. In addition

TABLE 11–1
Examples of Person-First Language

Use . . .	Do not use . . .
• person with a disability	• disabled or handicapped person
• individual without speech	• mute, dumb
• child who is blind or visually impaired	• blind child or "the blind"
• student who is deaf or hearing impaired	• deaf student or "the deaf"
• boy with paraplegia	• paraplegic
• girl who is paralyzed	• paralyzed girl
• individual with epilepsy	• epileptic
• student who has a learning disability or specific learning disability	• slow learner, retarded, learning disabled
• person with a mental disability, cognitive impairment	• crazy, demented, insane
• child with a developmental disability	• mentally retarded
• child with a congenital disability	• birth defect
• child who uses a wheelchair	• wheelchair-bound child

to expanding and clarifying special education services for children with disabilities, the legislation replaced the title of PL 94–142 (*Education for All Handicapped Children Act*) with a new one (*Individuals With Disabilities Education Act—IDEA*). Although the change may seem insignificant to some, IDEA communicated a monumental message. By replacing the term *handicapped* with *individuals with disabilities,* Congress declared that professionals should think of children with special needs as *children* first rather than centering on their disabilities. In addition, IDEA was noted for its use of "person-first" language; this means that the person is emphasized first, the disability second. Examples of appropriate person-first usage appear in Table 11–1.

To extend this idea of appropriate terminology, persons without disabilities should be referred to as *nondisabled* rather than *normal* or *able-bodied*. The word *handicap* should be used only in reference to a condition or physical barrier ("The stairs are a *handicap* for Nina," or "Larry is *handicapped* by the inaccessible bus").

Teaching Children With Disabilities

Although an awareness of appropriate terminology helps, our desire to include children with disabilities in all aspects of social studies classroom life must also be based on our own feelings toward and understandings of children with disabilities. How would you feel, and what would you do, for example, in each of the following situations?

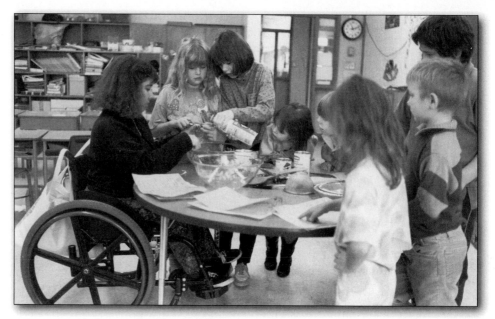

Social studies classrooms must be among the most honorable of all places to live in—a small society filled with understanding and appreciation for all.

- Sarah has a convulsion and you are the only adult around.
- David is lost and cannot hear you calling him.
- Armand seems unable to sit still; he constantly interrupts other children in class.

It might surprise you that the way most people choose to "deal" with problems like these is to avoid them. How many of us tend to steer clear of children with disabilities because we feel inadequate or insecure? You cannot take this approach as a teacher of elementary school children today. You must replace your feelings of inadequacy by confronting your uncertainties and replacing them with confidence based upon accurate knowledge. To effectively implement the spirit of *inclusion*, all professionals must learn something about how it operates; doing so may alleviate many fears and make those involved in the process feel more secure. The following suggestions are general and should be adjusted in consideration of each unique situation.

1. *Learn something about each specific disability.* You have a good start toward understanding children with disabilities if you know about child development. After all, children with disabilities are, first of all, children. It is important to know that children with disabilities are more like other children than they are different from them. Therefore, your first step in working with children with disabilities is to establish a framework with a solid understanding of child development.

When a child with developmental disabilities enters your classroom, take time to meet and get to know something about him or her. You might invite the family to visit your classroom, or find it instructive to visit the child at home. Whatever the choice, you will need a great deal of background information about the child. Other sources of background information include past teachers or other specialists who have previously cared for the child.

Certainly, it is not possible to know everything about each exceptionality you will meet during your teaching career, but you will have to learn a lot about each as you encounter it. That is why the relevant public laws stipulate that a team of specialists be involved in the formulation of each IEP. However, you should become familiar with the ways one can accept, understand, and become sensitive to the needs of every child. To help in this regard, search through many professional journals, books, and videotapes available through professional organizations or publishers of special education materials. Get to know each child well.

Once you gather basic information about a specific disability, you have taken the first step in working with a child. Solit (1993) uses the case of Marie to describe how this knowledge background fits into the total scheme of planning a program for children with disabilities:

> The teacher learns that Marie has a moderate hearing loss, with no developmental or cognitive delays. Marie wears hearing aids. The audiologist taught the teacher how to check the hearing aid to ensure it is working. The teacher learns that the hearing aid will make sounds louder, but it will not necessarily clarify speech. The parents explain that Marie uses American Sign Language to communicate. The [principal] decides to find a volunteer who can sign to Marie, communicate with the teacher, and also be a role model for Marie. The teacher also receives release time to attend sign language classes.
>
> The audiologist explains how to adapt the classroom environment so there are less auditory distractions for Marie. The teacher learns that many aspects of the program do not need to change because Marie will benefit from the high quality . . . classroom that is already in place. (p. 133)

2. *Maximize interactions between children with disabilities and nondisabled children.* Give simple explanations about a child's disability when needed. Youngsters are curious; they want to know about a new child and will be satisfied with an open, honest explanation. ("Russell's legs don't work well, so he needs a wheelchair.") Encourage the children with disabilities to share their strengths. For example, Russell can help another child in a project that involves the use of his hands (such as building a diorama or drawing a picture) while nondisabled learners may assist Russell with his special needs. In his classroom, for example, Russell regularly joined his classmates on the playground for recess. One of their favorite games was kickball. To play, Russell selected a "designated kicker" to kick the ball for him; after it was kicked, he wheeled around from base to base in his wheelchair. Social acceptance and cooperation help support students with diverse abilities.

3. *Individualize your program.* Start where the child is and plan a sequential program to encourage him or her to build one skill upon another. Visit classrooms where children with disabilities have been successfully included. Look for ways teachers individualize their instruction. How is peer interaction stimulated? Are parents involved in the classroom activities? Are peer questions about a child's disability answered openly and honestly?

4. *Assess your classroom environment.* Helping children with special needs feel comfortable in your classroom involves some very critical considerations. Overall, the inclusive classroom should contain the same materials and activities suggested for general social studies programs, but enhance these offerings with opportunities to meet the needs of children with disabilities. It helps to include photographs or pictures of people with disabilities participating with nondisabled people on the job or in a variety of other activities. Be sure the learning materials are accessible to all the children. Some children will need Braille labels to help them locate things while others may require ramps to move from one area to another. Whatever the case, be sure to explain to the other children why these special adaptations have been made: "This ramp helps Francine get to the top level when she is in her wheelchair." Invite adults with disabilities to share their special talents and interests with your children. In short, the classroom should offer a safe environment where all children feel accepted, whatever their capabilities or limitations. Despite the fact that some adjustments must be made, each child should be enabled to gain skills and understandings in all areas and to reach his or her full potential.

5. *Choose books that help children learn about and appreciate diversity.* In recent years, many good children's books have offered information about disabilities, explained difficulties youngsters with disabilities often encounter, and told stories about people who serve as positive role models for children with disabilities. Marc Brown's *Arthur's Eyes* (Little, Brown), for example, tells of how a little boy learns to cope with teasing about his new eyeglasses. Ada B. Litchfield's *A Button in Her Ear* (Whitman) explains deafness and how hearing aids help children with hearing losses. Lucille Clifton's *My Friend Jacob* (Lothrop) portrays a relationship between a young boy and his older friend with a learning disability. Maxine B. Rosenberg's *My Friend Leslie* (Lothrop) is a photographic essay of a young girl with multiple disabilities.

Literature can be one very important path to understanding and acceptance. This point can be illustrated clearly through this episode from the life of Helen Keller (1920), who lost her sight and hearing after a fever at the age of 19 months. The following high point in Keller's life occurs when her teacher, Anne Sullivan, places the hand of her then-7-year-old pupil under the spout of a pump:

> We walked down the path to the well-house, attracted by the fragrance of the honeysuckle with which it was covered. Someone was drawing water and my teacher placed my hand under the spout. As the cool stream gushed over one hand she spelled into the other the word *water,* first slowly, then rapidly. I stood still, my whole attention fixed upon the motions of her fin-

gers. Suddenly I felt a misty consciousness as of something forgotten—a thrill of returning thought; and somehow the mystery of language was revealed to me. I knew then that "W-A-T-E-R" meant the wonderful cool something that was flowing over my hand. That living word awakened my soul, gave it light, hope, joy, set it free! (pp. 23–24)

Keep many types of stories available and use them to promote questions, conversations, and empathy for children with developmental disabilities.

Inclusion involves changes in attitudes, behaviors, and teaching styles. Plan your inclusive social studies program to fit your children's needs. No single chapter in a textbook can hope to give you a complete idea of the responsibilities involved in doing so, but if you truly want to be a standout teacher, you must begin with sensitivity to the world of all children.

Gifted Children

Exceptionally talented or intelligent children were given little special attention in our elementary schools prior to the 1970s. The general consensus was that these youngsters could make good with very little or no help; their superior intelligence and advanced skills guaranteed success in whatever they chose to do. Sparse information, at best, described the specific needs of gifted students. To single out and offer special planning for these youngsters was considered by many to be elitist: "Why," people wondered, "should we channel extra money and resources into education for the gifted when they can learn so well on their own?" However, educators have now begun to realize the importance of accommodating gifted children's needs. Wolfle (1989) underscores this awareness effectively: "Every child deserves a developmentally appropriate education, not just 'average' children and children who are behind" (p. 42).

How will you know when you have an exceptionally gifted or talented youngster in your class? What are some of the gifts and talents that characterize this unique population? One way to begin answering these questions is to examine your children for exceptional characteristics:

1. *Verbal skills.* Uses advanced vocabulary; spontaneously creates stories; modifies language to the level of the person being spoken to; explains complex processes; influences the behavior of others; exchanges ideas and information fluently.
2. *Abstractions.* Retains easily what he or she has heard or read.
3. *Power of concentration.* Attentive to features of a new environment or experience; becomes totally absorbed in an activity; is alert, observant, and responds quickly.
4. *Intellect.* Carries out complex instructions; focuses on problems and deliberately seeks solutions; stores and recalls information easily; memorizes well and learns rapidly; explains ideas in novel ways; is curious; asks questions;

masters academics at an earlier age; has multiple interests; knows about many things of which other children are unaware.

5. *Behavior.* Sensitive to the needs and feelings of other children and adults; has strong feelings of self-confidence; influences others.

It makes sense for teachers of gifted children to follow the guidelines for establishing a personalized curriculum as described for children with disabilities. No single method will work with all gifted children; individual strengths and interests dictate varieties of approaches. As we consider special approaches, however, above all else, we must not remove childhood from the lives of gifted elementary school youngsters. The gifted youngster is a child first and should be treated like other children. Most 8-year-olds, for example, have similar interests whether they are gifted or not, but gifted children often show greater interest in learning *more* about a topic or material. So, like their peers, gifted first graders will enjoy playing with airplanes and trucks but, while their peers may stop there, gifted youngsters will want to investigate them in more detail (e.g., how high airplanes fly or how diesel engines differ from gasoline engines).

The goal of working with gifted students is not to separate them from their classmates with the materials and activities only they can use; rather, it is to provide enrichment and acceleration experiences within the classroom for all children. We must allow gifted students to grow to their fullest by building on their interests and talents in developmentally appropriate ways. Some general suggestions follow:

1. Gifted students require faster-paced instruction for skills- and content-based learning so they can move more rapidly through the curriculum.
2. Gifted students require more frequent use of inquiry and independent research projects that encourage independent learning.
3. Gifted students require more advanced materials—higher-level reading materials, computers, self-directed learning packets, and other more highly complex sources that allow students to explore topics in depth.
4. Gifted students require the reorganization of subject matter content so it allows them to explore issues across curricular areas and promotes higher-order thinking skills.

CULTURAL AND ETHNIC DIVERSITY

As a nation, we pride ourselves on our cultural, ethnic, and religious diversity. We are of many colors, speak many languages, and observe many different customs and traditions. All of us contribute to our nation's diversity, whether our backgrounds are Ukrainian, Polish, German, African, Swedish, Inuit, Sioux, Jewish, Korean, English, Puerto Rican, Mexican, Irish, Algonquin, French, Brazilian, Japanese, Italian, Russian, Cuban, Seminole, Lithuanian, Spanish, Scottish, Australian, or Chinese. In the past, the term *melting pot* (or crucible) was popularly used to characterize how to deal with this diversity, especially as it was applied to "Americanizing" foreign immigrants.

The Melting Pot Concept

The term *melting pot* was a metaphor that characterized how varied groups should be blended together into a new American culture:

> America is God's great Crucible, a great Melting Pot where all races of Europe are melting and reforming! . . . Germans and Frenchmen, Irishmen and Englishmen, Jews and Russians—into the Crucible with you all! God is making the American. . . . The real American has not yet arrived. He is only in the Crucible, I tell you—he will be the fusion of all races, the coming superman. (Zangwell, 1909, p. 37)

Proponents of the melting pot idea believed that all ethnic groups possessed strengths and that, as the "Crucible of America" amalgamated them into a single mass, a new, superior culture would be cast. The melting pot, then, was not meant to destroy cultural diversity per se, but to combine the strengths of many cultures into something new and unique: ". . . the new emerging American culture must be built not on the destruction of the cultural values and mores of the various immigrant groups but on their fusion with the existing American civilization. . . . In the burning fires of the melting pot, all races were equal—all were reshaped, and molded into a new entity" (Krug, 1976, p. 12).

The public school system of the time was reasonably tolerant of students from diverse backgrounds, but mirrored the melting pot ideals to the extent that diversity in the educational process was not tolerated. Stearns (1996) comments:

> It remains true that the American educational tradition . . . opted largely for a single cultural standard, to which all minority groups were expected to bow. The standard did not embrace all beliefs; religious diversity, most obviously, was still tolerated and to some degree honored. Yet growing emphasis on political conformity, . . . increasingly detailed rules about sex and hygiene (amid germ theories and residual Victorianism), and pervasive assumptions about proper economic discipline drove the single standard home. (p. 23)

A student cast as a singular mold of middle-class values was the fundamental goal of the educational tradition. Immigrants as well as members of the working class were instilled with "appropriate American middle-class standards" about such habits as personal grooming, punctuality and diligence at work, and political correctness.

Cultural Pluralism

Today, our nation as well as its schools has dismissed the idea of a single "American" culture emerging from a great melting pot. *Cultural pluralism* is the philosophy that now describes how all the parts of society contribute to an American whole; the United States is viewed as a multitude of cultures, each with unique, rich characteristics that contribute to the larger culture. Instead of a melting pot, our society has been likened to a salad bowl, pizza, mosaic, or patchwork quilt where each culture retains its own distinctiveness but yet contributes to the design of the whole.

The foundation of social studies instruction is based on this idea of cultural pluralism—becoming sensitive to and respecting the contributions of each group to society in general.

Cultural and Ethnic Minorities

Although a majority of Americans are of European descent, since 1980 the number of Americans identified as *ethnic minorities* (a term frequently used to refer to people receiving unequal or discriminatory treatment) has been steadily increasing. Woolfolk (1995) suggests that by the year 2020, almost half the population of the United States will come from African-American, Asian American, Native American, and Latino ethnic minority groups. As this trend continues to develop, it is obvious that significant changes in the makeup of our school population will follow. Our nation's growing diversity has important ramifications for all educators, but should be of special significance for those concerned with social studies; that is where the foundation of tolerance and understanding begins.

Cultural Identification (Ethnicity)

All individuals belonging to a group—whether an ethnic group, religion, peer group, or family—have a culture, or a system of behaviors, beliefs, customs, and attitudes. Culture is reflected in the group's artwork, literature, language, inventions, and traditions. Cultural differences are widespread, in both the overt (clothing, hairstyle, language, naming ceremonies) and the subtle (such as how one greets an elder). Culture consists of all the accepted and patterned ways of behavior of a group of people. It is a body of common understandings. It is the sum total of the group's ways of thinking, feeling, and acting. It is exhibited in the objects groups make—the clothing, shelter, tools, weapons, implements, utensils, and so on.

Different cultures manifest similar needs, but may choose to satisfy them in different ways. Gollnick and Chinn (1986) illustrate this fact by looking into a need shared by all cultures—food. All groups must obtain food in order to survive. However, different cultures have different ideas of just what items might be used as food:

> Many Americans reject foods, such as horses, dogs, cats, rice, mice, snakes, snails, grasshoppers, caterpillars, and numerous insects, consumed by other cultural groups in different areas of the world. At the same time, other cultural groups reject foods that are normal to many Americans. Muslims and Orthodox Jews do not eat pork. Hindus do not eat beef, some East Africans find eggs impalatable, and some Chinese do not drink milk. Do you remember the foods included on the . . . charts learned in elementary school? Often we find it difficult to believe that not everyone has a diet that includes the basic . . . food groups seen on those charts. (p. 6)

Whether we choose to compare and contrast food, religion, holiday customs, clothing styles, or any other of the array of cultural traits, the unique beliefs and be-

FIGURE 11–1
Cultural Diversity

Source: Turner, M. (1980). *Social studies and the young learner, 5,* 2. © National Council for the Social Studies. Reprinted by permission.

haviors of any distinct culture provide its members with a feeling of group identity (ethnicity) and offer a sense of continuity (belonging) that gives meaning to the life of an individual. Ethnicity is a feeling of continuity with the past, a feeling that one is an essential part of a group's collective history. Individuals bound by such strong ties feel their own survival is threatened if their group's existence is endangered. Ethnicity, then, involves a deep feeling of personal attachment in the historical continuity of a group (see Figure 11–1).

Strong feelings of ethnicity may have both a positive and a negative consequence. It is an asset, for example, for the culture to be viewed by its members as the natural and correct way of thinking, acting, and behaving. However, those feelings often evoke a sense of superiority over any other culture. The inability to view another culture objectively through its "cultural lens" prevents one culture from understanding or appreciating a second culture. This cultural insensitivity can lead to an inability to function effectively in a second culture.

When cultural differences bump into each other, misunderstandings often rise to the surface and members of one culture may misperceive others as "dangerous" or "strange." Those who ridicule a culture because its members eat grasshoppers rather than steak, for example, often do so in an effort to preserve their own long-held beliefs and values. This is not an uncommon reaction; all groups want to instinctively safeguard what they deeply value. If this protective instinct runs so deep, however, that a group becomes hostile toward any variation of its long-held beliefs, then the danger of excluding and alienating other groups is a strong possibility. Cultural insularity can result, a condition especially worrisome today in light of the need for interdependence among countries and the importance of establishing positive ties among all cultural groups. For this reason, an essential ingredient of effective teaching in social studies is to regard all children and their families with dignity and respect, regardless of ethnic, racial, or religious differences.

Teaching in Culturally Diverse Settings

In the not-too-distant past, many elementary schools attempted to meet the goals of multicultural education by including the study of holiday customs as separate topics not related to the regular social studies curriculum. This was an honest attempt to bring other cultures into the classroom, but our current professional responsibilities go far beyond that. Clark, DeWolf, and Clark (1992) warn that narrowly focused approaches to multicultural programs can often become *culturally assaultive*. To explain the nature of culturally assaultive programs, the authors ask you to pretend you are a young child, the only "non-Indian" child in your classroom, who will be learning some very interesting things about non-Indians during Thanksgiving time. As you timidly come into the room and seat yourself in the circle with the others, your teacher leads the following discussion:

> "Who knows what kind of houses non-Indians live in? Yes, that's right. They live in square houses with red tile roofs. Who lives in these houses? Mother and father and sister and brother. Yes, that's right. Grandmother? No; they don't live with their grandmothers, like we do. They send their grandmothers away to special places called retirement homes. Why? I don't know.
>
> "Next week, during Thanksgiving, we'll have a unit on non-Indians. We'll all make a non-Indian town out of clay. It's called a 'suburb.' Can you say 'suburb?' Non-Indians sleep in separate rooms, and they have little houses to keep their cars in.
>
> "Now this is a non-Indian hat." The teacher pulls out a Pilgrim's hat. "Non-Indians wore these when they first came to our land." (p. 5)

Clark, DeWolf, and Clark (1992) go on to explain that culturally assaultive classrooms perpetuate biases and stereotypes, usually incorporating the following elements into their practices:

- Discussion of cultures only as they existed in the past, such as that of the "Indians" who helped the Pilgrims on the first Thanksgiving Day.
- An incorrect or stereotypical version of how those people live or lived, such as characterizing "Indians" as wearing next to nothing and scalping people.
- Emphasis on differences from (rather than similarities to) other groups; for example, focusing on the kinds of houses "Indians" lived in rather than the fact that in the Indian cultures, as in other cultures, grandmothers tell stories.
- Use of songs, stories, and other devices that objectify the group and emphasize the group characteristics, ignoring the fact that individuals exist within the group; for example, using the song "Ten Little Indians" [Do not present any cultures as objects to count] instead of having an "Indian" from the community come in and tell a story from his particular tribe or nation.
- Token representations of the group in the classroom; for example, one "Indian"-looking doll among many Caucasian-looking ones.

- "Holiday units" on minority groups instead of saturation of the year-round curriculum with cultural diversity; for example, a unit on Mexican Americans on Cinco de Mayo and exclusion of that culture on all of the other . . . days of the year (p. 6).

Multicultural education, then, is not something we limit to special events or celebrations, such as observing African-American history only during the month of February, playing the dreidel game only during Hanukkah, tasting pork fried rice only during the Chinese New Year, or breaking a piñata only on Los Posados. Certainly, those events are important and should be rendered a special place in our dynamic social studies programs, but limiting a multicultural program to such singular, isolated events does not make the issue of ethnicity an important segment of the entire program. We must explore more deeply the likenesses and differences among cultures so that children learn ways to be more tolerant of people unlike themselves, and to recognize that it is healthier to accept difference rather than to steer clear of it. Teachers who believe that cultural awareness should be part of the child's education must plan to incorporate multicultural activities into the dynamic social studies curriculum.

Approaches to Teaching Culturally Diverse Content

To incorporate a culturally diverse approach to social studies education, a "cultural connectiveness" method is necessary. What this means is that instruction is focused on infusing cultural diversity into their daily learning experiences. This approach changes the structure of the curriculum by introducing various perspectives, frames of reference, and content from different ethnic groups into the study of the nature and complexity of U.S. society. It requires students to work toward a society that is more equitable by critically examining concepts, issues, or problems related to social conditions and taking action to relieve them.

The overall goal of multicultural education is to gradually and cumulatively empower students with the knowledge, skills, and attitudes needed to understand and appreciate racial, cultural, and ethnic diversity. The following steps are suggested as ways to infuse multicultural education into existing social studies programs:

1. *Know your community.* If you plan to turn your classroom into a place where diversity is a goal, start with a focus on the cultural groups represented by the school population. The students, community, and families your school is serving should be the primary starting point for culturally relevant programming, but this is a daunting prospect for many teachers. It is no easy task to incorporate cultural knowledge into one's teaching. However, Gollnick and Chinn (1994) explain that the first step is to know your students' cultural backgrounds:

> You should approach teaching multiculturally as an enthusiastic learner with much to learn from students and community members who have cultural backgrounds different from your own. You may need to remind yourself that your way of looking at the world evolved from the personal experiences

Culturally responsive social studies classrooms celebrate the diverse cultures of the school community.

you've lived through, which may vary greatly from the experiences of the students in your school. You will need to listen to the histories and experiences of students and their families and integrate them into your teaching. You will need to validate students' values within both school and their out-of-school realities—a process that is not authentic if you have feelings of superiority. . . . To make our classrooms multicultural we need to learn the cultures of our students, especially when the students are members of oppressed groups. (p. 296)

2. *Seek family support.* A prerequisite for meeting the needs of all families is the belief in their dignity and worth. Researchers have found that "to the extent that the home culture's practices and values are not acknowledged or incorporated by the school, parents may find that they are not able to support children in their academic pursuits even when it is their fervent wish to do so" (Florio-Ruane, 1989, p. 169). Be especially willing to listen as well as talk to the parents of your students—make sure that they understand your program's goals. Find out what they would like their children to learn about their own culture and other cultures. To teach multiculturally requires starting *where students are.* Teachers must know the families in their community. In school, Islamic parents could be upset with the uniforms that their daughters are expected to wear to gym class, and the

thought of co-ed gym classes could be quite disturbing. Islamic students might also be puzzled about why the school celebrates Christian holidays and never Islamic holidays.

You may find that the values and expectations of some families may differ markedly from your own. But rather than insist that the children mimic the behaviors and beliefs of *your* culture, gain a better understanding of the families making up the school culture. Although families may resist some of the content and activities you plan for your social studies curriculum, it does not rule out a multicultural approach to teaching. Celebrate diversity by incorporating appropriate content into the curriculum, but first know the viewpoints of the families served by the school.

3. *Give equal attention to all groups.* Social studies classrooms must reflect the diversity of cultures, whether or not the school population itself is diverse. Instructionally, the curriculum must incorporate information about many cultures and about intergroup relations. Gollnick and Chinn (1994) advise: "The amount of specific content about various microcultures will vary . . . but an awareness and recognition of the culturally pluralistic nature of the nation can be reflected in all classroom experiences. No matter how assimilated students in a classroom are, it is the teacher's responsibility to ensure that they understand cultural diversity, know the contributions of members of oppressed as well as dominant groups, and have heard the voices of individuals and groups who are from a different cultural background than that of the majority of students" (p. 309).

Teachers with a multicultural perspective know that, because they cannot possibly offer equal treatment to the hundreds of microcultures in this country, they must begin planning by developing an understanding of and sensitivity and respect for the various cultures of the children in their school community. In urban Chicago, a teacher would include activities and information from African-American culture, and in Lancaster County, Pennsylvania, a teacher should address the Amish culture. In other areas of the country, schools should focus on the character of the groups represented in the community. These cultures should become an integral part of social studies, expanding the standard curriculum with diversity and multiple perspectives. As the students begin to realize that they are important members of the school and that diversity is valued, the curriculum can examine sensitive issues and topics from the perspective of various ethnic and cultural groups—Chinese Americans, Irish Americans, African Americans, Puerto Ricans, Catholics, Southern Baptists, Caucasian males, or Jewish women. Cultural diversity must be infused into the social studies program and become the lens through which the pluralistic nature of our nation can be focused.

4. *Fill your room with fascinating things.* Think about all the curriculum materials in your room. Multiethnic dolls, pictures, and study prints from different parts of the world, crayons that match in degree of skin tone, examples of Japanese calligraphy, tortilla presses, kimonos, cowboy boots, nesting dolls, chopsticks, bongo drums, serapes, and tie-dyed cloth from Africa fascinate children and encourage interest in people.

5. *Invite visitors into your room.* Having people from the community who are willing to come to your room and share something of their culture is a splendid addition to a multicultural program. Resource people can demonstrate a special craft or talent, read or tell a story, display and talk about an interesting artifact or process of doing something, share a special food or recipe, teach a simple song or dance, or help children count or speak in another language. If you arrange for visits from different people throughout the year, then your children will begin to respect and value all cultures.

6. *Draw from the vast resources of the arts.* The arts offer one of the most valuable sources from which to draw suitable multicultural experiences, for music, art, and literature know no cultural boundaries. The common expressions of human feeling found in these art forms can be used effectively to develop children's capacities to identify with other groups and other societies—indeed, the totality of human civilization. The arts provide ways of giving the individual an opportunity to try on a situation—to know the logic and feeling of theirs—even though these others are remote to the lives of the school population.

It is never too early to introduce young children to the arts of various cultures. Songs, rhymes, and chants evoke pleasure and enthusiasm from the very young in kindergarten and first-grade classrooms. Stories, pictures, books, arts and crafts, stage plays, puppets, dance, and other forms of creative expression add zest to the early grades. Younger children as well as middle- and upper-graders can visit museums or displays, especially those associated with specific ethnic groups. Seeing the beautiful handmade crafts (pottery, silver and turquoise jewelry, and baskets) of the Hopi and Zuni, for example, helps students understand important aspects of these cultures. Take your children to musical events having distinct cultural characteristics—African chants, Yiddish folk tunes, sailor chanteys, Scottish bagpipe music, or Eastern European polkas. Invite guest speakers to demonstrate special arts techniques—Amish quilt making, Inuit soapstone carving, Cajun music, or Plains Indian pictographs. Read, tell, or dramatize stories of various cultures. For instance, *The Miracle of the Potato Latkes* by Malka Penn (Holiday House), *The Angel of Olvera Street* by Leo Politi (Scribner's), *The Gifts of Kwanzaa* by Synthia Saint James (Albert Whitman and Company), and *Christmas Around the World* by Emily Kelly (Carolrhoda Books) provide outstanding examples of the many ways people from various cultures celebrate December holidays. Children should be helped to understand that the arts reflect culture, and that one cannot fully appreciate the value of any art without some understanding of the cultural matrix from which it grew. Conversely, one cannot fully appreciate a culture unless one values the creative efforts of its members.

All social studies programs have the responsibility to provide quality educational experiences that help children become compassionate individuals who feel comfortable with their identities and sense their unity with other people. We must create positive environments where children learn to accept others with cultural differences and begin to develop the skills of living cooperatively in a culturally diverse nation.

GENDER

The image each of us acquires about our masculine or feminine characteristics and the various behaviors and attitudes normally associated with them is called *gender identity,* or *gender typing.* Like most other aspects of child development, gender identity emerges from dynamic interactions of biological and environmental forces. There is no question, for example, that there are basic genetic differences between males and females; biology sets the stage for gender identification. However, biology alone does not determine gender-specific behavior. From birth, our families begin to show us in subtle ways exactly what it means to be masculine or feminine. For example, little girls are most often dressed in something pink and frilly, whereas boys are routinely clothed in blue. Boy babies are commonly referred to by such terms as "big" or "tough," whereas girls are quite often described as "pretty" or "sweet." From these early days on, choices of toys, clothing, and hairstyles supplement verbal messages to influence gender identity.

Through such environmental influences, children unconsciously acquire a gender *schema,* or frame of reference, delineating what it means to be a boy or girl. This happens at about the age of 2; from that point on, children work hard to fit into their gender roles. By age 5 or 6, they have already learned much of the stereotypical behavior of their gender.

"Appropriate" behaviors are reinforced throughout the early years of life by internalizing the attitudes and responses of such environmental influences as family, relatives, peers, and the media. In other words, little girls who are rewarded for playing with dolls, read books about girls playing with dolls, and see girls playing with dolls on television will be more likely to play with dolls than with trucks. Likewise, little boys who have similar experiences with trucks will be more likely to play with trucks than with dolls. It would seem, then, that if gender-specific behaviors are influenced by such environmental phenomena, children raised in bias-free environments would not exhibit a preference for stereotypically gender-specific toys. However, a phenomenon referred to as "developmental sexism" seems to occur despite our most systematic attempts to shape a nonsexist environment. This means that young children grow to be enormously sexist in their perception of gender roles and choice of play activities (most boys choose to play firefighter and girls play house) even if they have been brought up in a nonbiased environment.

This concept of "developmental sexism" is supported by Kohlberg's (1992) idea of "gender constancy." That is, children soon learn that they permanently belong to a category called "boy" or "girl"—their gender cannot change. Once children grasp the concept that they cannot be transformed from girl to boy and back again, they organize the world into "girl" or "boy" categories and become powerfully attached to their gender. (I remember watching one warm spring day as Teddy, a kindergartner, proudly skidded around the playground on his sister's outgrown pink roller skates. "You're a-a gir-l! You're a-a gir-l!" chanted his friend Johnny in sing-song fashion. Each time Johnny finished his melodic rhyme, Teddy angrily retorted, "No, I'm not! No, I'm not!" but Johnny persisted: "You're a girl 'cause you have pink roller skates. You're a-a gir-l!" Johnny continued his taunts until Teddy was reduced to tears.) Such a strong attachment to one's gender continues to grow through the early elementary grades, cementing the peer solidarity that influences behaviors compatible with society's expectations for males and females.

Gender-role stereotypes seem to be decreasing in our elementary schools, but they continue to be a problem. Although books appear to be more fair and inclusive than in the past, Sadker, Sadker, and Klein (1991) believe that teachers prefer to read the books they grew up with; many of these older books represent highly traditional gender roles. Therefore, even today, the influence of sexist books can be found in some classrooms.

In addition to problems with books, Sadker and Sadker (1986) suggest that girls are shortchanged during classroom interactions. Elementary school teachers ask boys more questions, give them more precise feedback, criticize them more frequently, and give them more time to respond. The teacher's reaction may be positive, negative, or neutral, but the golden rule appears to be that boys get the most attention from teachers in elementary school classrooms.

Additionally, the American Association of University Women (AAUW) has reported "[t]here is clear evidence that the educational system is not meeting girls' needs. Girls and boys enter school roughly equal in measured ability. In some measures of school readiness, such as fine motor control, girls are ahead of boys. Twelve years later, girls have fallen behind their male classmates in key areas such as higher-level mathematics and measures of self-esteem (1992, p. 2).

Although most of the research into gender bias in schools has been centered on the unfair treatment of girls, Campbell (1996) cautions that, "It is boys who lack role models for the first six years of schooling, particularly African American, Latino, and Asian boys. While young, European American girls benefit from their female-centered primary school experience, children of color—particularly boys—fail. It is boys who encounter the most conflicts and receive the most punishments in school and most often get placed in special education and remedial programs" (p. 113).

Gender stereotyping can be tied to many influences and conditions, but teachers must take a positive role in recognizing bias and replacing it with equitable expectations for all children. This means eliminating one's own biases of gender-associated behavior and stereotyped notions about gender roles. This can be done by providing males and females with appropriate instruction and by avoiding gender-role stereotyping. Some guidelines to avoid sexism in teaching follow.

1. *Avoid stereotyping masculine and feminine roles.* Examine ways you might be limiting the options open to boys and girls. During class discussions, for example, many teachers attempt to reason with young children in order to create more objective attitudes about gender roles. When a child says, "Only boys can grow up to be truck drivers," teachers are tempted to reply, "That's not true. Women can be truck drivers, too." This approach often fails. The young child's way of classifying the world into male and female is new and not open to exceptions. A child may even become upset that the teacher fails to see the world in the same light and defend his

or her case even more strongly. We can compound the problem, therefore, by trying to reason with a child. This presents us with an interesting dilemma: We want children to experience a nonsexist world, but they tend to resist our efforts of objectivity. What can we do?

First, let the children know you understand and accept their unique system of trying to make sense of the world. Their willingness to come to you and share their excitement about new discoveries should always be accepted with openness and sincerity. You do, however, have a responsibility to help them understand that choices should be open to each person, regardless of gender. In responding to the "truck driver" comment, you might say, "I know you've never seen a woman truck driver before, so it's hard to understand that women can drive large trucks, too." However, trying to reason with a child through comments such as, "It's okay for women to be truck drivers, too. Many women are very good at driving large trucks," often elicits a response such as, "Well, they shouldn't be!"

Stereotyping should be avoided at all costs. This advice extends not only into how females are featured, but also into whether men are depicted in traditionally male roles and careers.

2. *Use gender-free language whenever possible.* Through words and actions, teachers assume the position of a positive role model. Be sensitive to your choice of masculine terms to refer to all people; for example, *police officer* replaces *policeman,* *firefighter* replaces *fireman,* and *mail carrier* replaces *mailman.* If your children use labels such as fireman frequently, begin a discussion with a comment such as, "Saying the word *fireman* makes it sound like only men fight fires. Do you think that's true?" Then introduce the word *firefighter* and point out that men and women (and boys and girls) can do the same kinds of jobs. Additionally, be aware of how actions can convey ideas of gender-role coequality. For example, the children may learn to interpret gender roles less rigidly if they see their teachers, male and female, displaying characteristics typically associated as either masculine or feminine; for example, being assertive and forceful, sensitive and warm, depending on the situation.

3. *Make sure your classroom materials present an honest view of males and females.* Just as you lead young children toward understanding the idea of equality through your words and actions, the activities and materials you choose for your classroom should resist gender stereotyping. Books like *Heather Hits Her First Home Run* by Phyllis Hacken Johnson (Lollipop Power Books) and Charlotte Zolotow's *William's Doll* (Harper & Row) are sensitive books that address stereotypes. *William's Doll* takes a look at a situation many little boys face:

> William would like a doll so he could play with it like his friend Nancy does with her doll. At the very thought of a little boy with a doll, his brother and friend call him creep and sissy. William's father buys him a basketball and a train, instead of a doll. William becomes a very good basketball player and he enjoys the train set, but he still longs for a doll. Finally, when William's grandmother comes for a visit, she buys him a doll and explains to his father that having a doll will be good practice for him when he grows up and has a real baby to love. (Raines & Canady, 1989, p. 50)

Teachers who wish to build a good classroom collection of gender-fair books will need to look for books that show children and other people engaged in a variety of activities, regardless of gender.

A program offering opportunities for both sexes to participate in positive classroom experiences should transcend obsolete sex-role expectations such as boys taking the lead when mathematics skills are required, and girls when sewing or cooking are needed for a project. Encourage the boys to wash the art table after completing a salt-and-flour relief map and the girls to hammer the nails needed to hold together a model clipper ship. If this doesn't happen freely, discuss the situation with your children. Say, "I notice that in most social studies projects the boys build the model. This seems to exclude the girls. Why do you think that is happening?" Invite equal access to activities by encouraging children to engage in a wide range of experiences that are free of gender stereotypes.

4. *Balance the contributions of men and women in the social studies program.* All students should be exposed to the contributions of women as well as men throughout history. Gollnick and Chinn (1994) point out that social studies courses that "focus primarily on wars and political power will almost totally focus on men; . . . courses that focus on the family and the arts will more equitably include both genders. . . . Students are being cheated of a wealth of information about the majority of the world's population when women are not included as an integral part of the curriculum" (p. 139). Banks (1994) suggests that women can be virtually ignored in written history. Citing the Montgomery, Alabama, bus boycott of 1955 as an example, Banks maintains that most textbook accounts emphasize the work of men such as Martin Luther King, Jr. and Ralph D. Abernathy, or organizations headed by men, but virtually ignore the work of women. He uses the memoirs of Jo Ann Gibson Robinson, president of the Women's Political Council of Montgomery, as an example. The Council was started in 1946 to "provide leadership, support, and improvement in the black community and to work for voting rights for African Americans" (Banks, 1994, p. 6). The Council received numerous complaints concerning bus driver offenses against African Americans who were asked to give up their seats on crowded buses to whites. On December 1, 1955, Rosa Parks was arrested for refusing to give up her seat. Disgusted by such hostile encounters with bus drivers, the Council distributed leaflets that called for a boycott of city buses. Referring to Rosa Parks, Robinson's leaflet read in part:

> This woman's case will come up on Monday. We are, therefore, asking every Negro to stay off the buses Monday in protest of the arrest and trial. Don't ride the buses to work, to town, to school, or anywhere else on Monday. (Garrow, 1987)

Although most textbook accounts credit King and Abernathy for the Montgomery bus boycott, plans to end bus segregation with a boycott were actually instituted 2 years earlier, in 1953, by Robinson's Council. The Parks case in 1955 just happened to be the "right time" to implement the boycott. The situation seems to be improving in recent years, but the work of historically significant females such as Jo

Ann Gibson Robinson still must find its way into our nation's textbooks. This does not mean we must sit back and wait for that day; it will take a great deal of scholarly effort to uncover their stories, but the experiences of women from all walks of life must be highlighted in the social studies curriculum.

Schools that foster positive gender roles will help children value the likenesses and differences in themselves, thereby taking an important step toward alleviating the damage resulting from long-ingrained patterns of sexism in our society. It is this unconditional positive regard for children that lies at the heart of social studies education.

SOCIAL CLASS

The term used by the U.S. Bureau of the Census and by sociologists to describe the variations of wealth and power among individuals and families is *socioeconomic status,* or SES. SES is normally determined by studying such economic factors as occupation, income, and level of education. Of all the forms of inequality affecting our children's education, SES could be the most powerful; it frequently surmounts the effects of race and gender. For example, although upper-SES Hispanics share customs and traditions with low-SES Hispanic families, they will be more likely to interact with upper-SES families of other ethnic groups than they will with Hispanic families at other class levels.

Research over the past 20 years has indicated a number of strong relationships between SES and school performance. A consistent connection is that low-SES students of all ethnic groups exhibit lower average standardized test scores, receive lower grades in school, and leave school earlier than high-SES students. Several factors explain the lower school achievement of low-SES students, such as poor health care for mother and child, limited resources, family stress, interruptions in schooling, and discrimination. Garcia (1991) cautions that research in this area is meager, but lists other explanations for lower achievement among low-SES children:

1. *Low expectations—low self-esteem.* Low-SES students often speak ungrammatically, come to school in old or dirty clothing, frequently are ungroomed and malodorous, are unfamiliar with the themes of mainstream children's literature, and are confused by the punishment and reward systems of the school. Since most teachers find it difficult to identify with these characteristics, they often conclude that low-SES students are not very good at schoolwork. They have reduced expectations for low-SES students, thereby contributing to phenomena called the *self-fulfilling prophecy* (tending to behave as others expect) and the *looking-glass self* ("I am what I think you think I am"). When teachers associate such reduced expectancies with socioeconomic status, discriminatory practices surface and low-SES students are denied access to equal educational opportunity.

2. *Learned helplessness.* Some children from low-SES homes come to school from communities where dropout rates of 50 percent are not uncommon. Since their

relatives and friends leave school early, many low-SES students are not motivated to go on. However, poor but stable families often value education and prepare their students well for school, viewing school as the best place to end their cycle of poverty. Teachers must understand that poverty is not the fault of the child and to think of teaching not in a negative or indifferent way, but as a challenge to help low-SES students overcome the effects of poverty.

3. *Resistance cultures.* Teachers, especially in inner-city schools serving poor Hispanic American, African-American, or Native American families, will find that some children come to school as part of a "resistance culture," members of which oppose upper- and middle-SES values, including school. This opposition can take the form of willfully breaking school rules, minimizing the value of achievement, and attaching more importance to manual rather than mental work. Students who accept any characteristics considered "middle class" or "white," including behaviors that would make them successful in school, are thought of as "selling out" their minority or peer group. To address this challenge, teachers must interact with the minority groups represented in the school population to determine the most effective instructional experiences for their children.

4. *Tracking.* A significant factor contributing to poor academic performance among low-SES students is that they are usually placed in low-ability groups or classes, where they are taught differently. Teachers see these children as less able academically and often use teacher-dominated strategies calling for lots of worksheets, rote memorization, and passivity. Additionally, less-experienced or less-successful teachers are generally assigned to the low-ability groups. After reviewing the research on the effects of tracking, Gamoran (1992) has concluded that "grouping and tracking rarely add to overall achievement in a school, but they often contribute to inequality. . . . Typically, it means that high-track students are gaining and low-track students are falling further behind" (p. 13). What can be done? Slavin (1987) suggests that tracking in elementary schools must be stopped because teachers often fail to match students' needs with instruction. Tracking will continue to have no positive effect on achievement until teachers use it to provide specialized instruction to children having specific needs: "For ability grouping to be effective at the elementary level, it must create true homogeneity on the specific skill being taught, and instruction must be closely tailored to students' levels of performance" (Slavin, 1987, p. 323).

Therefore, since ability grouping rarely contributes to positive academic achievement, it should be eliminated or curtailed. If it is used, students should be grouped according to skills need, and, when instruction is completed, the group should be disbanded.

LANGUAGE DIVERSITY

Hand-in-hand with the rich diversity of cultures enjoyed in the United States is a grand assortment of languages and dialects. Changing populations and an influx of immigrants from Asian and Hispanic nations have produced a situation where there are now over 200 languages spoken in the United States (Arends, 1991). Although this situation has created a fascinating cultural state of affairs, it has resulted in significant challenges for the education of our nation's youth. These challenges, however, are not new. Concerns about the goals and purposes of educating non-English-speaking students date back to the Revolutionary War, when school was taught in any of 18 languages spoken by the colonists—English, German, Scottish, Irish, Dutch, French, Swedish, Spanish, Portuguese, and others. As the colonies eventually blended into a new nation, however, English became the dominant language in public education. The freedom to use other languages has become a matter of unrelenting conflict since that time.

Although there has been evidence of the sporadic use of languages other than English in our nation's schools, the widespread use of bilingualism (teaching in two languages) did not occur until the 1960s. The civil rights movement brought attention to discriminatory practices throughout society, including children whose first language was not English. From the 1960s into the 1980s, federal court decisions have insisted that schools offer quality programs to enable students to participate successfully in all-English classrooms.

Currently, our nation's changing demographics require a new look at the needs of children from homes where English is not the first language. Campbell (1996) reports that since 1970 the United States has experienced massive immigration similar to the levels that occurred from 1890 to 1910. At least 10 million immigrants have come to live in this country during the past three decades. Well over 70 percent of these new immigrants are Latinos (from Mexico, El Salvador, Guatemala, Nicaragua, and Honduras) and a great number are Southeast Asians (from Vietnam, Cambodia, Laos, Korea, Japan, and China). Settlement patterns indicate that 80 percent of these immigrants settled mainly in 10 states, primarily California, Texas, and Florida. As teachers, we must be aware of how these changing demographics will influence the student composition of our classrooms into the next millennium: "By the year 2050, the present percentage of Latinos in the U.S. population will almost triple; the Asian population will more than triple; African Americans will increase about 3 percent; the non-Hispanic white population will drop from 76 percent to 53 percent. (U.S. Census Bureau, 1990)

Students coming to school speaking a main language other than English are referred to as *English as a Second Language* (ESL) students, and children who are not yet fluent enough in English to perform school tasks successfully are sometimes called *Limited English Proficiency* (LEP) students. Some educators have argued that because many people connect negative attitudes about linguistically diverse students with the term LEP, a more positive term such as *Potentially English Proficient* (PEP) should be used (Freeman & Freeman, 1993). Children who speak fluently in English at school and a native language at home are called *bilingual*.

Bilingual education is one of the most fiercely debated topics in education today: Is it better to teach social studies to LEP (or PEP) students in the primary (home) language until the children are fluent in English, or do these children need lessons in English before social studies instruction can be effective? Educators have attempted to answer that question in two primary ways. One proposed solution is *submersion,* in which students are assimilated into the English-speaking classroom and exposed only to English. In their haste to become "Americanized," submersed students quickly learn to value English, often avoiding the use of their native language, even at home.

Proponents of this approach advise teachers to introduce English as early as possible, to speak only in English, and to ask students to give up their family's language. They argue that valuable learning time is lost if children are taught in their native language. Woolfolk (1995) cites three important issues raised by critics of this approach. First, children who are forced to learn a subject in an unfamiliar language are bound to have trouble. Think of how successful you would have been if you had been asked to learn American history or economics using a language that you had studied for only a few short months. Second, students may get the message that their native language (and therefore their culture) is inferior and less important. This influences both language acquisition and self-esteem:

> To devalue a minority child's language is to devalue the child—at least, that's how it feels on the receiving end. The longtime policy of punishing Chicano students for speaking Spanish is an obvious example. While such practices are now frowned upon, more subtle stigmas remain. Children are quick to read the messages in adult behavior, such as a preference for English on ceremonial occasions or a failure to stock the school library with books in Chinese. . . . Whatever the cause, minority students frequently exhibit an alienation from both worlds. Joe Cummins calls it bicultural ambivalence: hostility toward the dominant culture and shame toward one's own. (Crawford, 1992, pp. 212–213)

Third, ironically, is that students who master standard English and let their home language deteriorate are often encouraged to learn a second language when they reach middle school. Had their native language been allowed to flourish in the first place, they would already be fluent in two languages without having to take Spanish I or similar introductory courses.

An alternative to submersion is *bilingual teaching;* that is, using two languages as vehicles of instruction. The primary goal of bilingual programs is not to offer direct instruction in English per se, but to teach children in the language they know best and to reinforce their understandings through the use of English. Grant and Gomez (1996) explain:

> The core curriculum in public school bilingual classrooms is the same as that for any other classroom. The only significant curricular difference is the focus on language development (ESL and the appropriate native language) and attention to the cultural heritage of the targeted language minority group. Besides teaching language, science, math, social studies, art, and music, bilingual teachers must facilitate language learning in everything they do. They are concerned with how best to teach non-English speaking stu-

dents the full range of subjects while developing native and [English language] skills. (p. 118)

If you visited an elementary school social studies classroom with a bilingual program, you would likely see an English-speaking teacher and another teacher or aide fluent in a native language using language experience or whole language approaches to literature-based instruction. Their preferred teaching approach follows Scarcella's (1990) preview/teach/review format. In this design, the content of the lesson is previewed in English, the body of the lesson is taught in the student's native language, and then the lesson is reviewed in English. This approach is often used when two teachers—one English-speaking teacher and one fluent in the native language—collaborate in a team-teaching effort. In addition to Scarcella's preview/teach/review format, Freeman and Freeman (1993) recommend the following guidelines for bilingual instruction:

1. *Environmental print.* Children learn to recognize words written in both English and their native language when they see print in a number of environmental contexts—magazines, newspapers, telephone books, menus, food packages, street signs, days of the week, classroom posters, labels, nametags, charts, bulletin boards, and other interesting sources. Words should be printed both in English and the children's native language.

2. *Culturally conscious literature.* Classroom use of multicultural literature written in the students' native language helps strengthen cultural values and beliefs. Quality books are now being written for children in a number of languages and are becoming increasingly available throughout the United States. For example, Carmen Lomas Garza's bilingual book *Family Pictures: Cuadros de Familia* (Children's Book Press) is an authentic portrayal of what it is like to grow up in a Mexican American family in South Texas (Rosalma Zubizerreta authored the Spanish version). If you cannot afford to purchase a number of like books, parents or other members of the community might be willing to lend books written in the children's native language. Having a parent or other volunteer come to school and read from these books adds respect and appreciation for the native language.

3. *Literacy instruction.* Traditional models of literacy instruction that are teacher dominated and skills driven—a lot of teacher talk and student listening—are not very effective in helping ESL students acquire literacy skills. To replace such "hands-off" learning, advocates for changing the nature of literacy instruction in the content areas such as social studies recommend constructivist practices. Constructivist practices assume that students will learn literacy by engaging in the full processes of reading and writing in a purposeful manner. In other words, the best method for teaching reading and writing to English language learners is through the natural use of these skills throughout the entire school curriculum.

4. *Language buddies.* Learning a second language is enhanced greatly when students are paired up with English-speaking classmates who speak the native language fluently. English proficiency is promoted by the classmate's careful explanations, modeling, and assistance with new words.

Bilingual education, like all dimensions of multicultural education, is based on a commitment to school success for all of our nation's children. A bilingual curriculum should provide students with educational opportunities that are meaningful, compassionate, and challenging—to develop the full range of oral and written language necessary to function in school and as a citizen in our democratic society.

AFTERWORD

Good social studies teachers always impress me with the affection they associate with the special moments they experience with all of their students: The expressions on children's faces when they learn something new; the excitement shown by parents as their children make progress during the year; just being with children and knowing that there is a common bond of affection; the children's unspoiled enthusiasm—such experiences revitalize teachers and can strengthen their commitment to the profession. Many teachers find great joy in the candid individual expressions that mark each child's uniqueness: "Henry came up to me holding his finger as if it were hurt. When I asked him what was the matter, he replied, 'An elephant bit my finger,' and then turned and walked away!"

These are the special sources of satisfaction awaiting a teacher of elementary school children. You will find extraordinary joy, affection, excitement, and personal satisfaction as you meet challenges each day.

Children thrive under good teachers who delight in children being who they are. These teachers adapt the social studies classroom to meet every child's cultural, linguistic, and individual needs. This includes providing the child with the time, opportunities, resources, understanding, and affection to achieve the important goals of social studies education. To affirm individual differences, teachers must eliminate bias from the elementary school environment. Every child must know he or she is appreciated and respected by the teacher and needs experiences that reflect an understanding and appreciation for individual and cultural differences. These experiences are not only memorable and pleasurable, but they also last a lifetime—they help make our world.

References

Alwin, D., & Thornton, A. (1984). Family origins and schooling processes: Early versus late influence of parental characteristics. *American Sociological Review, 49,* 784–802.

American Association of University Women. (1992). *How schools shortchange girls.* Washington, DC: Author.

Arends, R. I. (1991). *Learning to teach.* New York: McGraw-Hill.

Banks, J. A. (1993). Multicultural education: Characteristics and goals. In J. Banks & C. McGee Banks (Eds.), *Multicultural education: Issues and perspectives* (pp. 2–26). Boston: Allyn & Bacon.

Banks, J. A. (1994). Transforming the mainstream curriculum. *Educational Leadership, 51,* 4–8.

Campbell, D. E. (1996). *Choosing democracy: A practical guide to multicultural education.* Upper Saddle River, NJ: Merrill/Prentice Hall.

Chandler, P. A. (1994). *A place for me: Including children with special needs in early care and education settings.* Washington, DC: National Association for the Education of Young Children.

Clark, L., DeWolf, S., & Clark, C. (1992). Teaching teachers to avoid having culturally assaultive classrooms. *Young Children, 47,* 5.

Crawford, J. (1992). *Hold your tongue: Bilingualism and the politics of English only.* Reading, MA: Addison-Wesley.

Deiner, P. L. (1983). *Resources for teaching young children with special needs.* New York: Harcourt Brace Jovanovich.

Florio-Ruane, S. (1989). Social organization of classes and schools. In M. Reynolds (Ed.), *Knowledge base for beginning teachers* (pp. 163–172). Oxford: Pergamon.

Freeman, D. E., & Freeman, Y. S. (1993). Strategies for promoting the primary languages of all students. *The Reading Teacher, 46,* 552–558.

Gamoran, A. (1992). Is ability grouping equitable? *Education Leadership, 50,* 11–17.

Garcia, R. L. (1991). *Teaching in a pluralistic society: Concepts, models, and strategies.* New York: HarperCollins.

Garrow, D. J. (1987). *The Montgomery bus boycott and the women who started it: The memoir of Jo Ann Gibson Robinson.* Knoxville: The University of Tennessee Press.

Gollnick, D. M., & Chinn, P. C. (1986). *Multicultural education in a pluralistic society,* (2nd ed.). New York: Merrill/Macmillan.

Gollnick, D. M., & Chinn, P. C. (1994). *Multicultural education in a pluralistic society,* (4th ed.). New York: Macmillan.

Grant, C. A., & Gomez, M. L. (1996). *Making schooling multicultural: Campus and classroom.* Upper Saddle River, NJ: Prentice Hall.

Jones, E., & Derman-Sparks, L. (1992). Meeting the challenge of diversity. *Young Children, 47,* 12–17.

Keller, H. (1920). *The story of my life.* Garden City, NY: Doubleday.

Kohlberg, L. (1992). In Lefrancois, G. R., *Of children.* Belmont, CA: Wadsworth.

Krug, M. (1976). *The melting of the ethnics.* Bloomington, IL: Phi Delta Kappa.

Perez, J. (1993). Viva la differencia. *First Teacher, 14,* 24–25.

Perez, S. A. (1994). Responding differently to diversity. *Childhood Education, 70,* 151–153.

Public Law 101–476, October 30, 1990, Stat. 1103.

Raines, S. C., & Canady, R. J. (1992). *Story s-t-r-e-t-c-h-e-r-s: Activities to expand children's favorite books.* Mt. Ranier, MD: Gryphon House.

Sadker, D., & Sadker, M. (1986). Sexism in the classroom: From grade school to graduate school. *Phi Delta Kappan, 68,* 512.

Sadker, M., Sadker, D., & Klein, S. (1991). The issue of gender in elementary and secondary education. In G. Grant (Ed.), *Review of research in education.* Washington, DC: American Educational Research Association.

Sadker, M., Sadker, D., & Steindam, S. (1989). Gender equity and educational reform. *Educational Leadership, 46,* 44–47.

Scarcella, R. (1990). *Teaching language minority students in the multicultural classroom.* Upper Saddle River, NJ: Prentice Hall.

Slavin, R. E. (1987). Ability grouping and achievement in elementary schools: A best-evidence synthesis. *Review of Educational Research, 57,* 293–336.

Solit, G. (1993). A place for Marie: Guidelines for the integration process. In K. M. Paciorek (Ed.), *Early childhood education 94/95.* Guilford, CT: Dushkin Publishing.

Stearns, P. N. (1996). Multiculturalism and the American educational tradition. In C. A. Grant & M. L. Gomez (Eds.), *Making schooling multicultural: Campus and classroom* (pp. 17–33). Upper Saddle River, NJ: Merrill.

U.S. Bureau of the Census. (1990). *Current population reports. Series P-20.* Washington, DC: U.S. Government Printing Office.

U.S. Office of Education. (1977). Education of handicapped children. *Federal Register (part 2).* Washington, DC: Department of Health, Education and Welfare.

Wolery, M., Strain, P. S., & Bailey, D. B. (1992). Reaching potentials of children with special needs. In S. Bredekamp & T. Rosegrant, *Reaching potentials: Appropriate curriculum and assessment for young children, Vol. 1.* Washington, DC: National Association for the Education of Young Children.

Wolery, M. & Wilbers, J. S. (Eds.) (1994). *Including children with special needs in early childhood programs.* Washington, DC: National Association for the Education of Young Children.

Wolfle, J. (1989). The gifted preschooler: Developmentally different but still 3 or 4 years old. *Young Children, 44,* 42.

Woolfolk, A. E. (1995). *Educational psychology.* Boston: Allyn & Bacon.

Zangwell, I. (1909). *The melting pot* (A Play). Quoted in D. M. Gollnick & P. C. Chinn (1983), *Multicultural education in a pluralistic society.* St. Louis: Mosby.

Author Index

Subject Index